Law, Regulation, and Governance

Law, Regulation, and Governance

Edited by
Michael Mac Neil, Neil Sargent, and Peter Swan

OXFORD
UNIVERSITY PRESS

70 Wynford Drive, Don Mills, Ontario M3C 1J9
www.oup.com/ca

Oxford University Press is a department of the University of Oxford.
It furthers the University's objective of excellence in research, scholarship,
and education by publishing worldwide in

Oxford New York
Auckland Bangkok Buenos Aires Cape Town Chennai
Dar es Salaam Delhi Hong Kong Istanbul Karachi Kolkata
Kuala Lumpur Madrid Melbourne Mexico City Mumbai Nairobi
São Paulo Shanghai Singapore Taipei Tokyo Toronto

with an associated company in Berlin

Oxford is a trade mark of Oxford University Press
in the UK and in certain other countries

Published in Canada
by Oxford University Press

Copyright © Oxford University Press Canada 2002

The moral rights of the author have been asserted

Database right Oxford University Press (maker)

First published 2002

All rights reserved. No part of this publication may be reproduced,
stored in a retrieval system, or transmitted, in any form or by any means,
without the prior permission in writing of Oxford University Press,
or as expressly permitted by law, or under terms agreed with the appropriate
reprographics rights organization. Enquiries concerning reproduction
outside the scope of the above should be sent to the Rights Department,
Oxford University Press, at the address above.

You must not circulate this book in any other binding or cover
and you must impose this same condition on any acquirer.

National Library of Canada Cataloguing in Publication Data

Law, regulation, and governance/
edited by Michael Mac Neil, Neil Sargent, and Peter Swan

Includes bibliographical references and index.
ISBN 0-19-540765-2

1. Administrative law. 2. Rule of law. 3. Globalization. 4. State,
The. I. Mac Neil, M. (Michael) II. Sargent, N. (Neil) III. Swan,
Peter D. (Peter Daniel), 1948-
JZ1318.L39 2002   340'.11   C2002-901359-3

Cover design: Brett J. Miller

1 2 3 4 - 05 04 03 02

This book is printed on permanent (acid-free) paper å.
Printed in Canada

Contents

	Acknowledgements	vii
Chapter 1	Governing at a Distance: An Introduction to Law, Regulation, and Governance *Peter Swan*	1

Part I **Theoretical Dimensions**

Chapter 2	Regulation, Governance, and the State: Reflections on the Transformation of Regulatory Practices in Late-Modern Liberal Democracies *Trevor Purvis*	28
Chapter 3	Legal Governance and Social Relations: Empowering Agents and the Limits of Law *Alan Hunt*	54
Chapter 4	The Legal Regulation of Politics: Governance, National Security, and the Rule of Law in the Modern State *Barry Wright*	78
Chapter 5	Radical or Rational? Reflexive Law as *Res Novo* in the Canadian Environmental Regulatory Regime *David J. Schneider*	97
Chapter 6	Democratic Environmental Governance and Environmental Justice *Peter Swan*	121

Part II **Labour, Capital, and State**

Chapter 7	Corporate Governance *R. Lynn Campbell*	144

Chapter 8	Governing Employment Michael Mac Neil	171
Chapter 9	Law, Regulation, and Becoming 'Uncivil': Contestation and Reconstruction within the Federal Administrative State Rosemary Warskett	188

Part III   Dispute Resolution

Chapter 10	Is There Any Justice in Alternative Justice? Neil C. Sargent	204
Chapter 11	The Regulation of Mediation Cheryl A. Picard and R.P. Saunders	223

Part IV   Governing Self-Regulation

Chapter 12	Government, Private Regulation, and the Role of the Market Kernaghan Webb	240
Chapter 13	Cyberspace Governance: Canadian Reflections Michael Mac Neil	264

Part V   Historical and Cultural Perspectives on Regulation and Governance

Chapter 14	Extending La Longue Durée: Commercial Impact in the Reform and Use of the Law of Quarantine Logan Atkinson	288
Chapter 15	Escaping Interventionism: Negotiating Regulation and Self-Governance in the Wake of the Indian Act Jane Dickson-Gilmore	317
	Index	343

Acknowledgements

This book results from an extensive collaborative effort by authors associated with the Department of Law at Carleton University. Each of the contributors teaches in the department either as a full-time faculty member or as a sessional or adjunct professor. In addition, two of the authors have completed the Master of Arts in Legal Studies at Carleton. The project started with an initial plan to write an issue-oriented text on legal studies. As a result of ongoing discussion by the participants, it developed into a project that explores a plurality of perspectives on the interaction of legal and non-legal modes of regulation within structures of modern governance. Our intention is to provide theoretical frameworks on law, regulation, and governance and to explore the application of those frameworks to concrete areas such as national security, the environment, employment, cyberspace, conflict resolution, Aboriginal governance, and health policy.

It is not surprising that this type of text would emerge from Carleton University. The Law Department at Carleton offers the largest Bachelor of Arts program in Legal Studies in Canada and the only Master of Arts program in Legal Studies. Unconstrained by the requirements of professional legal education, the program allows for a great deal of flexibility in curriculum development and promotes contextual and interdisciplinary approaches to the study of law and law's relation to the social, the political, and the economic.

The editors' job has been a pleasure as a result of the co-operation we have received in the conception and execution of this project. Our first thanks must go to our colleagues who have written the chapters of the book. They have demonstrated a great deal of patience as the undertaking moved along, and a great deal of flexibility in responding to the many requirements that have been made of them as they have adapted their work to better integrate it into the project as it has evolved. We would like to thank the three anonymous reviewers who provided extremely helpful advice on how to better structure the text and to improve various chapters. We are also grateful for their words of encouragement. Various people at Oxford University Press have nudged this effort towards completion over the years, including Euan White, Laura Macleod, Len Husband, and Phyllis Wilson. We are particularly indebted to Richard Tallman, who did an outstanding job in copy-editing the work. Catharinah Faux, a student in the Master of Legal Studies program at Carleton, provided helpful research support along the way. We also want to thank Brettel Dawson, one of the initiators of this project, whose enthusiastic support provided a great deal of motivation.

Finally, we wish to thank the students in the Carleton program, who have stimulated and challenged us, and for whom we hope this text does the same.

Michael Mac Neil
Neil Sargent
Peter Swan

Chapter 1

Governing at a Distance: An Introduction to Law, Regulation, and Governance

PETER SWAN

Introduction

Does the globalization of economic production and exchange mean that economic production has become autonomous from political controls through legal regulation? Does the increasing ease with which money, technology, people,[1] and goods move across borders mean that the state has less and less power to regulate these flows, and does it signify that political sovereignty has declined? How has the power of the state been affected by its inability to control new technologies such as the Internet? Have the demands of ethnic or cultural groups in increasingly pluralist societies undermined the validity of the rule of law? These questions are just a few of the many that have led legal and political scholars to question the continued viability of both the nation-state and legal regulation.

According to John P. McCormick, such issues raise the possibility of a 'crisis of the state' and call into question the role of law within that state. Arguing from the perspective of contemporary political theory and legal theory, he suggests that such perceptions of crisis can lead to a fundamental interrogation and self-examination of the role of law by theorists.[2] According to this analysis, forms of law are embedded within specific socio-economic complexes that must provide a ground for critical reflection on the possible transformations in the role of the law and the state.[3] Consequently, the analysis of specific forms of law is closely associated with an understanding of specific social and economic conditions that reflect and constitute, and are in turn constituted by, those legal forms. This social-historical approach to legal analysis informs many of the most significant contemporary assessments of the changing role of law and the state.

Many present-day critical reflections on law focus on the transformation of the interventionist law of the welfare state as a result of such socio-economic forces as economic globalization, technological change, and waves of immigration to the wealthy nations of North America and Europe. Conceptually and historically, legal interventionism transformed the liberal state with its formal laws into the modern welfare state. This interventionism imposed particular egalitarian conceptions of the public interest on society and the economy. It also represented the takeover of social distribution and allocation by modern governments. One of the major purposes of legal interventionism was to provide a 'social safety net' by guaranteeing minimal conditions of life for the weaker members of society. At the same time, it also aimed to correct 'market failures' by imposing political and social goals on the economy. This legal interventionism has been widely regarded as the dominant form of law and mode of social regulation during the last century.

Interventionist law was purposive law. Its substantive goals included a better allocation of resources, the protection of those who were marginalized and excluded from societal decision-making, and a more just distribution of wealth. As articulated in the vision of citizenship of T.H. Marshall, the goal of legal interventionism was to put a human face on capitalism by providing for a degree of equality that would allow all social classes to participate in social life.[4]

Within the last 30 years, the limits of an interventionist legal regulation of economy and society have become more apparent. The welfare state and legal interventionism have been assailed from a variety of directions. From the political right, commentators and theorists have argued that the purposive interventions associated with the welfare state have hindered both economic productivity and innovation.[5] Neo-liberals have advocated deregulation in areas where there has been direct government intervention in market activities.[6] From the left, there has been concern that excessive bureaucratization associated with interventionism has depoliticized society and undermined challenges to existing power relations.

Interventionism has also been assailed by contemporary forces of modernization. Changes associated with economic, technological, and political globalization have led to many of the most significant decisions about social life being made at the transnational or international level. When governments have difficulty regulating transnational financial transactions or the content of an apparently boundless Internet, questions about the continuing viability of national sovereignty and domestic law are raised.

In some quarters the attacks on legal interventionism and the changes at the global level have led to a perception that there is a crisis of interventionist law or a crisis of regulation. This perception of crisis has been articulated in a variety of ways. Commentators have posed a number of questions. After the retreat from the welfare state can public law be used to promote public purposes or social justice? What role can law play in emerging forms of governance that rely more on co-operation among government, industry, and the associations of civil society than on top-down directives from government? At the level of social and political theory one can ask whether the state is being displaced by emerging regulatory roles for local, regional, and global institutions and forces.

Not everyone accepts that there is a crisis of the state or even a crisis of legal regulation. Nonetheless, it is widely agreed that fundamental changes taking place in the regulatory terrain require us to re-evaluate the roles of both the state and the law itself. These changes necessitate a reformulation of the concept of regulation not only to take account of the use of law to prescribe behaviour but also to consider a range of self-governing practices by non-state organizations that also have a regulatory effect.

The rest of this introductory chapter considers the relationship between the regulatory crises of the state and of law and the development of legal reflections on how the role of law is changing. I will begin by examining the relationship between perceptions of such crises and the development of legal thought during the 1920s and 1930s, when there was a shift from a liberal emphasis on a minimalist state to the legal interventionism of the welfare state. Different responses to such crises by German legal theorists and by 'legal realism' in North America help us to understand what is presently occurring in the relationship between law and the state. Contemporary scholars are attempting to re-evaluate legal and non-legal forms of regulation, and this re-examination is reflected in ideas about the changing role of law in relation to regulation and to the idea of multiple sites of governance. Finally, as I will outline here, the various chapters in this book contribute to this reflection on the changing role of law. While most of the authors in this collection insist on a continuing role for the state and for legal forms of regulation within that state, they also recognize that the traditional role of law is changing and that increasingly legal forms of regulation must be examined in relationship to the non-legal rules and practices that contribute to the development of new projects of governance. The contributors to this volume examine various aspects of this broader conceptualization of regulation while exploring the tension that remains between legal rules and various degrees of economic and social self-regulation.

Legal Theory and Crisis

The contemporary situation of law and the state and the questions that it conjures have a parallel in a similar questioning of the state and law in the earlier part of the twentieth century. Some of the most decisive developments in contemporary legal thought took place during the 1920s and 1930s as a result of considerations of the limits of laissez-faire liberalism by academics in North America and Europe. In particular, these scholars wondered whether new regulatory functions for law had become necessary. These seminal inquiries in Europe began with Max Weber's sociology of law and the work of a politically diverse group of German legal thinkers in the Weimar Republic. At roughly the same time in North America a group of scholars who became known as the 'legal realists' were asking similar questions. The German 'jurisprudence of crisis'[7] addressed the issue of crisis in terms of broad, historically informed theorizing about the changing role of law and state regulation. The concept of 'regulatory crisis' is illustrated clearly in the work of Robert L. Hale, who formulated a more expansive idea of regulation by demonstrating the role of private power in tasks of governance.

The primary objective of the legal paradigm of classical liberalism was to limit the authority of the state. It sought to regulate and contain state power through formal and codified legal norms that in theory would ensure the security and freedom of individuals against a potentially dangerous state. According to Max Weber's conceptual and historical critique of this paradigm, 'formal law' achieved its culmination under conditions of nineteenth-century laissez-faire capitalism in North America and Western Europe.[8] This formal mode of law was characterized by general and supposedly neutral rules that were to apply equally to everyone. Yet as Weber suggested, fundamental contradictions in legal rules emphasized the formal equality of private actors while leaving real social inequalities in areas such as employment relations unaddressed.[9] Around the same time, legal theorists such as Carl Schmitt pointed to other internal inconsistencies besetting this liberal paradigm of formal law. According to Schmitt the generality of such rules left them vulnerable to both discretionary interpretation by judges and arbitrary intervention by the state to promote specific political purposes.[10] There thus was a perception of an internal crisis of law in the sense that the formal law paradigm was plagued with contradictions.

This perception of crisis was enhanced by external social and political developments such as the spread of a mass-based politics and the replacement of the nineteenth-century model of a non-interventionist state by a twentieth-century welfare-state one. For Weber, 'formal law' is replaced with what he calls 'deformalized law' wherein law serves the purposes of an interventionist welfare state.[11]

Those who participated in the legal debates over the 'crisis of law' in the Weimar Republic tended to develop broader theoretical assessments of the consequences of the paradigmatic shift from formal to material law. From a conservative perspective, Schmitt emphasized the increasing influence of discretionary and arbitrary forms of adjudication. Above all, he drew attention to the close relationship between law and politics and to the attempt to deny such a connection in the ideology of liberal parliamentarism and liberal legalism. From the perspective of the social democratic left, Hermann Heller also underlined the nature of the relationship between law and politics by developing a democratic theory of law in which situated ethical values provide criteria for evaluating the legitimacy of law.[12]

For Heller one of the primary ethical values shaping the legitimacy of law was the promotion of social equality. Yet this relationship between law and the promotion of equality was also greeted with ambivalence by other leftist legal scholars. Franz Neumann and Otto Kirchheimer were particularly wary of the possible effect of the transition to interventionist law on the rule of law and regulation.[13] They contended that any analysis of the rule of law needed to address the social, economic, and political imperatives that have transformed regulation and political administration since the nineteenth century. For Neumann and Kirchheimer, the rule of law, defined as governing through abstract general laws that apply equally to all persons, is specific to the development of competitive capitalism. If the conditions for the rule of law changed, then a new concept of the rule of law would be needed. Accordingly, the conception of the rule of law as generality worked under conditions of competitive capitalism, when all those with access to power were relatively equal in power. With the sharing of political power with the working class in the Weimar Republic, the generality of the law was undermined as regulations specific to the working class were needed to promote the material equality necessary for political participation. Consequently, such purposive regulation rendered the classical definition of the rule of law obsolete.

If for many German legal theorists the 'crisis of law' was construed theoretically in terms of the dangers presented by growing purposive law and regulation, the legal realists in North America were more concerned with clarifying how legal regulation actually took place and with showing how power relations were implicated in legal rules. The origins of legal realism were based on hostility to laissez-faire economics.[14] Accordingly, their concept of legal crisis was presented more in terms of how the categories of formal law that emphasized individual freedom actually created and concealed real inequalities in society. In the United States and Canada, this was expressed in a questioning of the rigid demarcation between a public law associated with the constitutional and administrative power of the state and a private law that defined and facilitated an area of individual economic freedom. Through this critique of the public/private distinction, the legal realists presented their understanding of the relation of power to law in which they uncovered the role of private economic power in regulating everyday social relations.

The sophistication of the realist approach to regulation is given its clearest expression in Robert L. Hale's analysis of wider conceptions of governance by non-state actors. Hale rejected the idea of natural liberty that was attributed to the market and argued that all economic relations are forms of regulation or control of individual conduct. He argued that legal coercion or compulsion was implicitly derived from the coercive power of economic relations.[15] The legal expression of these power relations, defined as the ability of a person or organization to control the behaviour or conduct of another, according to Hale, could be located in the exercise of property rights and the ideal of freedom of contract. Consequently, Hale contended that coercive power relations were not confined to government by the state and apparatus of public administration but were also asserted by a private regulatory form of government. Thus, the conduct of an individual may be controlled by the state, by an individual, or by a private body or association.

One of his central concepts is the idea of 'private government'. According to Hale, government is a private as well as a public phenomenon. Personal control and public regulation often went hand in hand.[16] However, the exact location of this private government in Hale's work remains somewhat ambiguous. In his unpublished papers, he states that: 'The invisible government is not a single, coherent unit but rather a cluster of different groups and persons who hold sway in different fields.'[17] This suggests a plurality of sites of 'government' or 'governance'. However, in other aspects of his work, Hale seems to be proposing that the primary source of private coercion comes from the minority who possess the monopoly of bargaining power in the economic system.

In this emphasis on wider notions of power and the regulatory role of private as well as public bodies, one can notice a similarity between Hale's work and the more contemporary analysis of the concept of governmentality by the French social theorist Michel Foucault.[18] Suggestive ideas for such a conceptual linkage can be found in Hale's assertion that the exercise of choice does not necessarily indicate a lack of coercion or power imbalance.[19] Just as Foucault is often ambivalent about the effects of power, Hale is ambivalent about the effects of coercion. Although Hale recognizes that authoritarian aspects in the private realm must be controlled by the state, he does not regard all coercion as bad and he also acknowledges that the owners of capital do not hold all the power; employment is an example of mutual coercion where employees also are able to exercise power. Like Foucault, Hale also claims that the authoritarian aspects of power can be found as much in the private realm as in the public. Accordingly, he argues that by focusing only on coercion in the public realm, the legal discourse on the public/private split ends up veiling the actual extent of private governing power and leaving it unchecked.[20]

Just as the approaches to legal crisis took different forms in Europe and North America, so did the practical consequences of this theorizing in the real world. In Germany, the crisis of regulatory power had a parallel in a constitutional crisis. The unstable mixture of traditional liberal legalism and social democratic social justice concerns in the Weimar Constitution ultimately proved to be untenable. Kirchheimer and Neumann's warnings about the inability of liberal legalism to address rising demands for social justice proved to be correct. In the end, however, the authoritarian legal theory of Schmitt proved to be most influential and the Weimar Constitution collapsed when the Reich president invoked his emergency powers to prevent a civil war. In the United States, the consequences were not so dire. The work of legal realists such as Hale also proved to be influential in shaping the direction of public policy until the 1970s. Their ideas about strengthening public law in order to remedy the inequalities associated with private law had a profound impact on President Roosevelt's New Deal legislation. Although this did contribute to a constitutional crisis in the 1930s, realist ideas about the public and private continued to inform the development of the welfare state in North America for over a generation.

Theorizing the Contemporary Crisis of Regulation

Hale's work remains important for pointing out the regulatory effects of the activities of non-state actors. Similarly, the most important German legal theory from the 1920s and 1930s retains a contemporary relevance because it illuminates the dynamic relationship between law and politics more effectively than most recent theory. However, both of these approaches are based on conceptions of power that are too narrow to explain the complexity and range of contemporary regulatory practices and discourses. The crisis of the state and the role of law within it that contemporary legal thinkers confront are qualitatively different from the one that the legal realists and the German critics of Weimar faced. Instead of dealing with an ascendant welfare state, these thinkers must confront a state that outwardly at least appears to be in decline. As in the last century, explanations of contemporary regulatory practices must address both the internal contradictions of the state and crises derived from external sources. Critics on the right and on the left point to the contradictory effects of state intervention. Neo-liberals argue for deregulation because they believe that purposive welfare interventions prioritize equality over economic liberty and thus act as obstacles to economic productivity and innovation.[21] Critics on the left have been concerned with a lack of bureaucratic accountability and suggest that excessive reliance on legal regulation by the state may lead to a citizenry who are mainly passive clients of the state rather than active participants and deliberators in processes of social, political, and economic decision-making.[22]

The regulative capacity of the contemporary state may also be threatened by external developments. According to John P. McCormick, contemporary legal theorists and sociologists of law must recognize the diminishing capacity of the state to address questions of administrative accountability, economic regulation, and social justice 'under emerging conditions of what has variously been referred to as internationalization, regionalization, multilateralism, Post-Fordism and, most fashionably, globalization.'[23] Although this idea of a 'diminishing capacity' of the state to govern is contentious, contemporary legal thought must confront challenges to a declining welfare state that may be attributed to the growing interdependence of nation-states, the increasingly transnational character of economic activity, and an acceleration of global cultural interactions.[24] This raises the question about whether globalization will transform structures of governance or the way that rules are made or the ways that authority is exercised in the world. Indeed, many would argue that the national governments are no longer in a position to determine the policies that most directly affect their citizens.

In addition to the forces of globalization, the state encounters demands for recognition by members of cultural groups and other groups that historically have been excluded from participation within the institutions of the dominant culture. No longer are these groups satisfied with the claims of inclusion through universal citizenship rights; instead, they demand special rights that will value their own cultures and allow them to participate on their own terms both within and without the dominant culture. Aboriginal groups in particular are demanding self-government rights that will allow them self-determination on some issues and an ability to participate in the dominant society in others. Such demands again challenge the traditional ideals of rule of law and state neutrality while limiting the authority of the nation-state to regulate relations between itself and reconstituted First Nations.

Although assessments of the decline of state power are being contested both theoretically and empirically, there seems to be agreement that the role of the state and of law is being transformed by contemporary social, economic, and political forces. While under conditions of globalization the locus of sovereignty may be shifting away from the nation-state, the specification of a new locus for sovereign decision-making remains difficult. One of the tasks of a critical reflection on the changing role of law and regulation should be to explore the nature of this shift and to try to identify new institutions and practices in which 'sovereign' decisions are made.

An increasing skepticism about the capacity of the state to regulate and co-ordinate socio-economic concerns has led to a growing interest in interrogating the techniques and practices of regulation. This reassessment has focused on both the institutions and the broader practices through which the regulation of social life is carried out. Like the intellectual examination of the earlier crisis of the state, this re-evaluation of the regulatory role of law has taken the form of both theoretical conceptualizations and empirical studies. Albeit that many explorations of the changing role of the regulatory state continue to be informed by the theory and techniques of political economy, two alternatives have recently emerged as a response to perceptions of regulatory crisis. The first approach advocates the proceduralization of law as a response to the inadequacy of the formal law of traditional liberalism and of the material law of the welfare state. The second method suggests that legal regulation must be examined within the context of broader projects of governance.

At its most general level, the emphasis on procedural law represents a shift to more indirect means to co-ordinate socio-economic activity. Procedural law entails a more limited and focused use of state power to achieve goals that are often implemented by other subsystems, such as the economy.[25] Thus, procedural law is not a form of deregulation but rather a means of indirectly realizing the goals of the regulatory state by shifting the implementation of political and social tasks to other systems. According to its advocates, the development of a procedural rationality is necessary to deal with the conditions of uncertainty that result from the complexity of interactions between different aspects of the socio-economic and political systems in contemporary society. They argue that a 'new emphasis must be placed on flexibility and reversibility of decision-making procedures' that would allow for a reflexive revision of earlier decisions.[26]

One of the key articulations of procedural law is found in the 'reflexive law' of Gunther Teubner, which was a specific response to the use of law to intervene directly in social processes. Legal intervention, according to Teubner, fails to account for the social complexity caused by the increasing differentiation of society into semi-autonomous spheres or systems that are too dispersed and powerful to be readily controlled by law.[27] He suggests that instead of trying to control directly the specific character or direction of social change under conditions of growing social complexity, law should attempt to use the self-reflective and self-governing capacities of other social institutions by providing the procedures to structure their decision-making processes.[28]

This emphasis on the proceduralization of regulation is an example of only one type of proceduralization. Within the context of competing theories of democracy, Julia Black suggests that Teubner's theory of reflexive law can be regarded as a form of 'thin proceduralism'.[29] According to this formulation, reflexive law is concerned primarily with procedure as 'a technique of regulation'. It attempts to address the best way to implement a policy while downplaying more normative concerns about the content of policy or about who should make that policy.[30] Thus it can be associated with a liberal view of democracy that emphasizes law-making through the technical aggregation of individual views rather than a formulation of views in an equal exchange of opinion. Although such a thin proceduralism pays lip service to ideals of participation and democracy, it is mainly concerned with techniques that can ensure that social complexity will be respected and that no one of the social, economic, or political subsystems can impose its will on the others through legal regulation. Its main effect, according to Black, is therefore to ensure that regulation remains neutral with respect to traditional interest group politics.

The alternative conceptualization is a 'thick proceduralism' that focuses on more substantive values and emphasizes the moral and political dimensions of regulation as opposed to regulation as technique. Participation as an ideal is even more central to the 'thick' variation of proceduralism because it focuses on nurturing more direct democratic virtues by promoting conditions that will allow all individuals and groups affected by regulatory decisions to deliberate upon and make the decisions that affect their lives. Accordingly, participation is more closely associated with forms of deliberative democracy that emphasize an expansion of the range of possible non-state and non-economy sites for participation in projects of regulation and governance. With the inclusion of previously marginalized groups in regulatory processes, forms of 'thick proceduralism' are intended as a means to enhance the epistemic character of regulation as a greater range of interests and perspectives are taken into account in the formulation of regulations.[31] The role of legal procedures within this model is to provide a framework for the deliberative co-operation of all who are affected by regulatory decisions.

However, the critical analysis associated with this form of democratic proceduralism must be pushed even further. Although the deliberative model of proceduralism emphasizes the need to further democratize the regulatory process by including traditionally marginalized groups, the analysis of such processes must necessarily involve a recognition of the role of power relations in deliberative processes. Not only must imbalances in economic resources and political opportunities be recognized, but also it is necessary to examine the effects of differentials in knowledge and of the traditional privileging of technical discourse in areas such as the regulation of environmental risks. Thus, this model of proceduralism may also need to emphasize the role of the contestation of norms and knowledge as well as the discussion of their merits through the rational use of public reasoning.[32]

A potentially more radical theoretical response to the perception of crisis in law and regulation can be found in recent explorations of broader concepts of governance. While the term 'governance' frequently refers to the investigation of a plurality of sites of non-state regulatory activity, it also refers to a specific social theoretical approach that has its point of departure in the later work of the French social theorist and philosopher, Michel Foucault. Although Foucault's own work on social control and discipline tends to downplay the role of law in contemporary society in favour of the mechanisms of discipline, social and legal theorists influenced by Foucault have suggested that the concept of 'governmentality' from Foucault's later work may provide a way to reconceptualize the relationship between legal and non-legal forms of regulation or governance.[33]

For Foucault, governmentality or the 'rationality of government' refers to specific ways for making the practices or activities of government intelligible to 'its practitioners and to those upon whom it is practiced'.[34] More specifically, it points out ways of thinking about what government is, about who can govern, and about who or what can be governed that are often taken for granted and not considered by those who govern.[35] Conceptualized in this manner, the study of governmentality focuses on the techniques and practices of governing as deliberate attempts to shape conduct in accordance with specific social, political, or economic objectives.[36] In addition, the focus of such studies is not so much on the state itself but rather on how the fiscal, organizational, and juridical functions of the state can be related to non-state undertakings for the management of various aspects of socio-economic life of the population.

However, despite presenting a broad conception of the complexities of the rationality of government, Foucault's account of the 'art of government' and of the role of law within it remains fragmentary. This project his been taken up and furthered by a number of thinkers who have been influenced by the analysis of governmentality. Mitchell Dean[37] in Australia, Nikolas Rose[38] in England, and Alan Hunt[39] in Canada, as the leading figures of the emerging field of the sociology of governance, remain committed to the basic thrust of Foucault's analysis while continuing to elaborate on its implications.

The practitioners of the sociology of governance insist that it is necessary to examine the interactions of a range of political actors, of which the state is only one. Thus governance, defined as any strategy, process, procedure, or program for controlling, regulating, or exercising authority over either animate or inanimate objects or populations, is regarded as being much broader than the traditional conception of state-centred regulation.[40] At the same time, there is an attempt to address Foucault's relative inattention to the contemporary role of law. Using insights derived from legal pluralism, Alan Hunt brings law back into the analysis of governance by acknowledging the persistence of a plurality of forms of law as new forms of rights are developed and regulatory laws become more focused and particularistic. According to Hunt, this recognition allows us to develop an expanded conception of the role of law as one of a plurality of ways of providing the mechanisms through which processes of governance are practised.[41] This view allows a re-emphasis and reformulation of public law. As a reflection of the multiple sites of governance, public law now focuses not only on the regulatory activities of the territorial state but also on the activities of quasi-state institutions, professional bodies, and regulatory agencies.

Just as the work of the legal realists led to the questioning of the public/private distinction, this idea about the expanded role of law leads to a further *problematization* or reflection on the complexity of the contemporary distinction. The emphasis on the role of governance by theorists such as Hunt, Dean, and Rose is more attuned to the sociological complexities of contemporary governance than the earlier work of the realists. This recognition forces us to reconsider the role of public law and to reconceptualize the complexity of the public/private distinction as it relates to the governing of a wide range of activities in contemporary society.

The Decentred State

Despite significant conceptual differences, a common theme in the above theoretical approaches is the questioning of the assumption of the autonomy of the political that has characterized modernity. Profound skepticism increasingly greets the ideal of autonomy as expressed through the state as a 'sovereign decision-making unit' with a monopoly on the organization and regulation of socio-economic affairs.[42] Most of these theoretical responses reject the idea of the state as the 'centre of action' from which the whole can be observed, known, and controlled. At the same time, however, aware of the expansionary character of the economic power of capital and its resulting inequalities, they also reject simple alternatives that would relocate regulatory competencies to the market. As a result, both as description and as prescription, these recent approaches emphasize a 'decentring' of a state that is but one of a plurality of sites of governance or regulation.[43]

An exact definition of the term 'decentring' in relation to the state is somewhat elusive and its exact character varies with the different approaches to regulation. For some, the concept of decentring questions the possibility of a unified sense of identity on the part of the state. The state, according to this version, does not form an 'effective unity' *pace* Hermann Heller[44] but rather is fragmented and inherently unstable. Claus Offe, arguing from the perspective of political economy, reflects on the impossibility of the state acting as a coherent unit of control and co-ordination. According to his analysis, internal pluralization and the fragmentation of perspectives and rivalries within the administration, along with increasing demands on the state from different sectors of the public, have undermined the ability of the state on its own to organize and regulate social and economic life effectively.[45]

However, the idea of a decentred state does not simply refer to its internal fragmentation in a structural sense. It also refers to the fragmentation of the nation-state as a result of cultural factors within the traditional territorial boundaries of the nation-state. As a result of different groups asserting claims for the recognition of the value of their cultures, citizenship is increasingly regarded as plural rather than unitary.[46] As Aboriginal nations begin to assert their desire for cultural and legal recognition and demand at least the right to limited self-government, the ineffectiveness of paternalistic state regulation is exposed. In Canada, recent concessions have created space for traditional Aboriginal justice techniques in the criminal justice system. Demands for self-government have led to the possibility of alternative forms of regulation in other areas as well. With increases in immigration from various parts of the world, various ethnic groups have expressed a desire to protect their cultures through forms of special rights that can protect them from being absorbed in the dominant culture.[47] Increasingly, it becomes more difficult for the state to be considered a 'unified entity' in which all differences disappear or cease to matter.[48]

The idea of a decentred state also acknowledges that many of the most important decisions affecting our daily lives are not made at the level of the nation-state but rather at either the local or global level. For some, therefore, the continued privileging of the state as a focus for political analysis deflects attention from the need to analyze the real loci of power and decision.[49] They insist that we must reconceptualize the political in such a way as to address local and global politics. This has not meant that the analysis of governance should ignore the state but rather that the activities of the state be examined in relation to the global organization of the economy and to the activities of a variety of social movements at both the domestic and transnational levels.[50]

Those analysts influenced by the work of Michel Foucault interpret the decentring of the state in a different manner. While insisting on the depriviliging of the state as a locus of power in society and as an organ of social control, their approach tends to be more descriptive than prescriptive. Following Foucault, the sociologists of governance acknowledge that the modern state, from its formation in the sixteenth century, has never been the sole locus of the activities of governance. However, they also recognize that the multiplication of sites of regulatory practices is central to the pursuit of a neo-liberal agenda whose projects of governance increasingly involve interactions between state and non-state actors.[51]

From the perspective of the theorists of proceduralization, the emphasis on the decentring of the state can be either descriptive or prescriptive. Theorists such as Teubner, who are influenced by systems theory, argue that because of the complexity of modern social, economic, and political systems and their interactions, there can be no privileged vantage point such as the state from which the whole can be known and co-ordinated. Thus, they contend that it would overburden law to assume that the state on its own could effectively provide such co-ordination. Their concept of decentring therefore emphasizes the importance of affirming a more indirect role for state law by using procedures to help other subsystems to employ their self-directing capacities in order to provide more effective forms of regulation.

For the advocates of the democratization associated with 'thick proceduralism', the decentring of the state is articulated as a form of decentralization of state authority. Although recognizing the need to continue to ameliorate the effects of social inequality, they question whether or not this function should be left to the state alone. Acknowledging that the bureaucracy of the welfare state has contributed to the creation of forms of passive citizenship,[52] they argue for a decentralization and democratization of the welfare state. Accordingly, they support the devolution of authority and accountability to local authorities and to non-governmental organizations through laws that specify the procedures through which decisions are to be made but that do not specify the content of these decisions. These agencies are given power to negotiate with both the private sector and the regional and national state in order to ensure the continued delivery of services and to guarantee that their formerly marginalized clientele have a voice in the process. To reconceptualize contemporary governance and regulation, the advocates of 'thick proceduralism' focus on pluralistic and deliberative processes of norm- and will-formation by the state in conjunction with a variety of non-state actors. The role of the state, then, is to establish legal procedures that allow for truly democratic regulatory decision-making in which those most affected are assured of the political equality necessary for more meaningful participation.[53] While this process emphasizes a need to 'empower' those at the margins of society, the advocates of proceduralism do not abandon the state but attempt to find ways for participants in informally organized public spheres to influence the formally organized public sphere in the state. The role of law becomes less one of directly co-ordinating social activity through regulations and more one of structuring the participatory input of societal groups to ensure the representativeness of state policy and the accountability of the experts who co-ordinate its regulatory projects.

The Decentring of Canadian Legal Regulation?

Despite a common focus on a decentring of the state and a corresponding emphasis on a decreasing role for interventionist law, most of the above approaches to understanding the contemporary crisis of law do not assume that state power is necessarily in decline. They all point out that the reality of contemporary regulation is much more complex. If anything, there is strong evidence to suggest that states today have as much power if not more than the welfare states that were their predecessors. The power of the regulatory state has not come to an end but is being transformed. This transformation means that we can no longer think of such power as being indivisible and confined to the territory of the nation-state. The locus of power is shifting and with this shift there is a growing concern about the most effective ways to realize goals relating to a plurality of conceptions of the 'public interest'. Concern also grows about how to subject seemingly unaccountable transnational political and economic power to democratic control.

The recognition of this alteration in the loci of power necessitates a reconceptualization of regulation. In part this reformulation should acknowledge a transition from 'strategies of command' to 'strategies of inducement'.[54] The role of regulators as rulers who prescribe the activities of regulated groups remains, but increasingly regulators also become supervisors or overseers who promote the self-regulatory activity of actors in the economy and in social life. The scope of regulation as command shrinks while the parameters of regulation as self-governance unfold. Laws, therefore, do not specify the substantive ends to be achieved but rather encourage the informational and governance capacities of organizations. And yet there is no abdication of the realization of the common ends of regulation. Rather, the tasks of achieving the objectives of projects of governance are increasingly shared by the state with business interests and civil society. At the same time, as many of the contributors to this volume suggest, a tension remains between regulation as command or compulsion and regulation as self-regulation.

Such concerns for both the multiplication of sites of governance and an expanded concept of regulation underlie many contemporary approaches to the crisis of law and also inform the work of the contributors to this book. Not everyone agrees that there is a crisis of law or of the regulatory state, or that there is significantly less room for legal compulsion. The contributors to this book are not unanimous in accepting that the state is being decentred or that its regulatory authority is being usurped by new sites of governance. Some suggest that the state, which was once in retreat, is beginning to find new legal ways to structure the activity of economic and social actors. Even among those who subscribe to the idea of the decentring of law and the state, there is no consensus on how that decentring is taking place. Others, while recognizing the devolution of governance capacity to non-state actors, suggest that this phenomenon has been occurring for some time and is by no means new. Most insist that state intervention is still needed to ensure that the norms that emerge from decentralized decision-making meet specified standards. The authors' conceptual approaches differ in regard to their perspective on the exact character of the interaction between legal compulsion and the rule-making practices of self-governing agencies, institutions, and organizations. Despite these disagreements, all of the contributors recognize that investigations of contemporary patterns of governance must account for the regulatory contributions of a growing range of societal actors. They all attempt to provide explanations of new patterns of the relationship between the contemporary state and these actors and of the role of law within the emerging regulatory regime.

The first section of this book approaches the issue of governance and regulation from an explicitly theoretical perspective. Chapters 2 and 3 provide theoretical overviews of the contemporary multiplication of sites of regulation and governance from perspectives that relate to the 'sociology of governance'. Chapter 4 contests the notion of the dispersion of sites of regulation by using a broad historical-political framework to show how politics is subjected to a range of forms of legal regulation, including repressive state security law. Chapters 5 and 6 articulate variants of procedural legal theory in the context of suggestions for the theoretical and practical reform of different aspects of environmental regulation.

Drawing on Foucauldian explorations of governance and recent political economy, Trevor Purvis in Chapter 2 assesses regulation in terms of wider theoretical debates over the role of the state in the economy and society. Purvis shows that, rather than leading to the demise of the state's regulatory role, such phenomena as privatization, deregulation, and globalization create a form of decentring in which new regulatory functions develop in relation to a 'diffusion of sites of regulation'. He then provides a theoretical explanation of this decentring that expands our understanding of regulation as societal governance. He contends that despite being assailed from within and without, the state remains central to contemporary regulatory projects by increasingly 'governing from a distance'. The state, in other words, provides rules to guide the self-organizing activity of the disparate forms of associations and agencies that contribute to social and economic governance. Law thus plays a more indirect role by facilitating the rule of non-state agencies and practices.

In Chapter 3, Alan Hunt addresses concerns about a contemporary crisis of law. He suggests that while greater demands have recently been put on law to reform inequitable social relations, the failure of law to deliver on these demands has led to increased frustration and anxiety from both theoretical and practical perspectives. Hunt contends that reforms based on recent critical responses to the crisis of interventionist law, such as legal pluralism and reflexive law, may actually result in increased legalization. He is also critical of attempts to empower legal subjects through the articulation of legal rights. Noting the complexities of processes of 'legal subjectification' through which individuals are formed as selves and as ethical subjects, he raises doubts about the capacity of law reform on its own to deliver social justice. Drawing on Foucault's concept of governmentality and on the sociology of governance, Hunt's analysis of the decentring of law is articulated as a rejection of the privileging of the state as the locus of governance. He insists on viewing legal regulation in relation to non-legal modes of governance. By implication, all legal strategies, including rights strategies, can only be considered in terms of their interaction with other modes and practices of governance.

Barry Wright, in Chapter 4, examines the role of national security laws in the regulation of politics. Wright considers a range of historical and contemporary approaches to the regulation of politics, focusing on political trials, surveillance, and state security law, and shows how an examination of such repressive laws can be interpreted as part of a larger discourse about the regulatory role of law. A central premise of this investigation of the relationship between state and law is that states will seek to preserve themselves through the law itself. In assessing modern developments, the chapter rejects any Foucauldian claim that the emergence of new, more subtle forms of governance is matched by an abandonment of national security laws. Rather, such laws persist and are renewed within a more complex deployment of legal and social regulatory practices. The chapter thus serves as a counterpoint to much of the literature, which identifies the dispersal of power away from the state into a multiplicity of more or less privatized sites. Although there is increased emphasis on the prosecution of political offences, executive enabling legislation, and official secrets law, new institutions of surveillance and for assessing threats to security remain as important subcomponents of the legal regulation of politics.

This chapter explores some of the means by which states use law to accomplish this goal. It takes into account the complexities of power and the emergence of new forms of political rationality, but urges against the neglect of the 'real remnants of repressive state-centred power'.

In Chapter 5, David Schneider provides a framework for the legal regulation of the environment through the proceduralization of law in the form of 'reflexive regulation'. Schneider argues that while law still has a role to play in regulation, it is incapable of directly controlling the activities of other autonomous social systems. Borrowing from the systems-theoretical work of Gunther Teubner and the idea of 'responsive regulation' from Ian Ayres and John Braithwaite, Schneider suggests that law should assume the less ambitious and more indirect regulatory function of co-ordination in order to influence the development of self-regulatory processes within other social systems, such as the economy. He argues that rather than continuing to assume a role of command at the centre, the state should focus on helping firms to generate their own environmental rules that are consistent with the public policy objectives of regulators. The role of legal regulation thus becomes the 'regulation of regulation', that is, promoting the self-regulation of other social subsystems.

Where Schneider's analysis represents a form of 'thin proceduralism' that focuses on the technical improvement of regulatory instruments, Peter Swan, in Chapter 6, critically evaluates the application of 'thick proceduralism' to environmental regulation. The specific focus of the chapter is legal proceduralism, in the form of 'information regulation' or performance-based regulation, promoted by Charles Sabel and his colleagues. While agreeing with the expressed aim of achieving more effective and more democratic environmental decision-making through a legal framework that allows a plurality of actors to exchange information and to realize goals established co-operatively in processes of deliberation, Swan questions whether the information-based regulation (IR) approach adequately addresses issues of power in public deliberations over environmental risks. He suggests that the theories of legal proceduralism and deliberative democracy developed by Sabel and his colleagues need to be supplemented by a theory of environmental justice that emphasizes equality of access to participation in decision-making processes. Swan shows how legal proceduralism fails to address issues such as the asymmetrical distribution of discursive power in deliberative processes. He contends that law has a role to play in addressing these power differentials by allowing for claims of political equality and rights to participation in the definition of environmental risks and in finding solutions to these risks. Law thus continues to play an important role in environmental governance by establishing the conditions under which decentralized decision-making by a plurality of social actors can be rendered more effective, accountable, and democratic.

Part II, 'Labour, Capital, and the State', looks at contemporary changes to specific areas of regulation and governance within the traditional triad of the corporation, labour, and state that played such a central role in shaping modern governance during the last century. Chapter 7 examines recent changes in formal legal modes of regulation of the corporation. Chapter 8 explores changes in the relationship between the external legal regulation of employment and the internal modes of the governance in employer-employee relations. Chapter 9, on the other hand, suggests that recent conflicts within the federal public service have undermined ideological claims about a politically neutral, bureaucratic state.

R. Lynn Campbell, in Chapter 7, explores the relationship between internal governance issues involving corporate structure and responsibility and the external imposition of legal duties and liabilities by courts and legislatures. The chapter focuses on the shift that has taken place in the formal legal modes of regulating corporate governance. According to Campbell, there has been a shift from a laissez-faire, hands-off approach to corporate governance to one where courts and the state are more willing to extend the liability of corporate officers. This is accomplished through the imposition of regulatory controls backed by penal sanctions in such areas as consumer packaging, worker health and safety, environmental controls; through legislative reforms designed to protect minority shareholders; and through the judicial development of directors' and officers' fiduciary duties and duties of care and skill. Against those who would question the capacity of the contemporary state to make corporations accountable for their actions, Campbell emphasizes the continuing predominance of formal legal modes of regulation over informal regulation through voluntary guidelines or internal, non-legal models of accountability.

In Chapter 8, Michael Mac Neil emphasizes a distinction between 'regulation' and 'governance' in the employment relationship. He is primarily concerned with the extent to which the external regulatory role of law has shaped or can shape or change the dominant, internal governance of the employment relationship. According to Mac Neil, the employment relationship has its own particular governance regime, which is determined by the economic character of the capitalist economy. Law provides the external regulatory framework through which this regime of governance can be exercised, and also limited. However, according to Mac Neil, despite the fact that there have been fundamental regulatory shifts in an attempt to equalize the relationship between employer and employees, the power imbalance that characterizes the dominant regime of governance remains fundamentally unchanged. Mac Neil maintains that the regulatory role of law in this context is complex: on one hand, the state through legislation provides more rights for employees and more opportunities for bargaining over the terms and conditions of work; on the other hand, these regulatory changes have not altered the fundamental power imbalance that dominates the internal regime of governance of the employment relationship. In part this has occurred because courts have interpreted legislative standards and contract law provisions and collective agreements in ways that do not fundamentally challenge the traditional allocation of power and control. Accordingly, the state can be seen as decentred in two fundamental ways: its laws must compete with the customs and traditions that define the traditional rules of governance of the employment relationship; and the state itself is fragmented to the extent that its legislative, administrative, and judicial branches have historically functioned at cross-purposes within the external regulatory regime.

Rosemary Warskett, in Chapter 9, looks at the internal challenges to the ideology of regulation in which the 'civil service' is legitimized as a technically proficient, apolitical meritocracy that functions separately from both the political sphere and the economy. She focuses on the internal conflicts around issues of race and class within the federal civil service during the 1980s and 1990s, which demonstrate a significant challenge to this unifying and legitimating discourse. Warskett also considers how shifts in the mode of legal regulation of the civil service, particularly the creation of the Public Service Alliance of Canada and the impact of pay equity litigation, have provided some of the conditions necessary to challenge the traditional discursive construction of the state. Warskett looks at the civil service strike in 1980 and the ongoing pay equity complaint and litigation in the 1990s as key elements in the challenge to the traditional unitary discourse of the federal civil service. Law is seen as an arena of struggle in which competing ideological or discursive positions are in competition, even within the state. Thus she contends that theoretical understandings emphasizing the functional unity of the bureaucracy, and correspondingly de-emphasizing internal contradictions and conflicts within the civil service, are inadequate. The state apparatus, according to Warskett's perspective, is riven with internal contradictions and conflicts rather than being a unified regulatory centre.

The two chapters in Part III, 'Dispute Resolution', explore regulation, governance, and the role of law in relation to informal dispute resolution. Informal or alternative dispute resolution (ADR) was one of the earliest responses to perceptions of a crisis of law and the legal system. For more than three decades, it has been presented as an effective way to address the inadequacy of the formal judicial system in dealing with a growth in legal disputes and with the perception that the rigidity of the legal system prevented a more contextualized assessment of the interests of all parties involved in a dispute. ADR, initially intended to provide a less costly and more expeditious response to disputes, soon became popular in North America. The chapters in this section address whether or not informal dispute resolution has in fact displaced the state-centred judicial system as the main forum for settling disputes.

In Chapter 10, Neil Sargent deals with the changing character of the relationship between the formalized system of dispute resolution established by the state and an informal system based mainly on techniques of mediation. A central theme here is whether disputes are to be settled in a coercive fashion by centralized institutions established by the state, in accordance with legal rules developed by legislatures and courts, or on the basis of individual consent through bargaining processes that may be assisted by mediation. Sargent comes to the conclusion that the state and law are not likely to be displaced by alternative systems of justice, but rather will adapt and to some extent incorporate aspects of alternative systems into the formal processes. He also maintains that the alternative systems will not operate in isolation from the formal, state-based processes, and indeed, the choice of whether to engage in alternative mediation processes may well be attenuated by mandatory court processes. Hence, according to Sargent, the interaction of the formal process of adjudication and the informal alternative processes leads to a complex, mutually constitutive relationship. Although there may be a shift towards a greater scope for alternative justice, he argues that the two systems modify one another, without one displacing the other.

Cheryl Picard and Ron Saunders, in Chapter 11, also explore the changing nature of the relationship between formal and informal systems of dispute resolution. Like Sargent, they agree that the interaction between the two systems is likely to continue. Unlike Sargent, and contrary to much of the current literature on the decline of the state, they see a reassertion of the role of the state in the area of dispute resolution. They reach this conclusion in their exploration of the modes of regulation or governance applicable to informal dispute resolution such as mediation. They maintain that although mediation and other forms of alternative dispute resolution have emerged in response to a perceived 'crisis of legitimacy' within the formal legal system, the shift towards institutionalizing mediation within the judicial system and the regulation of the standards governing qualifications of mediators indicate renewed hegemony on the part of the state and the formal legal system. One of the major issues explored by the authors is the extent to which the standards of practice and qualifications of mediators are likely to be influenced by the incursion of lawyers into the field of mediation. Although the growth of mediation appears consistent with a shrinking role for the state and for law, and a form of institutional abdication by the legal system over certain types of disputes over which it formerly claimed a monopoly, the authors argue that in practice the legal system and the state are increasingly able to assert a form of hegemonic control over both the goals and the forms of practice in which this process of delegalization takes place. Rather than seeing a decentred state, Picard and Saunders see a state increasingly asserting its authority over informal systems by attempting to regulate their operations.

Over the last decade there has been an increasing emphasis on private, market-oriented forms of self-regulation. Part IV, 'Governing Self-Regulation', deals with the relationship between state law and forms of self-regulatory activity by economic actors. Chapter 12 explores this issue in the context of environmental law. Chapter 13 examines the challenge posed to regulation by the growth of the Internet and how the emerging regulatory regime may be based on a mixture of legal regulation and self-regulation. Both chapters in this section suggest that the state continues to play a role by attempting to govern the self-regulation of economic actors and by seeking to facilitate individual responsibility on the part of consumers.

In Chapter 12, Kernaghan Webb argues that limits to legislative approaches to environmental issues suggest that they need to be supplemented by private, market-oriented forms of self-regulation. This shift, however, does not mean that the state will be displaced. He argues that it is not a question of substituting a private approach for a public one but rather how to integrate the two effectively. The law, after all, still provides a framework for private regulation. It also can help to avoid problems such as 'free riders' by providing sanctions for those firms not participating in voluntary agreements whose performance does not meet the standards voluntarily agreed upon. Webb maintains that the development of this regulatory integration in Canada has been somewhat haphazard. He therefore argues for a more ambitious role for law wherein an explicit legal framework provides the parameters for voluntary programs such as in the Eco-Management and Auditing Scheme in the European Union.

Michael Mac Neil notes a similar regulatory approach in Chapter 13. He argues that new developments such as the Internet provide both challenges and opportunities to re-evaluate how we govern ourselves in a period of transition where older forms of regulation on their own are inadequate. In examining the relationship between state-centred law and the governance of cyberspace, he asks whether the unique character of cyberspace will necessitate the development of special rules and modes of governance. Mac Neil examines both the general debates over state involvement in the regulation of the Internet and the evolution of Canadian government policy in this area. He suggests that although policy documents since the mid-1990s appear to be moving towards the idea of self-regulation, they also indicate the need to retain a role for the state in the protection of the public interest by ensuring that private standards adequately protect the privacy of information gathered for the purposes of e-commerce. In addition, Mac Neil acknowledges the continued state involvement in multilateral regulation of the Internet. He maintains that in order to address problems of both domestic and international regulation, governments in Canada continue to use legal intervention to provide rules that govern voluntary standard-setting and enforcement. The state thus continues to govern, though at a distance.

Finally, the two chapters in Part V, 'Historical and Cultural Perspectives on Regulation and Governance', consider what lessons can be learned from historical and cultural forms of regulation. Chapter 14 explores the role of law and regulation in the battle against epidemic diseases in the eighteenth and nineteenth centuries and asks why the law of quarantine has remained relatively static since the seventeenth century. In the context of recent controversies around the determination of Mohawk citizenship, Chapter 15 looks at the issue of the replacement of state legal intervention by forms of local intervention as part of a move towards more autonomous rule in the community of Kahnawake.

Logan Atkinson, in Chapter 14, challenges the notion that the diffusion of regulatory power associated with the 'crisis of the state' is unique to the late twentieth and early twenty-first centuries. Instead, he suggests that, in the area of public health protection at least, state authorities have always been constrained in their attempt to enact and enforce regulation that meets policy objectives around the health of the citizenry. Through two historical case studies, one from Great Britain in the 1720s and the other from Upper Canada in the 1830s, Atkinson advances the position that the needs of those concerned with immigration and interstate trade have consistently usurped the regulatory power of the state to control the importation of dangerous goods. In effect, this chapter asks us to reconsider the schema that neatly traces the development of the modern state through laissez-faire, interventionist, and 'globalized' forms by reflecting on the historical complexity of the exercise of regulatory power.

Jane Dickson-Gilmore, in Chapter 15, reveals the ambiguity associated with legal interventionism. Referring to the paternalistic intervention associated with the Indian Act, she questions the assumption that the purpose of the interventionist state was to provide social justice to marginalized peoples. Dickson-Gilmore points out that this goal is particularly problematic when the state's notion of justice is not shared by the recipients of that legal intervention. She shows how resistance to intervention has provided part of the justification for the Mohawk community of Kahnawake to negotiate an agreement with the government of Canada to remove the Indian Act and allow a limited form of self-government. However, focusing on the community control of membership through the determination of Mohawk blood, Dickson-Gilmore questions whether local autonomy will necessarily lead to less interventionism from the outside and suggests that it may lead to more interventionism from within the community. More specifically, she argues that any agreement for reordering the relationship between the Canadian state and Mohawk government presents a strong potential for a proliferation of regulation and potentially more intrusion in the lives of individual Mohawks.

Conclusion

If we are to look for a common theme in these explorations of the role of law and the state in this contemporary period of rapid change, at an abstract level it is in the idea of a decentring of the state and regulation. Thus, we cannot and should not look for the principal location for regulatory impulses in the state alone. Increasingly, we must also locate such impulses in the codes of conduct of corporations or industrial associations, in the activities of domestic and transnational non-governmental organizations, and in the rules and practices of supranational organizations such as the European Union and international organizations such as the World Trade Organization. And yet the state cannot be ignored. What is also shared by the various theories of crisis and transformation, as well as by most of the authors in this book, is an acknowledgement that law, regulation, and governance are inherently relational activities. They are constructed primarily in the relationship between the state and the rest of society. As such, they confirm the project of legal studies, which insists that law is found not only in state laws and regulations and in court decisions but also in the practices of various social groups and institutions and in their relationship with a state that is not in decline but is being continually transformed.

Notes

1. The idea of the ease of movement of people across borders must be qualified. Despite recent waves of immigration to North America and Europe, Western nations, including Canada, have been selective in the areas of the world from which they allow immigration and grant refugee status. Restrictions have been placed on entry, immigration, and refugees from various nations in Africa, the Middle East, and Asia. Canada has maintained a policy of incarceration and deportation of illegal migrants, as illustrated by its handling of boatloads of Chinese migrants in British Columbia. In light of American pressure as a result of the terrorist attacks on the US, entry and immigration to Canada is likely to become even more restrictive.
2. J.P. McCormick, 'Max Weber and Jürgen Habermas: The Sociology and Philosophy of Law During Crises of the State', *Yale Journal of Law and the Humanities* 9 (1997): 298–9.
3. Ibid., 328.
4. T.H. Marshall and Tom Bottomore, *Citizenship and Social Class* (London: Pluto Press, 1991). Despite this emphasis on the promotion of social equality, the interventionist state does not only act benignly towards the less fortunate members of society. The promotion of Marshall's 'right to a modicum of welfare and security' is necessary to the post-war welfare-state compromise that maintains the stability of capitalism, and the regulatory state also uses legal intervention to control and coerce Aboriginal peoples, women, trade unions, and other groups marginalized from social, political, and economic decision-making.
5. For an overview of the neo-liberal critique of the welfare state, see W. Kymlicka and W. Norman, 'Return of the Citizen', in R. Beiner, ed., *Theorizing Citizenship* (Binghamton, NY: SUNY Press, 1995), 303–7.
6. See Chapter 2 in this volume, where Trevor Purvis suggests that even deregulation as promoted by neo-liberals should be regarded as a form of regulation to the extent that it is intended to promote specific practices through voluntary initiatives and self-regulation.

7. For the concept of the 'jurisprudence of crisis', see A.J. Jacobson and B. Schlink, *Weimar: A Jurisprudence of Crisis* (Berkeley: University of California Press, 2000).
8. M. Weber, *Economy and Society: An Outline of Interpretive Sociology*, G. Roth and C. Wittich, eds (Berkeley: University of California Press, 1978), vol. 2, 641.
9. See McCormick, 'Max Weber and Jürgen Habermas', 302.
10. A. Kalyvas, 'Carl Schmitt and Modern Law' (review of William Scheuerman, *Carl Schmitt, The End of Law*), *Telos* 116 (1999): 154.
11. Weber, *Economy and* Society, vol. 2, 880–9.
12. See H. Heller, 'The Essence and Structure of the State', in Jacobson and Schlink, *Weimar*, 265–79.
13. See the essays by Kirchheimer and Neumann in O. Kirchheimer and F. Neumann, *Social Democracy and the Rule of Law*, ed. Keith Tribe, trans. Leena Tanner and Keith Tribe (Boston: Allen & Unwin, 1987).
14. B. Fried, *The Progressive Assault on Laissez Faire: Robert Hale and the First Law and Economics Movement* (Cambridge, Mass.: Harvard University Press, 1998), ix.
15. R.L. Hale, 'Law making by Unofficial Minorities', *Columbia Law Review* (1920): 456.
16. R.L. Hale, 'The Constitution and the Price System: Some Reflections on *Nebia v. New York*', *Columbia Law Review* 34 (1934): 402.
17. R.L. Hale, Hale Papers, as quoted in N. Duxbury, 'Robert Hale and the Economy of Legal Force', *Modern Law Review* 53 (1990): 435.
18. M. Foucault, 'Governmentality', in C. Gordon, ed., *The Foucault Effect* (Hemel Hempstead: Harvester Wheatsheaf, 1991). This line of inquiry is also pursued in more recent discussions of a concept of 'governance'.
19. R.L. Hale, 'Bargaining, Duress, and Economic Liberty', *Columbia Law Review* 43 (1943): 606.
20. Duxbury, 'Robert Hale and the Economy of Legal Force', 436.
21. See Kymlicka and Norman, 'Return of the Citizen', 301.
22. See J. Habermas, *Theory of Communicative Rationality*, vol. 2 (Boston: Beacon Press, 1987), 357–63.
23. McCormick, 'Max Weber and Jürgen Habermas', 299.
24. Ibid.
25. Julia Black, 'Proceduralizing Regulation: Part I', *Oxford Journal of Legal Studies* 20 (2000): 603.
26. Ida Koppen and Karl Heinz Ladeur, 'Environmental Rights', in A. Cassese, A. Clapham, and J. Weiler, eds, *Human Rights and the European Community: The Substantive Law* (Baden-Baden: Nomos Verlagsgesellschaft, 1991), 43.
27. Gunther Teubner, 'Substantive and Reflexive Elements in Modern Law', *Law and Society Review* (1983): 263.
28. E. Orts, 'Reflexive Environmental Law', *Northwestern University Law Review* 89 (1995): 1254.
29. Black, 'Proceduralizing Regulation', 607.
30. Ibid., 598.
31. The most important theorist of 'thick proceduralism' is undoubtedly Jürgen Habermas, who contends that proceduralism has become a new paradigm of law replacing the 'material law' paradigm of the welfare state. See, generally, J. Habermas, *Between Facts and Norms: Contributions to Discourse Theory of Law and Democracy*, trans. W. Rehg (Cambridge, Mass.: MIT Press, 1996); and, more specifi-

cally, J. Habermas, 'Paradigms of Law', *Cardozo Law Review* (1996): 771–84. For an interpretation of the procedural theory of law that emphasizes the value of epistemic communities, see J. Cohen and C. Sabel, 'Directly-Deliberative Polyarchy', *European Law Journal 3* (1997): 313.
32. See J. Black, 'Proceduralizing Regulation: Part II', *Oxford Journal of Legal Studies* 21 (2001): 33–58; also, Chapter 6 in this volume.
33. Foucault, 'Governmentality'.
34. C. Gordon, 'Governmental Rationality: An Introduction', in Gordon, ed., *The Foucault Effect: Studies in Governmentality* (Hemel Hempstead: Harvester Wheatsheaf, 1991), 3.
35. Ibid.
36. Ibid., 4.
37. M. Dean, *Governmentality: Power and Rule in Modern Society* (London: Sage Publications, 1999).
38. N. Rose, *Powers of Freedom: Reframing Political Thought* (Cambridge: Cambridge University Press, 1999).
39. A. Hunt and G. Wickham, *Foucault and Law: Towards a Sociology of Law as Governance* (London: Pluto Press, 1994).
40. Ibid., 78; Rose, *Powers of Freedom*, 17.
41. Hunt and Wickham, *Foucault and Law*, 112.
42. My interpretation of the 'decentring' of the state draws on similar analysis by Black, 'Proceduralizing Regulation: Part I', 598–602.
43. As will become clearer below, 'decentring' does not imply an abandoning of the state as an appropriate locus of regulation but rather a recognition that governance/regulation does not simply take place in the state or through public law alone. Accordingly, the activities or practices associated with governance/regulation should not imply a place from which social unity can be constructed.
44. See Heller, 'The Essence and Structure of the State'.
45. C. Offe, *Modernity and the State: East, West* (Cambridge, Mass.: MIT Press, 1996), 63.
46. See Kymlicka and Norman, 'Return of the Citizen'. Kymlicka and Norman argue that multicultural states such as Canada must accommodate such demands for difference through 'differentiated rights' or 'special rights' that attach to individuals by virtue of their membership in certain 'disadvantaged' or 'excluded' groups. They argue that such accommodation can be justified within the confines of liberal political theory. Unfortunately, their critique of T.H. Marshall's classic conception of modern citizenship does not acknowledge the significance of the aspects of power and struggle that are involved in recognition. As a result, only after cultural groups have struggled for years has some degree of recognition been achieved. Cultural groups such as members of First Nations have only recently begun to have their demands heard. One could argue that this has only occurred after armed standoffs, such as at Oka in the last decade.
47. Ibid., 310–13.
48. The reference to the state as a unified entity refers to the concept of the modern state as a sovereign centre from which the political boundaries of society can be determined in relation to other sovereign communities and to non-political associations and individuals within that society. At the level of the empirical existence of

actual states, it can be argued that the Canadian state never has been such a unified entity both because of the contentious character of federal-provincial relations and because of historical assertions of Quebec nationalism and the more recent assertions of identity and of nation status by First Nations.
49. See W. Magnusson and R. Walker, 'De-Centring the State: Political Theory and Canadian Political Economy', *Studies in Political Economy* 26 (1988).
50. See especially the work of David Held and his colleagues on the politics of globalization. D. Held, *Democracy and the Global Order: From the Modern State to Cosmopolitan Governance* (Stanford, Calif.: Stanford University Press, 1996); D. Held et al., *Global Transformations: Politics, Economics and Culture* (Stanford, Calif.: Stanford University Press, 1999).
51. Despite this acknowledgement of the dispersal of power throughout society, the aim of the genealogical method that informs the sociology of governance is to 'destabilize and denaturalize the present' to aid in a more democratic shaping of knowledge and the contestation of authority and the practices of governance by those who are governed. Rose, *Powers of Freedom*, 282.
52. See Habermas, 'Paradigms of Law'.
53. The work of Charles Sabel and various colleagues represents one of the most significant bodies of work that investigates the application of forms of 'thick proceduralism' to practical decision-making processes. See Cohen and Sabel, 'Directly-Deliberative Polyarchy'; C. Sabel, A. Fung, and B. Karkainen, 'Beyond Backyard Environmentalism', *Boston Review* 24 (1999).
54. Black, 'Proceduralizing Regulation: Part I', 598.

PART I

Theoretical Perspectives

Chapter 2

Regulation, Governance, and the State: Reflections on the Transformation of Regulatory Practices in Late-Modern Liberal Democracies

TREVOR PURVIS

We live in a time of sweeping social, political, and economic transformation. While the twentieth century saw enormous changes on each of these fronts, there seems something qualitatively different about the past few decades. One of the most poignant characteristics distinguishing this period from the preceding one is the dramatic shift in thinking with respect to the appropriate place of states in the organization and regulation of life in the advanced capitalist societies of the West. In the heady days of the post-war era there had been a pervasive sense that states could harness the forces of politics, economy, and society, providing the ultimate point of reference for political action, mechanisms for the guidance of economic development, and assurances of the quality of life of their citizens. The optimism that drove this state-centric vision of development has all but vanished today. Indeed, the past few decades have seen an almost complete reversal on these issues. Today, themes of regulatory crisis, privatization, and deregulation have combined with those of globalization, local governance, and subsidiarity, dramatically reshaping the nature of debate over the state's role in the economy and society. At the same time, and relatedly, the provision of welfare services has moved markedly away from the universalist pretensions that had dominated welfare state expansion in the post-war years and towards a residual model aimed at providing bare subsistence services for those unable to fend for themselves. The state *appears*, in some important respects, to be very much in retreat.

The central aim of this chapter is to address critically a rather common perception that states are becoming increasingly impotent in the face of the sweeping social, political, and economic changes currently transforming life in the advanced capitalist societies of the West. I am particularly concerned to press our thinking about processes of social regulation and their relationship to the state and law beyond conventional conceptions of regulation as functional responses to market-driven imperatives or notions of 'social control'. Instead, I will cast contemporary transformations in regulatory activities against the backdrop of a diffusion of regulatory practices and their intersection with various modes of governance in liberal democracies.

The apparent decline of the nation-state in recent years has important implications for any effort to understand contemporary social regulatory practices and their transformation. To the extent that regulating is an important dimension of 'what states do', one of the features that points to the apparent impotence of the state is a perception that states can no longer effectively regulate economy and society in an effort to shape social and economic development. This chapter stresses the persistence and pertinence of states as crucial actors in the social, political, and economic transformations currently underway in the international political economy. This being said, I also want to stress the importance of recognizing the shifting role and nature of the state as it confronts these transformations. States are, in a very palpable sense, being 'decentred' in the face of sweeping social and economic changes. Any adequate discussion of the state, theoretical or practical, must endeavour to locate the state in the context of an array of new modalities of social governance. States, in short, are not disappearing, but their place in the grand regulatory scheme of things economic, political, and social is being modified dramatically.

The Regulatory State: Economic and Social

Any effort to address the transformation or reform of regulation and regulatory practices in contemporary liberal democracies immediately confronts a certain conceptual roadblock. There is a burgeoning literature on issues related to the role, nature, and likely future of the so-called 'regulatory state'. By and large, however, these discussions take a narrow view of just what sorts of regulatory activities and actors are pertinent to a discussion of the future of that state.[1] The principal focus of this literature is 'regulatory agencies' and 'regulatory law', that variable group of state institutions born of statutory enablement, whose *raison d'être* is the strategic deployment of a possible range of policy instruments aimed at intervention in narrowly conceived market sectors (e.g., financial, telecommunications, transportation, and utilities markets). In legal theory, political science, policy studies, and the sociology of law whole subdisciplines now are dedicated to analyzing the nature and limitations of such agencies' efforts at shaping and directing market development. Unfortunately, the narrowness of this focus keeps most of this discussion from addressing the broader regulatory nexus within which such efforts take place. Such accounts largely fail to, or refuse to, address the role of the state as a regulator, not only of economic life narrowly defined, but of social and political life more generally.[2]

While a narrow focus on economic regulation may provide practical insights for purposes of policy-making and macroeconomic analyses, outside of descriptive accounts of institutional change such an approach is unlikely to shed much light on the broader transformations currently reshaping the relationship between state and society in Western liberal democracies. It is precisely against the backdrop of the broader complex of regulatory practices (of which economic regulation is simply one aspect) that an adequate appreciation of the transformations in the role and nature of the contemporary state might be found. The apparent 'withdrawal' or 'retreat' of the state from a wide array of fields of regulation is perhaps better understood in terms of a rearticulation of the elements of regulation. The role and nature of the 'regulatory state' and its current transformations can be better appreciated against this broader backdrop, examining not only the state's role in regulating the economy but social relations more broadly speaking. Both the post-war and current transformations of the state have involved the (re)articulation of state regulatory endeavours vis-à-vis other sources and modes of regulation and governance in all spheres of social life.

Of course, the state's role in the regulation of other aspects of social life has not been wholly ignored. The lacunae of conventional regulatory studies are partially overcome by another vast literature that looks to the role of the state in broader processes of 'social control', a field of study typically the province of sociology, criminology, and anthropology.[3] These disciplines have offered numerous, often competing, and frequently very sophisticated accounts of the state as an agent of social control. It is not my intention to delve into this literature here, but rather to enter a few caveats regarding the relationship between these two seemingly distinct fields of study.

First, it is important that, in our effort to gain a clearer sense of the myriad regulatory forms and practices pervading contemporary societies, we abjure a retreat to simple accounts of social control that see the gradual colonization by the state of more and more of the tasks, processes, agencies, and sites through which control is exercised. If the dominant approaches to regulation and the regulatory state have failed to take account of the state's role as a regulator of social life more generally, a state-centric and teleological notion of social control that sees the state gradually assuming all aspects of the direction, ordering, and disciplining of society seems similarly restricted in its ability to provide a more balanced perspective on contemporary modes of societal regulation. The simple Orwellian vision conjured up by such formulations reduces the state to a vehicle for the ideological inculcation and/or coerced pacification of populations. Through this lens, the state's role as 'regulator' is reduced to the instrumental and functional reproduction of dominant relations of class, race, gender, sexuality, etc. There is little space in such an account for either viable resistance or meaningful agency.

So, too, must we reject accounts of social control that see the state's broader role in the regulation of social life as standing in a zero-sum relationship with 'extra-state' agencies and institutions. It is far too simple to see the apparent withdrawal of the state from, for instance, the field of welfare service provision as entailing a total abandonment of these regulatory fields to the 'private' sphere. Certainly the consolidation of neo-liberal hegemony in Western liberal democracies has ushered in a withdrawal from direct intervention in these fields, but the relationship is much more complex than one of abandonment or wholesale retreat. Clearly, a new relationship is emerging between the regulatory state and the objects of its social-regulatory efforts, but the state remains importantly implicated in the outcomes of these transformations. The techniques and practices deployed in pursuit of particular regulatory objectives are shifting, but the state frequently remains implicated in the broader regulatory scheme of things, now, however, 'governing from a distance' through a dispersion of expertise.

Thus, with its rather myopic focus on markets and private and public economic actors, the conventional view fails to elucidate the myriad techniques and practices that constitute the very conditions of existence of markets. If we accept this view we lose sight of the intrusion of regulatory practices in multiple sites that buttress a broader mode of governmental rationality. If we take a state-centred perspective on social control, we seem bound to look at state-based regulation as either a mechanical replicator of relations of domination/subordination, or as a zero-sum game in which the apparent withdrawal of the state is envisaged as a total abandonment of the field of regulation to the anarchy of a heterogeneous social realm. Either way we lose sight of the complex relationship that actually prevails between states and a burgeoning array of para-state agencies of social and economic regulation.

We need, somehow, to generate a perspective on regulation that encompasses the problems and issues identified in each of these perspectives and at the same time transcends their limitations. Each has important lessons to teach the other, but neither alone, nor in combination with the other, is adequate to a theoretical and analytical project aimed at understanding the transformations in regulatory forms currently at hand. Regulating economic life is something that has always gone hand in hand with the regulation, control, and governance of social life, broadly speaking. Economic regulation has always been conditioned by broader economies of moral, ethical, and political discourses and practices. Social regulation, by comparison, has also generally been much more than simply an instrumental and/or functional response to the exigencies of prevailing relations of domination/subordination. That is, social regulation has always occurred in specific contexts and has always stood in a constitutive relationship with economic, moral, ethical, and political discursive and practical contingencies.

The foundations for precisely such an approach can be located in a recent literature exploring the notion of 'governmentality', a conceptual innovation introduced in the later work of Michel Foucault.[4] Governmentality focuses our attention on governmental rationalities, highlighting the complex array of technologies and practices of rule, activated in and through plural, dispersed sites, through which individuals (both corporeal and corporate) 'place themselves under the control, guidance, sway and mastery of others, or seek to place other actors, organizations, entities or events under their own sway'.[5]

One of the strengths of this approach is that it moves us away from a conception of politics, political power, and government centred in the institutions of the state, instead focusing our attention on the exercise of political rule beyond the state. This being said, it is important to stress that this decentred notion of politics does not deny the existence, or even the importance, of the state. On the contrary. But it does force us to reconceive projects of government, cast in terms of the multiple forms and means whereby rule is effected, in terms other than the state, enabling us to link, for instance, the political project of producing healthy workers to the techniques and practices of social workers, doctors, and family counsellors. None of these actors has an immediate connection to the state. Indeed, the point is that we need not ultimately link these projects to the state. Instead, the state appears as simply one site, or a constellation of sites, tactics, practices, etc., among others engaged in governing the behaviour of others. It shifts our attention from the state to the techniques of rule exercised in a plurality of sites, linking what Foucault called the 'microphysics of power' to broader rationalities of government.

The Historical Emergence of Social and Economic Regulation

One might be forgiven if, having tackled the conventional literature on the regulatory state, he or she arrived at the conclusion that state regulation of the economy is something of the very recent past. Most contributors to this literature seem to date the emergence of the regulatory state to roughly the second half of the nineteenth century. In actuality, however, the roots of the state as regulator of economy and society are thoroughly entwined with those of the state itself.

The advanced capitalist states of the West have emerged out of a centuries-old set of social developments. But while each state has developed differently, with the specific shape of its institutions and practices determined by particular historical forces (place in the world economy, wars, internal economic development, specific national socio-cultural and political histories, etc.), there were nevertheless broad points of similarity in the nature of that development from case to case. The imperative to regulate has constituted an important example of such similarities. On the one hand, the forerunners of the modern state were implicated in the regulation and ordering of numerous aspects of economic life. Indeed, as Majone has recently suggested, 'Regulation, in the sense of rules issued for the purpose of controlling the manner in which private and public enterprises conduct their operations, is of course as old as government.'[6] Some of the earliest of these efforts saw pre-modern state and quasi-state forms engaged in the production and regulation of currencies, the implementation of restrictions on market concentration,[7] the administration and collection of taxes and tariffs, the establishment of weights and measures, and the planning and construction of various features of social infrastructure, including military fortifications, roads, and waterways. To these functions, the advent of the modern state brought the gradual consolidation and concentration of policing and military power in state institutions, a move that secured both the conditions of property and contract while providing a geopolitical framework within which market relations might proceed. In important respects, each of these was directly or indirectly related to the state's gradual monopolization of legal order, which provided an increasingly formal-rational framework within which economic enterprise might operate and flourish, a framework that facilitated the purpose-rational orientation of life in emergent capitalist societies.[8]

On the other hand, states and their predecessors have, for centuries, been involved in the regulation of a wide array of social relations irreducible to the economy, issuing and enforcing rules to control how individuals and groups conduct themselves. Some of these have obvious direct links to economic life. Thus, states have been actively engaged in creating and ordering relations of social class through such activities as policing, disciplining, and incarcerating indigent populations, promulgating and enforcing sumptuary laws, and gradually developing legislation aimed at regulating aspects of working life. But they have also been significantly implicated in the constitution of social relations irreducible to the economy, such as relations of culture, gender, sexuality, race, and nation. And while the actual practices employed on each of these fronts often varied dramatically from case to case (depending, for instance, on a state's relation to competing sources of moral discourse and its place in the colonial scheme of things), states have historically been implicated in the constitution and ordering of a plurality of social, political, and economic relations and identities.

It is important to stress this role of the state in the regulation of social relations beyond a narrowly conceived view of the economy, for it was precisely its simultaneous gradual intrusion into various economic *and* extra-economic social relations that paved the way for the emergence of the regulatory state as that term is conventionally understood. An understanding of the complex relationship between the state and the full panoply of processes of social regulation requires that we move beyond a narrow view of regulation as the state's role in, and impact on, the organization of economic life, to a view of the direct and indirect role the state has played in the organization of populations, the constitution of productive citizens, and the promulgation and privileging of particular conceptions of moral life. The foregoing suggests an image of an emerging state that entails two simultaneous and inextricably related regulatory thrusts. On the one hand, the state became actively engaged in the organization of certain aspects of the economy and involved in the production of social infrastructure, activities that bear directly upon economic relations in various ways, not the least of which is that they create some of the essential material conditions for the conduct of commerce. On the other hand, the state was increasingly involved in policing its territorial boundaries, policing and monitoring its domestic population, and forming productive citizens. While none of these functions are specific to capitalist state forms, each underwent gradual intensification and centralization with the rise of capitalist social relations, an intensification deepened by the onset of industrialism and the related dramatic shift from rural to urban life as an emerging working class sought out employment in the new industrial centres of Europe and North America.

From Discovery of 'the Social' to the Keynesian Welfare State

If states had long been implicated in various dimensions of social and economic regulation, the sweeping social transformations wrought by economic, technological, political, and social change in the nineteenth century saw the convergence of an array of social forces that provided an emerging capitalist state solid foundations for a significant expansion of its regulatory intervention in economy and society. Of crucial importance to this consolidation was the emergence of 'the social' as a distinct object of governance.[9] This era ushered in new techniques of social regulation, embedded in the knowledges produced by political economy, the newly consolidated medical sciences, and the rise of the social sciences (particularly sociology and criminology), along with their attendant methods of analysis, perhaps the most important of which was the development of social statistics. The development of these scientific discourses 'produced a new subject amenable to regulation, namely populations.'[10] These new scientific discourses and their attendant techniques and practices created the possibility not only of monitoring, enumerating, surveying, and analyzing social development, but also of regulating, guiding, shaping, mastering, and exerting authority over the social, thus raising for the first time the possibility of broad-based efforts at 'social engineering'. Together, and bolstered by their frequent articulation via juridical and quasi-juridical discourse, they came to constitute a new rationality of government.

It is no simple coincidence that the standard periodizations of the 'regulatory state' and the 'welfare state' bear such striking similarity. The emergence and consolidation of each are closely linked to this emergence of 'the social' as an object amenable to governance. The appearance of regulatory agencies (staffed by bureaucrats, administrators, and industry experts) whose task of monitoring and directing aspects of market development and workplace conditions paralleled that of new forms of expertise (social workers, philanthropists, doctors, psychologists, etc.) geared towards the production of effective workers, healthy bodies, hygienic homes, and functional families. On both fronts liberal states were relatively measured and gradual in terms of their willingness to engage in direct intervention in these fields.[11] Nevertheless, the last decades of the nineteenth century and the first decades of the twentieth saw the growth of state orchestration of, and increasing direct involvement in, the co-ordination and organization of all facets of society.

If slow off the mark, the rate and scope of state intrusion into economic matters and the proliferation of special agencies to direct specific features and/or sectors of market activity underwent a dramatic intensification in the years following World War II. A recent history of dramatic cyclical fluctuations in economic growth/stagnation in the West posed serious challenges to the stability of liberal democratic regimes. Political discontent with the unfettered market had been stirred in Europe through the great deflation of 1873-97. The Great Depression that preceded World War II had seen massive economic dislocation and posed the very real threat of political strife in both Europe and North America. Recent direct state involvement in the direction of economic output, which had been driven by pursuit of the war effort in two world wars, combined with these other factors to lay the groundwork for both popular and intellectual support for a new measure of state interventionism. The unprecedented economic growth, technological innovation, and urbanization that dominated the post-war period ushered in a new era of optimism regarding the capacities of states to shape the developmental trajectories of their respective societies. The conditions were ripe for the consolidation of a new rationality of government: welfarist liberalism.

The expansion of the post-war interventionist state advanced on both economic and social fronts simultaneously. First, there was a significant general consensus in the liberal democracies of the West that states had an important role to play in shaping the rate and direction of economic development, although the ways this was to be achieved would vary quite significantly from one country to another. This post-war era saw a wide variety of instruments and techniques deployed in varying combinations aimed at steering economic development. At one end of the spectrum many states got involved in direct forms of economic intervention by providing particular goods and services, either through state-run monopolies in some industries or as competitors with private enterprise in others. Another form of direct intervention was the use of regulatory and monetary inducements to influence the establishment and/or location of economic enterprises. Somewhat less direct were regulatory restrictions on entry into industries, controls on competition in particular fields, fiscal transfers between levels of government, tariff barriers to bolster the development prospects of particular domestic industries, and quality and output regulations in other industries. Finally, many states used various indirect instruments to promote specific forms of economic development, such as the strategic allocation of government contracts to promote economic development opportunities and the location of military installations as a mechanism to provide infusions of capital to less prosperous regions. These instruments are what are generally assumed in conventional discussions of 'regulatory law' or the 'regulatory state'.

This same era, however, witnessed a dramatic increase in direct state intervention aimed at regulating the welfare of its citizenry. An array of agencies developed to ameliorate economic disparities through income transfers and the provision of particular social services. The thinking behind these interventions was twofold. On the one hand, through the strategic use of particular policy instruments (i.e., unemployment insurance schemes, welfare programs, job retraining) states could help to offset some of the effects of cyclical economic downturns by transferring income to those most directly affected by periods of general or sectoral economic decline. On the other hand, through the provision of particular social services (i.e., health, education, workers' compensation, pensions, relief for single parents) states aimed to engender greater socio-economic equality and provide a buffer against the economic challenges an individual might encounter throughout life. As individuals move through their lives their capacities for economic self-maintenance vary, from the dependencies of childhood, through potential economic insecurities arising from responsibilities for dependants, job loss, or the possibility of injury and/or debilitating illness, to the declining capacities for economic productivity that frequently attend old age. Welfare states aimed to mitigate these contingencies in an effort to stabilize economic growth and provide the legitimatory foundations for continued general mass support from their democratic constituencies.

The actual mix of these instruments deployed in any given national instance varied, often dramatically. Some states, for instance those in the Scandinavian countries, tended to take a much more aggressive and direct role in their respective economies, promoting specific industrial sectors by legislating state monopolies in particular industries, government competition with the market in others, and extensive government investment in yet others. These states also developed the most extensive universal welfare state schemes, with the objective of creating greater economic equality across their societies. At the other end of the spectrum stood countries like Canada and the US, in which much more indirect forms of intervention were the norm, with welfare state development targeted more at those who 'fell through the cracks'. But despite the wide variety of configurations of policy instruments from country to country there was nevertheless a common point of influence underlying this general approach to states' efforts to steer the trajectory of macroeconomic development. That common point was the widespread influence of Keynesian macroeconomic theory, which suggested that dramatic fluctuations in national economic performance, particularly fluctuations in employment levels, could be mitigated by stabilizing economic demand through a judicious combination of redistributive policies and government management of markets aimed at securing conditions conducive to full employment in the economy. This consensus began to break down through the 1970s when, contrary to theoretical expectations, unemployment and inflation began to rise simultaneously. Keynesianism and its interventionist prescriptions began to fall out of favour.

The intrusive post-war state, which had such an important role in regulating and directing economic life in Western liberal democracies, was inextricably related to the hegemony of Keynesian-inspired macroeconomic policy in virtually all of these states. It is no mere coincidence that the interventionist regulatory state associated with Keynesianism is frequently referred to as the 'Keynesian welfare state'. Regulatory efforts aimed at guaranteeing the vitality of various economic enterprises went hand in hand with those aimed at securing the social security of a labour force and the conditions whereby that security would be guaranteed. But the most marked feature of the strategies of rule that dominated this phase of liberalism was the proliferation of expertise. The distinctive mechanism by which government of 'the social' would proceed under the Keynesian welfare state 'was through an advancing governance by experts orchestrated under the umbrella provided by the agencies of the welfare state.'[12] The crisis of Keynesianism brought with it a crisis of this mode of governance. The extended reach of the state into the social body was to come under critical fire in the face of a concerted challenge from neo-liberalism, an attack buttressed by the mounting fiscal and legitimation crises of the welfare state.

Neo-Liberalism: Deregulation, Privatization, and the Recalcitrant State

There are innumerable critiques of the regulatory state, but undoubtedly the most influential in recent years has been that advanced by neo-liberal thinkers. Since the late 1970s most Western liberal democracies have witnessed a significant shift in the dominant ways of thinking about the relationship between the state, the economy, and social and political life more generally. Neo-liberal thinking has enjoyed a significant measure of popular legitimacy, and while its analyses were introduced to the realm of public policy by conservative regimes in the 1970s and 1980s, its tenets have come to dominate the macroeconomic policies of most liberal democratic governments today.

These critics suggest that the regulatory state is in retreat today as a result of its dramatic 'overextension' in the post-war era.[13] On the one hand, they regard the state's efforts to guide the general trajectory of national economic development through extensive exercises in direct and indirect forms of regulation as having been fundamentally misguided, leading inevitably to regulatory crises and failures. One frequently cited criticism in this vein highlights the fact that regulatory agencies are frequently 'captured' or 'co-opted' by the industries they seek to regulate, thus calling into question their capacity to advance the 'public interest'. Another is that states' efforts to guide national economic development frequently have unforeseen negative consequences because states are incapable of effectively gaining and managing the complex and comprehensive knowledge that would make such efforts feasible. Societies are, by their very nature, so complex as to render the desired outcomes unlikely, if not impossible. The result of such imperfect knowledge is imperfect policy-making, which can potentially lead to dramatic misallocations of social and economic resources. Still another focuses on the lack of accountability in much government bureaucracy and the tendency for bureaucrats to engage in exercises of 'empire-building'. Rather than attempting to advance the public interest, successful efforts at such empire-building threaten to concentrate scarce resources in suboptimal ways. Finally, it is argued that liberal democratic states, in their efforts to meet the many demands arising from a vast plurality of social and political interests, have become 'overloaded'. Confronted by a burgeoning array of overlapping and frequently contradictory interests, states that try to respond actively to each of these are eventually bound to crisis and failure. To the extent that the state might remain an actor in economic relations, neo-liberal critics generally see its role as a minimal one of providing guarantees for the security and protection of property and contracts.

On the other hand, neo-liberal critics argue that the massive expansion of welfare state services in the post-war era has led to a number of problems. One of the dominant criticisms is that the welfare state tends to undermine the natural capacities of markets to distribute social wealth optimally. State interference in the distribution of wealth and services is seen to be a disincentive to investment and productivity. An excessive tax burden, it is suggested, tends to sap resources that would otherwise be available for investment and the expansion of capital. Another criticism focuses on the tendency for welfare state provision of services to undermine the incentives for individual self-reliance that market-driven provision of these services would engender. State-sponsored welfare programs, such as those in the areas of health care, education, employment security, and pensions, are seen to create a disincentive for individuals to look after these affairs themselves. Were they to do so, it is suggested, something approaching an optimal distribution and provision of such services might be realized. Still another criticism highlights the impersonal service yielded by cumbersome bureaucratic structures that tend either to be unattentive to clients' needs or to intrude unnecessarily into their lives. Finally, it is frequently suggested, state provision of these services has led to gross overexpenditure on social services, which led to a fiscal crisis of the state. Profligate spending on such services removed from the self-regulating tendencies of markets, it is claimed, causes inefficiencies that stifle economic growth and are bound to rack up huge deficits. To the extent that a welfare state is necessary, it is concluded, it should be a residual or minimal welfare state shorn of the excesses of its past and driven by significantly humbler objectives.[14]

To all of these criticisms (and others) the neo-liberal solution is a dramatic reorganization of the relationship between state and society. But we would do well at this point to engage in a process of separating the clichés of New Right rhetoric from more theoretically informed approaches to the supposed crisis of the post-war regulatory state. In the case of the former there is a distinct sense that necessity dictates withdrawal of the state from most areas in which it has been an active intervenor in the post-war years. A rhetoric of laissez-faire, translated in this instance to imply an absence of government, urges abandoning the hubris of regulation for the self-correcting mechanisms of free markets. Extensive retreat of the state is seen to be the only appropriate antidote for a past of mismanagement and over-management. Markets, it is suggested, will naturally tend to an equilibrium that will ensure a maximal allocation of resources and wealth. Only markets, under these formulations, have the built-in capacity to reach such an equilibrium. Government interference in these processes is merely bound to produce suboptimal, and possibly disastrous, results. The prescription: deregulate those areas in which government has made an active effort to guide economic development, and pare down the welfare state to an extent that it poses minimal interference to market imperatives. The apparent demise of the regulatory state is seen, from this vantage point, to be an unqualified good thing.

But, as Rottleuthner has suggested (albeit in a somewhat different context), 'The clichés of the political arena—as far as the critique of the welfare state is concerned—contribute to the acceptance of the theory of a regulatory crisis, not its systematic empirical truth.'[15] Indeed, despite much rhetoric to the contrary, neo-liberal theory works a significantly different seam. The vision it proposes of this new relationship is not one of the minimal state of classical laissez-faire liberalism, in which laissez-faire connoted a distinctly passive resignation to the superior capacities of natural markets to secure the greatest common good. Thus, while the discourse of laissez-faire abounds in the theoretical discourses of neo-liberalism, gone is the passivity of the classical formulations. Markets are no longer seen as spontaneous natural entities fettered by the unnatural forces of state regulation, but rather as a field in which the state plays a crucial and active role in creating conditions conducive to market development and freedom of property. And a crucial dimension of this shift is a reconstitution of classical liberalism's *Homo economicus*, the wholly unfettered individual free of all state intrusions. In 'his' stead arises a new *Homo economicus*, the individual who must be shaped, manipulated, regulated, in short, governed, while given the opportunity, through the medium of choice, to engage as a reflexive rational actor in a world infused with the logic of enterprise culture.[16] As Gordon suggests:

> *Laissez-faire* is a way of acting, as well as a way of not acting. It implies, in Foucault's words, an injunction 'not to impede the course of things, but to ensure the play of natural and necessary modes of regulation, to make regulations which permit natural regulation to operate': 'manipuler, susciter, faciliter, laissez-faire'. The permissive meaning of *laissez-faire* needs to be understood in an activist, enabling sense no less than in its character of passive abstentionism.[17]

The state remains crucially implicated in these processes, as facilitator and enabler. As Alan Hunt and I have suggested elsewhere, 'we see it as better to understand the role of the state within neo-liberalism as no longer being the conductor of the orchestra, but rather as providing the conditions for the exercise of rule by many other institutions and agents that operate by inciting and stimulating the active choices of enterprising individuals.'[18]

With this reworking of the fundamentals of classical liberalism, neo-liberalism has brought the extension of market rationality—the techniques, practices, and principles of enterprise—into virtually all spheres of life. Its objective is to further 'the game of enterprise as a pervasive style of conduct, diffusing the enterprise-form throughout the social fabric as its generalized principle of functioning.'[19] The analytics of neo-liberal economics are extended to:

> all purposive conduct entailing strategic choices between alternative paths, means and instruments; or, yet more broadly, *all rational conduct* (including rational thought, as a variety of rational conduct); or again, finally, all conduct, rational or irrational, which responds to its environment in a nonrandom fashion, or 'recognizes reality'. . . . Economics thus becomes an 'approach' capable in principle of addressing the totality of human behaviour, and, consequently, of envisaging a coherent, purely economic method of programming the totality of governmental action.[20]

The rationality of the market infuses a wider and wider realm of social discourses and practices, as rational actors exercising their reflexive capacity for choice maximize their opportunities as self-governing *responsibilized* and autonomized individuals.

While announcements of the death of the welfare state seem premature, it is abundantly clear that the general direction of welfare state development has changed and that a significant shift has occurred in the intended scope of state provision of services. The political hegemony of neo-liberalism has ushered in a 'new ethics of the subject' that reformulates the relationship of government, expertise, and subjectivity, recasting that relation through strategies of autonomization, pluralization, and responsibilization. Increasingly, we are subjected to exhortations that we are responsible for ensuring the conditions of our own welfare. We are, in short, increasingly supposed to look after ourselves (and our dependants)—in childhood, in sickness and injury, in our educational pursuits, in the face of indeterminacies engendered by market changes (e.g., unemployment and the need for retraining), in old age, etc. But what is truly remarkable about this new ethics of the subject—a responsibilized, autonomized self—is the place the residual welfare state is to play in this new scheme of things.

Welfare states are not disappearing (although it might be reasonable to refer to them as 'workfare states'), but they are being significantly transformed with the new objective of subordinating 'social policy to the demands of labour market flexibility and structural competitiveness'.[21] This new dynamic of welfare state provision has meant that while state-directed benefits and services are being hollowed out in favour of promoting an autonomized and responsibilized citizenry, those unable to make it in the new scheme of things are increasingly subjected to more intrusive surveillance, monitoring, and means testing, all of which have implied an intensification of direct and increasingly draconian intervention in the lives of a burgeoning disenfranchised underclass. Coupled with imperatives to 'top up' diminished welfare benefits by way of reliance on private charities and philanthropic agencies, the disciplinary dimensions of the new modalities of governance of the marginalized poor appear increasingly ominous.

Cast against this backdrop I believe we can find a plausible account for a seeming contradiction between the shallow rhetoric of the New Right and the actuality of neo-liberal governmental rationality 'in action'. Despite the triumphalism informing a good deal of that rhetoric there is, nevertheless, mounting empirical evidence to suggest that 'deregulation' is at best a misleading moniker for some very complex processes. Although the New Right frequently accounts for the failures of 'deregulation' as stemming from a failure to do so rapidly or thoroughly enough,[22] what in actuality seems to be happening is a shift to 'freer markets and more rules'.[23] In practice, *de*regulation frequently seems to mean *re*-regulation, while privatization seems to imply new forms of regulation.[24] This should not be surprising. As Polanyi pointed out some time ago: 'The introduction of free markets, far from doing away with the need for control, regulation, and intervention, enormously increased their range.'[25] For present purposes two features of this shift are particularly significant. First, if the state can meaningfully be said to be withdrawing or in retreat, then that withdrawal or retreat is qualitative, not absolute. One of the great ironies of the apparent withdrawal of the state is that the state itself has been the most active participant in its own apparent demise. As Santos has suggested, the irony is that increasingly 'the state must intervene in order not to intervene.'[26] Second, insofar as neo-liberal hegemony has decentred the state, it has not been an instance of institutional suicide. While 'much of the decentring of the state is indeed conducted by the state itself',[27] upon close inspection it becomes clear that 'This "retreat from the state" is also itself a positive technique of government; we are perhaps witnessing a "de-governmentalization of the State" but surely not a "de-governmentalization" *per se*.'[28]

One of the consequences of this practical decentring has been a blurring of the boundaries of the state.[29] Thus, while it has always been impossible to establish a definitive and transcendent concept of the state and to demarcate its boundaries with any finality, the processes of decentring have seen the proliferation of quasi- and para-state agencies working at arm's-length from the institutions and bureaucracies so long associated with 'the state'. These agencies, many of which have moved in to fill areas of social regulation vacated by the state, do not act wholly autonomously of state influences, but, rather, are frequently subjected to broad regulatory oversight (albeit now at arm's-length, and on criteria based less on social than on market principles) or are beholden, at least in part, to state funding or state-sponsored financial arrangements. As Dean suggests, the prevailing neo-liberal hegemony once again poses the theoretical problem of locating 'a division between "the state" and its outside. This can be instanced by the development of non-profit community and social services which are funded at least partially by the state, but run by citizen associations, and the neo-liberal use of corporations, and families, to achieve governmental objectives (e.g., the provision of welfare, the establishment of prisons, job-centres, etc.).'[30]

Rather than an absence of regulation, deregulation actually suggests the emergence of a form of regulatory renewal or transformation that 'leads to a redistribution of legal advantages and disadvantages'.[31] And insofar as privatization marks an important dimension of the regulatory state's supposed retreat, there are good reasons to question whether or not this form of 'deregulation' should be cast in this terminology at all. In at least some jurisdictions there has been a significant increase in the use of legal regulatory mechanisms to oversee the conduct of enterprise in fields of economic activity previously dominated by public ownership. As one author has suggested in a discussion of the implications of privatization in Britain, 'the requirements of public information, overview of privatized companies, and the prevention of monopolistic abuses suggest that regulators are likely to look to a greater reliance on the courts to develop legal techniques to supervise the privatized industries.'[32] There is good reason, therefore, to question the very notion of *de*regulation. Much more fruitful would be an emphasis on the new modalities of regulation arising as *re*-regulation proceeds with the explicit intention of providing for greater liberalization of markets and capital flows, accompanied by new forms of social regulation facilitative of the imperatives of the new order.[33]

The Challenge of Globalization and the New World Disorder

It is unfortunate that political rhetoric has such power to obfuscate the true complexities of life. In the case of the future of the state a good deal of political rhetoric does just this, frequently overlooking some very significant complicating factors that might help us to understand better the shifting role and nature of the contemporary state rather than simply assuming its inevitable withdrawal and/or disappearance. Too frequently we are told that the crises that have befallen the advanced liberal democratic states of the West have been engendered by the inherent flaws of regulation. That is, the problems of the regulatory state are seen to be endemic to regulation generally. If we are to appreciate their inherent complexities and gain an adequate appreciation of the possibilities and potential pitfalls of regulatory projects, any assessment of the relative success or failure of regulatory endeavours (and 'relative' is an important qualifier to keep in mind here, something a ubiquitous rhetoric of 'regulatory failure' frequently fails to acknowledge) must take full account of the contexts within which such efforts proceed.

The past few decades have ushered in an array of social, economic, and political transformations, changes that transcend and challenge the sovereignty and regulatory capacities of the nation-state in dramatic fashion. These changes are frequently treated as indicators of an epochal shift of some sort— the notion that we are moving out of 'modernity' into a new 'late-modern' or 'postmodern' era. The transformations generally associated with this perceived shift are many, but perhaps the most widely cited of these is the apparent 'globalization' or (perhaps more appropriately) transnationalization of social relations. It is worth detailing the various dimensions of these changes, for to the extent that globalization is occurring, it is 'progressing' on a number of interrelated fronts simultaneously, and each of these has its own specific implications for the future shape of regulatory practices in late-modern societies. Moreover, to the extent that the state is being decentred, that process is occurring in a context in which new global modalities of governance intersect local, regional, and national governmental practices.

First, and perhaps most importantly, the past few decades have witnessed a dramatic internationalization of economic relations. This has taken a number of forms. Thus, in recent years a burgeoning array of regional pan-national trading blocs and agreements have emerged, including the European Union, the North American Free Trade Agreement, the ASEAN (Association of Southeast Asian Nations) Free Trade Area, and MERCOSUR (Mercado Común del Sur) in the southern part of South America. The post-war era has also seen the development of a number of international economic institutions such as the General Agreement on Tariffs and Trade (now the more formally institutionalized World Trade Organization), the Organization for Economic Co-operation and Development, the World Bank, and the International Monetary Fund. In addition to these institutional changes there has been a general internationalization of capitalism with an increasingly international division of labour. These latter changes have had profound implications for both the economies of Western liberal democracies and those of the 'Second' and 'Third Worlds'. I will return to these below.

Second, and closely related, has been the impressive growth of regional and global international political and juridical institutions. Organizations such as the United Nations, the Organization of American States, the Council of Europe, the Organization for Security and Co-operation in Europe, and the Organization of African Unity have added a new dimension to political activity above the level of the nation-state. These bodies have combined with an array of international legal institutions, such as the World Court, the Court of Justice of the European Union, and the International War Crimes Tribunal in the Hague, creating new networks of political and legal interaction that traverse the boundaries of nation-states.

Third, a veritable revolution in information and communications technologies in recent years has, in a very palpable sense, made the world a much smaller place. These changes, of course, are closely related to the economic transformations just mentioned, for capitalism has been a principal generator of many of these innovations, and they have in turn facilitated a marked change in the nature of international capital markets. Among the most recent changes are the rapid expansion of telephone and satellite communications networks, the advent of fibre optics technology, the development of the personal computer, and the birth and explosion of the Internet.

Fourth, there has been a revolutionary transformation in transportation technologies. The impact of the automobile on urban life and the commercial development of air transport in the latter half of the twentieth century have combined with the new information and communications technologies to make the world a much smaller place than it seemed a mere half-century ago.

Fifth, the changes mentioned so far have also lent themselves to a significant intermingling or 'hybridization' of cultures. This appears to be taking place in two ways. In the first instance, global capitalism, new information technologies, and new transportation technologies have seen the intermingling of cultural influences. Consequently, today the youth of Ivory Coast, Nunavut, and New York can conceivably watch MTV, aspire to wear the same clothing, and even converse in cyberspace. But, as I will suggest, this often tends to an overly simplistic view of globalization as some sort of broad levelling force at the expense of a recognition of the profoundly uneven nature of these processes and their implications for peoples occupying very different life circumstances around the globe. This unevenness is at the heart of the second source of the recent intermingling or hybridization, that is, the dramatic increase in Second and Third World population migrations as peoples attempt, or are forced, to flee the effects of poverty, war, famine, natural disasters, and environmental degradation.

Sixth, the world seems to be shrinking in the face of technological developments that in some instances severely threaten the quality of life in some regions and in other instances threaten the very existence of life on this planet. The implications and interrelation between such environmental changes as those signalled by the buildup of greenhouse gases, global warming, ozone depletion, and the possible gradual intensification of cyclical climate and weather events such as tropical storms and El Niño are not always clear. A growing body of evidence suggests that however much these environmental changes might be put down to natural shifts in the earth's biosphere, they are almost certainly largely rooted in environmental degradation stemming from human activities.[34] To this we can add the horrific prospects of nuclear holocaust in which just a few belligerents could conceivably put an end to all life on the planet.

Seventh, all of the foregoing have extended the scope of political action beyond the boundaries of the nation-state. New networks of social movements operate outside of the traditional state-centred conception of political action that had dominated political development from the end of the nineteenth century through the first three-quarters of the twentieth century. Feminist, gay, and indigenous peoples' rights groups, for instance, have concentrated much of their political focus on 'grassroots' community organization and change rather than restricting themselves to the hierarchical organization of political life implied by traditional state-centred approaches to political action. But these and other groups (such as the international anti-nuclear movement) have also created international political action networks whose organization and activities significantly circumvent the institutions of the state, formerly the principal actors in the sphere of international relations. We see in these activities the consolidation of new axes of political action, a new dynamic of local/global that is to some extent displacing the old hourglass model of political action, which saw all political action funnelled upward and inward to the state and from there out into the broad sphere of international relations.

It is not difficult to see how the regulatory state, broadly conceived, is implicated in each of these features of late-modern life. We have here a picture of a world being rapidly integrated on some fronts, while the fallout from those processes is giving rise to new social, political, and economic tensions. Confronted by the internationalization of markets, states are finding it increasingly difficult to regulate the movement of capital within and across their borders. The rise of international political and legal institutions poses a potential challenge to the legitimacy of states, which had long operated with relative impunity vis-à-vis their own internal affairs. The globalization of communications and information technologies presents great difficulties for states that would endeavour to regulate information flows, an issue perhaps most poignantly highlighted by efforts to regulate the Internet. The cultural intermingling and hybridization precipitated by the new information, communications, and transportation technologies, in tandem with new levels of international population migrations, pose real challenges to the sense of unity of identity and collective solidarity so frequently associated with the idea of the nation, the constitution and reproduction of which has been, contrary to the myths of primordialism, both an object and outcome of various state regulatory practices.[35] New transnational threats to ontological security stemming from environmental degradation and nuclear and biological weapons proliferation have been addressed, with varying degrees of (un)success, through the sustained regulatory efforts of states, various non-governmental organizations, and a variety of transnational agencies and institutions. But precisely their inherent transnational qualities seem to defy efforts on the part of single states to address or meet adequately the concerns of such groups.[36] Finally, the identities and practices associated with the new social movements have long been the object of social regulation on the part of the state (think here of the policing of homosexual practices and the concerted state-led efforts to undermine the credibility of anti-nuclear activists). But precisely their non-correspondence with national identities and those practices associated with nation formation and reproduction mean that they pose distinct challenges to the unity or primacy so long assumed to be associated with the identity of the nation.

In each of these thrusts of globalization we can locate elements of new modalities of governance. The characteristics of these are such that one might conclude the state is indeed gradually fading into obscurity, assailed on all sides by the twin forces of globalization and localization, its sovereignty transgressed by a growing plurality of social forces and relations, and its regulatory capacities diminished by the pluralization of dispersed sites and practices of governance emerging in late-modern life. Clearly the forces of 'globalization' have impinged on the capacity of states to act autonomously in a world that has been described as simultaneously shrinking and fragmenting.[37] The dramatic changes brought about by this panoply of forces has undoubtedly further contributed to the practical decentring of the state. Two things must, however, be stressed: first, *states have played a principal role in this decentring, and there is no reason to assume they will not continue do so*; and second, *decentring should not be interpreted as necessarily constituting either the disappearance of the state or a diminution of its capacity to govern*, contrary to far too many triumphalist interpretations.[38]

But at the level of 'the global', what does this dispersion of sites and practices of governance imply for how we understand the regulatory transformations of late modern life? And where do these transformations leave the state? As was mentioned earlier, the notion of governmentality takes as its premise the dispersion of sites and practices of regulation and governance. The state then is seen as a strategic condensation of such dispersed practices operating through a plurality of sites and institutions and employing various techniques in its pursuit of regulatory objectives. Thus, we need not change our approach to appreciate the specific implications of late-modern life. We can track the transformations in the regulatory state as a rearticulation of the agents, techniques, methods, technologies, spatial organization, etc. of governance. States, in other words, have always been located in a relatively dispersed field of governmental practices, and as such have always simply contemplated the strategic articulation of such practices in specific conjunctures. This, of course, does not mean that states are not indeed being displaced or becoming less important. But insofar as we might be led to think that the forces of globalization are having just such an effect, we may want to have a closer look at the evidence.

The dissolution of the state's power to regulate or govern economic life and the general trajectory of social and economic development, as we have seen above, suggest that the strategic terrain has indeed shifted, but that states remain formidable actors in the regulatory scheme of things. Economic globalization constitutes a part of that terrain, and there is no reason to suspect that globalizing forces are going to change that. Indeed, states remain far more crucial actors in the organization of an international political economy than many are willing to admit. And, for the First World at least, they remain far more powerful agents of governance than, say, multinational corporations. In regard to the rise of trading blocs and alliances, states remain the principal agents of and parties to negotiation of the terms of such agreements, and, perhaps more importantly, the agents of their implementation and enforcement. Similar points could be raised vis-à-vis the negotiation and/or enforcement of agreements on environmental protection and the mitigation and/or amelioration of environmental damage. So, too, with the emergence of any number of international juridical forms and the enforcement of their edicts.

Insofar as cultural hybridization is concerned, there is undoubtedly some truth to the idea that the unprecedented migrations of populations create the possibility of the pluralization of the cultural, ethnic, and racial constitution of nation-states. But national identity has always been a construction of unity premised on a denial or minimization of the natural heterogeneity of social life. National identities may be threatened by such processes, but to the extent they are, their moral foundations were profoundly dubious in the first place. Indeed, precisely such dubious foundations are only too manifest in the anti-immigration movements afoot in the First World in recent years. And these movements have far less to do with international migration among nationals of the First World than they do with stemming the threat of massive migrations from the Second and Third Worlds and the infusion of racial and ethnic 'undesirables' that such migrations might portend. Indeed, the capacity to withhold access to citizenship rights remains one of the most powerful manifestations of the contemporary state's sovereignty.[39] And in these terms sovereignty becomes a profoundly odious mechanism for the regulation of populations and their racial, ethnic, cultural, and religious composition, not to mention sexual morality.

New social movements are another new and important manifestation of the pluralization of political action and of the sites in which political struggle is played out in late-modern life. But social movements remain policed by states, are often reliant on state-funding, support, or recognition, and find states to be very important focal points for the articulation and substantive actualization of their claims.

Finally, there are disturbing parallels between what is happening at the level of infra-state and supra-state governmental transformations. Just as in the case of the decentred state governing the economy and social life 'from a distance', at the level of transnational relations states seem to be the principal facilitators and enablers of global markets. The implications for Second and Third World countries parallel, in startling fashion, those that confront the marginalized and disenfranchised underclasses of the First World. The new ethics of the subject seem to be replicated at this level, with the ethos of autonomization and responsibilization applied to national and regional collectivities. Just as the new ethics of the subject ushered in by neo-liberal thought has seen the increasing marginalization of a burgeoning domestic underclass in First World nation-states, a similar process seems to be taking place in the processes of globalization. The neo-liberal extension of market rationality—the techniques, practices, and principles of enterprise—into virtually all spheres of life has been extended to relations between nation-states, albeit mediated by the mechanisms of governance that stand at arm's-length from the states of the First World. The growing divide between North and South is regulated, managed, and governed by an array of new para-state international agencies, almost all of which operate, however, with the express blessing and complicity of First World states. Perhaps nowhere is the perversity of New Right triumphalism so blatantly manifest as in the face of the deepening immiseration, or misery, of the majority of the planet's inhabitants and in the growing gap between the First World, on the one hand, and the Second and Third Worlds, on the other. States remain profoundly implicated in the regulation, governance, and reproduction of this immiseration.[40]

All the talk of the demise of the state seems terribly shallow when cast against the reality of brutal repression and enforced destitution confronted by the poorest nations of the world. And what makes this so outrageous is that it is precisely the power of the state to regulate and govern by means of direct and systematic violence when confronted by the imperatives of transnational capital and the new techniques of governance deployed by such 'non-state' agencies as the IMF and the World Bank. As Rose has pointed out, 'even terror can be a calculated instrument of government, as can naked violence',[41] a point that resonates in the life experience of the immiserated populations of the world when they confront the dictates of these new transnational actors. As Santos suggests, 'Only the nation-state can formulate credible nationalist ideologies to compensate for impossible nationalist practices, and a strong repressive state may often be needed to facilitate the requirements of global capital.'[42] Unfortunately, nationalist ideology is seldom adequate in Second and Third World quests for economic justice in a world increasingly bifurcated along lines of life opportunities and wealth. Indeed, too often the state—in these instances, the national states '*of* these people'—acts as the henchman of global capital, and usually with the implicit, if not the explicit, support of First World states.

Conclusion

In this chapter I have suggested that we cast a wary eye at theses that portend the demise of the regulatory state. An adequate appreciation of the state's role in contemporary regulatory practices must escape the confines of a narrow focus on 'economic regulation', and must examine the broader regulatory nexus within which such activities proceed. The shift to so-called 'deregulation' in the 'purely economic' spheres of life has been abetted and supplemented by a dramatic reshaping of the state's role in the regulation of other features of social life. As Santos has suggested, 'the extent to which the regulatory functions of the nation-state have weakened . . . is a matter of debate. Unquestionable is only the fact that such functions have changed (or are changing) dramatically, and in ways that question the traditional dualism between national and international regulation.'[43] A focus on modes of governance and governmental rationality holds out significant promise as a vehicle for gaining a better understanding of these processes.

The colonization of multiple spheres of governance by liberal democratic states in the post-war era has given way to a dispersal of sites, practices, and technologies of governance contingently sutured by neo-liberal hegemony in an unstable equilibrium of compromise. But neo-liberalism has a provocative dimension: the appeal of autonomy and individual responsibility. It opens space for a very real (if somewhat narcissistic) actualization of autonomy for those fortunate enough to be on 'the inside' of the new rationalities of government. The important question, I would suggest, is whether or not there are ways we could harness the autonomization impulse in a new ethics of solidarity, one that transcends the boundaries of the territorial nation-state while remaining attentive to the limitations and benefits that can potentially arise from membership in the political communities that are nation-states.

Notes

1. A recent OECD document offers a typical example, defining regulation as 'the various instruments (both formal legal instruments and such informal tools as "guidance") used by government to control some aspect of the behaviour of a private economic actor. Regulation can also include rules issued by non-governmental bodies (e.g., self-regulatory bodies) to which governments have delegated regulatory powers. All regulations are supported by the explicit threat of punishment for non-compliance.' W. Helgard, *Regulation and Industrial Competitiveness: A Perspective for Regulatory Reform* (Paris: Organization for Economic Co-operation and Development, 1997), 12.
2. It is not that the contributors to this literature fail to acknowledge the role of the state as social regulator more broadly speaking, but rather that they seem to regard these broader issues of social regulation as constituting a distinct genus of regulatory activity, largely unrelated to their principal focus on the regulation of economic actors. The broader regulatory context is thus deliberately excluded from consideration as seemingly irrelevant to the concerns of regulatory studies. See, for instance, P. Selznick, 'Focusing Organizational Research on Regulation', in R.G. Noll, ed., *Regulatory Policy and the Social Sciences* (Berkeley: University of California Press, 1985), 363; G. Majone, 'Regulation and Its Modes', in Majone, ed., *Deregulation or Re-regulation: Regulatory Reform in Europe and the United States* (London: Pinter, 1990), 1; G. Majone, 'Regulation and Its Modes', in Majone, ed., *Regulating Europe* (New York: Routledge, 1996), 9.
3. Within the 'conventional' literature just discussed, one frequently encounters references to 'social regulation'. While I use this same term, my usage is much broader than that typically employed in that literature. In the latter, the focus of 'social regulation' is generally limited to 'fundamental aspects of the production process and on its negative externalities. . . . [S]ocial regulatory policies address the quality of the goods and services that are produced, and the by-products of the industrial economy which threaten human health and life, and the environment.' M.A. Eisner, *Regulatory Politics in Transition* (Baltimore: Johns Hopkins University Press, 1993), 119. In other words, as the term is generally applied in the literature it is taken to encompass those areas in which the state intervenes to shape the behaviour of economic actors involved in the production of goods, services, or negative externalities.
4. See especially, M. Foucault, 'Omnes et Singulatim: Towards a Criticism of Political Reason', in S. McMurrin, ed., *The Tanner Lectures on Human Values* (Salt Lake City: University of Utah Press, 1981), 225; M. Foucault, 'Governmentality', in G. Burchell, C. Gordon, and P. Miller, eds, *The Foucault Effect: Studies in Governmentality* (London: Harvester Wheatsheaf, 1991), 87.
5. N. Rose, *Powers of Freedom: Reframing Political Thought* (Cambridge: Cambridge University Press, 1999), 16.
6. Majone, 'Regulation and Its Modes' (1996), 9.

7. Malchup, for instance, traces the history of anti-monopoly regulation as far back as AD 483 when 'an edict of the Emperor Zeno . . . prohibit[ed] all monopolies, combinations and price agreements.' Ibid.
8. M. Weber, *Economy and Society*, 2 vols (Berkeley: University of California Press, 1968).
9. C. Gordon, 'Governmental Rationality: An Introduction', in G. Burchell, C. Gordon, and P. Miller, eds, *The Foucault Effect: Studies in Governmentality* (London: Harvester Wheatsheaf, 1991), 27 ff.; J. Donzelot, 'The Mobilization of Society', ibid., 169.
10. C. Smart, 'Disruptive Bodies and Unruly Sex: The Regulation of Reproduction and Sexuality in the Nineteenth Century', in Smart, ed., *Regulating Womanhood: Historical Essays on Marriage, Motherhood and Sexuality* (London: Routledge, 1992), 12.
11. This, of course, was not a uniform process. Not only were there marked differences from case to case, there was also a significant difference between the pace of growing state-led involvement in these activities on either side of the Atlantic. The economic and political turmoil that had shaken Europe through the latter decades of the nineteenth century had left Europeans much more suspicious than their North American counterparts of the merits of unfettered markets. Majone, 'Regulation and Its Modes' (1996).
12. T. Purvis and A. Hunt, 'Citizenship versus Identity: Transformations in the Discourses and Practices of Citizenship', *Social & Legal Studies* 8 (1999): 469.
13. Perhaps the most influential voices articulating the foundational principles of the New Right are those of F.A. Hayek and M. Friedman. See F. Hayek, *The Road to Serfdom* (Chicago: University of Chicago Press, 1944); Hayek, *The Constitution of Liberty* (London: Routledge, 1960); M. Friedman, *Free to Choose* (New York: Harcourt Brace Jovanovich, 1980). For a useful rejoinder to these criticisms, see B. Hindess, *Freedom, Equality and the Market: Arguments on Social Policy* (London: Tavistock, 1987), 127 ff.
14. The general tenor of dismissiveness common to these criticisms is nicely captured by Zank: 'In the name of environmental protection and other social goals, government has ignored or ridden roughshod over property rights; federal bureaucrats and regulators have deprived many individuals and small businesses of the exclusive rights to use and derive income from their assets or resources in order to protect the "wetland" created in the previous week's rainstorm and to ensure the "rights" of snail darters, owls, and a whole host of animals. In the name of fair employment, we had to endure years of race norming. Under the total barrage of regulation, we have also seen scores of industrial facilities close throughout the nation and move their operations and employment opportunities.' N.S. Zank, *Measuring the Employment Effects of Regulation: Where Did the Jobs Go?* (Westport, Conn.: Quorum Books, 1996), 1.
15. H. Rottleuthner, 'The Limits of Law—The Myth of a Regulatory Crisis', *International Journal of the Sociology of Law* 17 (1989): 274.
16. Gordon, 'Governmental Rationality', 43.
17. Ibid., 17.
18. Purvis and Hunt, 'Citizenship versus Identity', 469.
19. Gordon, 'Governmental Rationality', 42; cf. R. Keat and N. Abercrombie, eds,

Enterprise Culture (London and New York: Routledge, 1991); M. Dean, *Critical and Effective Histories: Foucault's Methods and Historical Sociology* (London: Routledge, 1994).
20. Gordon, 'Governmental Rationality', 43.
21. B. Jessop, 'Towards a Schumpeterian Workfare State? Preliminary Remarks on Post-Fordist Political Economy', *Studies in Political Economy* 40 (1993): 9.
22. Zank, *Measuring the Employment Effects of Regulation*; cf. M. Bienefeld, 'Financial Deregulation: Disarming the Nation-State', *Studies in Political Economy* 37 (1992): 31.
23. S.K. Vogel, *Freer Markets, More Rules: Regulatory Reform in Advanced Industrial Countries*, reprint edn (Ithaca, NY: Cornell University Press, 1998); Majone, 'Regulation and Its Modes' (1990); J. Kay and J. Vickers, 'Regulatory Reform: An Appraisal', in Majone, ed., *Deregulation or Re-regulation: Regulatory Reform in Europe and the United States* (London: Pinter, 1990), 223.
24. Moran and Prosser point out that while 'It would be natural to assume that deregulation meant a retreat by the State. . . . [i]n practice in many sectors the decade of deregulation [the 1980s] actually meant an increase in State control over markets.' M. Moran and T. Prosser, 'Introduction: Politics, Privatization and Constitutions', in Moran and Prosser, eds, *Privatization and Regulatory Change in Europe* (Buckingham: Open University Press, 1994), 9.
25. K. Polanyi, *The Great Transformation: The Political and Economic Origins of Our Time* (Boston: Beacon Press, 1944), 140.
26. B. de Sousa Santos, *Toward a New Common Sense: Law, Science and Politics in the Paradigmatic Transition* (London and New York: Routledge, 1995); cf. Vogel, *Freer Markets, More Rules*, 256–69.
27. Ibid., 279.
28. A. Barry, T. Osborne, and N. Rose, 'Introduction', in Barry, Osborne, and Rose, eds, *Foucault and Political Reason: Liberalism, Neo-liberalism and Rationalities of Government* (Chicago: University of Chicago Press, 1996), 11.
29. L. Salter and R. Salter, 'The New Infrastructure', *Studies in Political Economy* 53 (1997): 82.
30. Dean, *Critical and Effective Histories*, 151.
31. Rottleuthner, 'The Limits of Law', 274.
32. J. McEldowney, 'Law and Regulation: Current Issues and Future Directions', in M. Bishop, J. Kay, and C. Mayer, eds, *The Regulatory Challenge* (Oxford: Oxford University Press, 1995), 422.
33. Drawing on an examination of regulatory agencies in Canada, Salter and Salter provide an excellent illustration of the state's new role as facilitator, enabler, and animator of the conditions of possibility for the reproduction of neo-liberal rationalities of government. Challenging the assumption that 'deregulation' implies a simple withdrawal of regulatory agencies from the field of regulation, they point out that in the case of the 'deregulation' of telecommunications in Canada, changes in the statutory foundations of telecommunications regulation mean that now 'the regulator is instructed to be pro-active, not in the old-style way of developing and administering rules, but in policing and promoting competition, innovation and free markets. . . . [The] emphasis has shifted from managing firms to managing markets.' L. Salter and R. Salter, 'The New Infrastructure', *Studies in Political Economy* 53 (1997): 76–7.

34. However hackneyed the turn of phrase may seem, it remains true that pollution does not recognize borders. As I write, the Danube is in the midst, it seems, of becoming a river of death in the wake of an enormous cyanide spill originating from a gold-mining operation in Romania. The spill, the region's worst environmental disaster since Chernobyl, has flowed through and decimated the aquatic life of the Danube in Romania, Hungary, and Yugoslavia.
35. For excellent illustrations of the nature of the regulated constitution of the nation, see M. Valverde, *The Age of Light, Soap, and Water: Moral Reform in English Canada, 1885–1925* (Toronto: McClelland & Stewart, 1995); P. Corrigan and D. Sayer, *The Great Arch: English State Formation as Cultural Revolution* (Oxford: Blackwell, 1985); Smart, ed., *Regulating Womanhood*.
36. In the sense in which the notion of governance is being used here, the notion of 'non-governmental' organizations seems a distinct misnomer.
37. J.A. Camilleri and J. Falk, *The End of Sovereignty: The Politics of a Shrinking and Fragmenting World* (Aldershot: Edward Elgar, 1992).
38. M. Horsman and A. Marshall, *After the Nation State* (London: HarperCollins, 1994).
39. R. Brubaker, *Citizenship and Nationhood in France and Germany* (Cambridge, Mass.: Harvard University Press, 1992), 180.
40. This highlights the shallowness of the state-centric visions of some contemporary New Right thinkers, exemplified by John Gray's suggestion that 'poverty—at least the poverty of the underclass, rather than of those whom illness or disability has struck—is not a monetary but a cultural phenomenon, caused by a breakdown in moral and familial traditions.' J. Gray, *Beyond the New Right: Markets, Government and the Common Environment* (London: Routledge, 1993), xiii. While he clearly limits the scope of his analysis to the environment of the First World state, there seems little reason to believe that Gray's logic would not be extended to the problems of global poverty as well, finding equal fault with the immiserated populations of the world for their own destitution.
41. Rose, *Powers of Freedom*, 24.
42. Santos, *Toward a New Common Sense*, 280.
43. Ibid., 259.

Chapter 3

Legal Governance and Social Relations: Empowering Agents and the Limits of Law

ALAN HUNT

Introduction: Shadows over the Dream of Law

Modern societies have developed escalating expectations of law.[1] There has long been a recognition of an inescapable gap between what law promises and what it delivers. Today it is not only a matter of the effectiveness of law. There is now a widening gulf between two competing ideals. The first is that law should be a mechanism that helps secure due recognition to difference within a pluralistic social realm. The second is the desire to secure social stability through an acceptance of a framework of shared values. One major manifestation of this tension is exemplified in the 'dream of law', which envisages an expanded role for law as the solution to conflicting interests and identities. This vision of law as the solution to social problems is captured by the term 'juridification', which indicates a predisposition to tackle social problems by means of law-like rules, processes, and procedures.

This chapter identifies the nature of the doubts and anxieties that have come to beset the instrumental use of law as a vehicle of social policy. It explores the manifestations of this insecurity of law and reviews social theories that provide some insights into how this destabilization of law may be understood. Attention is focused on why the projects grouped under the label 'social law' and the related context of the welfare state, which have been the dominant trend since the early twentieth century, have faltered. The chapter then elaborates an account of a 'law as governance' analysis in order to understand the nature of these changes and their current trajectory. It ends with some suggestions for the future direction of research on law and governance.

Anxiety About Law

The greater the expectations invested in the legal system, the greater is the anxiety about law's capacity to deliver. As more demands are placed on law and more rights are extended, a mounting sense of dissatisfaction emerges among many constituencies. For example, the Canadian Charter of Rights and Freedoms holds out to feminists the promise of equality for women, but the delivery is both slow and ambiguous. Similarly, law promises a response to hate speech and racism, but these manifestations are stubbornly resistant and are barely touched by well-intentioned civil rights legislation and ponderous enforcement regimes. In general, an increasing number of social movements and single-issue campaigns make demands on law, but all too often they are disappointed and frustrated by the results. These 'delivery' and 'effectiveness' concerns become aggregated and linked together into a diffused dissatisfaction with law. This expresses itself particularly sharply in the prevailing 'common sense' that something is seriously wrong with the criminal justice system. This cumulative dissatisfaction has now reached such proportions that we should consider whether it is appropriate to speak of a 'crisis' of law.

Care is needed in speaking of crisis. If there is a crisis, it does not involve any sense of breakdown—there is no widespread lawlessness; the institutions of law continue to function; enforcement agencies complain (as always) about shortage of resources and personnel, but continue to make arrests, lock up prisoners, and supervise sanctions. In general, it is business as usual. Perhaps the problem is not so much a crisis as an uneasiness or pervasive sense of dissatisfaction, in which law is no longer regarded as a reliable source of legitimation that ensures respect for institutional authority. Nor does the normative content of law self-evidently embody a distillation of shared social values, thereby providing a framework for a consensual social order. But it is important to avoid painting an image of a golden age when law once guaranteed an open and transparent consensual social order. There have always been constituencies, most frequently class and ethnic blocks, who experienced law as the imposition of the interests of dominant others. What is perhaps new is the experience of suspicion and exclusion of wider and more diverse social groupings. The most important manifestation is a pervasive distrust of authority and a widespread cynicism towards the personnel and institutions of governance.

One distinct dimension of the current predicament of law is an intellectual crisis, which gives rise to a cynicism that destabilizes an expanding range of authorities; judges, politicians, and other traditional figures of authority no longer command respect. This weakening of authority relations is not simply the final unwinding of a social order grounded in hierarchical social respect; rather, it has its roots in an intellectual climate in which no social position or status has the prestige to underpin its prescriptions or truth-claims. While this mood of doubt and skepticism is experienced most intensely among intellectuals, it is also present among the personnel of the institutions of law and government and in wide sections of the population. The core of this intellectual discontent is the recognition of the fragility of truth and values that no longer can be buttressed by a shared confidence in progress, science, and technology. Law is beset by an intellectual destabilization that over the past decades has produced intellectual turmoil in the social sciences and the humanities.[2]

If the doubts that beset law are diverse, fluid, and complex, they are nonetheless real. The more law is pressed into service as a solution to diverse social problems, the more its limitations become apparent. If this set of anxieties that builds towards a crisis is to be addressed, it is important to examine more closely the dimensions of what Roscoe Pound famously once called the 'causes of popular dissatisfaction with the administration of justice'.[3]

The Impersonality of Law

The first important dimension of the current anxiety about the legal system is found in the belief that law is an impersonal and distant mechanism, one that is not 'user-friendly'. Law is perceived as being 'unresponsive', and this unresponsiveness involves a number of elements. First, there is something that goes beyond the older concern that law is technical, cumbersome, and slow. Law no longer seems able to meet the rising expectations among diverse categories of legal subjects, in particular the 'occasional users' of law.

Consider the example of the paradox of 'victim's rights'. Criminal law, traditionally organized around 'offences against the Crown' or 'society', offered little or no role for the 'victim' except that of being a witness. The attempts to meet the demand for the recognition of 'victim's rights' via the introduction of 'victim impact statements' seem not to have overcome the sense of exclusion and may have exacerbated it since the vengeance-oriented demands of many victim statements clash with the judicial tariff of proportionality. This, in turn, has fuelled the widespread ideology that 'law is soft on criminals', which has been amplified by politicians and media ever-prepared to play the 'law and order' card. For example, in the 1999 election in Ontario the Progressive Conservative campaign sought to show that the Liberals were 'soft on crime'; the evidence was that the Liberal leader, Dalton McGuinty, had once been a 'defence lawyer' and defence lawyers, as we all know, are a bad thing because they 'try to get criminals off'. Another example concerns the quest for justice in divorce; the law holds out the promise of justice, but divorcing couples seem to be left profoundly discontented. Ex-wives complain that the law seems incapable of making 'deadbeat dads' pay up; ex-husbands complain that the law seems to take no account of the complexity of serial monogamy, which results in them having to meet the costs of new (often 'acquired') families while continuing to make payments to an ex-wife; and while the law promises a 'clean break', it never comes.

One important task is to explore whether law can better serve those to whom it seems to offer the appearance of 'help'. It must be insisted that this is not just a question that much preoccupied the 1960s, namely, the 'effectiveness of law' that could be addressed by either more and better law, or by bigger and better enforcement. Today there is a sense that the problem significantly revolves around the very 'form of law' itself.[4] Therefore it is necessary to consider accounts that have proposed theses suggesting that there have been significant shifts in the forms of law.

Theorizing the Contemporary Crisis of Law

It will be helpful to distinguish between two different types of theorization of the predicament of contemporary law. The first involves 'internal' theories, which provide accounts of changes that have taken place *within* the legal form. For example, Nonet and Selznick advance a three-stage model in which law undergoes a developmental process from repressive law, to autonomous law, and then to responsive law.[5] The second type of theory proposes 'external' accounts of changing legal forms that are responses to changes in general societal developments. The classic example is Durkheim's account of the transition from 'repressive' to 'restitutive law' as a manifestation of changes in the social division of labour.

At this point a proviso needs to be made. I will be concerned to avoid any suggestion of 'evolutionism' or 'progress' that posits some unilinear sequence of stages in which one stage replaces another. Rather, I will assume as a methodological protocol that legal forms can and do coexist and that at any particular point in time there are likely to be fluctuating and reversible changes. Further, I will avoid any normative assumption that different forms are 'better' or 'more advanced' than others. Instead, my strategy will be to identify two recent influential accounts that offer an overview of current trends by comparing the views of Michel Foucault and Jürgen Habermas.

Foucault, it should be stressed, never took law as a primary object of analysis; nevertheless, he provides three important lines of inquiry. First, governance by law is increasingly supplanted by disciplinary mechanisms exercised by a wide range of experts, professionals, and disciplinary agents. The hospital, the workplace, and the school are key instances of disciplinary institutions. Second, the law, while not disappearing, increasingly takes the form of 'norms' that specify general policy objectives rather than imposing imperative rules of conduct; for example, employment legislation requires that employers treat their employees 'fairly' without stipulating what fairness is in the diverse contexts to which it is to be applied. Third, law loses its autonomy and becomes increasing interpenetrated with other forms of expert knowledge (medical, psychological, etc.).

In contrast, Habermas envisages the following linked processes. First, law is implicated in the 'colonization of the life-world' in which the everyday mechanisms of co-operative existence become subject to processes of 'juridification'; for example, many organizations replace informal means of dispute resolution by formalized law-like systems of rules and procedures. Second, law is complexly involved in this colonization of the life-world, but while law is part of the problem of juridification, it also has potential as part of the solution, as a means of importing considerations of 'justice' into everyday processes of handling disputes. For example, over the last century employers have become less able to 'hire and fire' at will, but must provide a statement of grounds for dismissal and a forum for adjudication and rights of representation. Third, there has been a marked trend towards 'proceduralism'; law is less directed to supply rules for decision than it is to introduce procedures through which substantive decisions may be reached. For example, in pay equity processes legislation does not seek to determine rates of pay, but introduces requirements as to the procedures by which these are determined. Finally, this proceduralization provides a framework within which an expanding diversity of conflicts can be regularized through procedures that open up the possibility of 'dialogue' between participants. In an increasingly stratified and pluralist social world—an agonistic society in which agreement on substantive issues is less and less possible—identifying procedures that require participants to take some steps towards the recognition of 'the other' is the best than we can aspire to.

These sketches will provide a theoretical framework within which to locate a discussion of the tensions that arise in contemporary instrumental projects to employ legal techniques to address social problems.

The Crisis of 'Social Law'

Implicit in the views of both Foucault and Habermas is a recognition that law increasingly has come to be seen as a purposive project, part of a general strategy of governance directed at sustaining political stability by ensuring the social conditions for individual well-being. This teleological conception of law is most significant in the extent to which during the twentieth century there has been an expansion of 'social law' or 'welfare law'. Social law corresponds to the final stage of T.H. Marshall's influential 'stages of citizenship' (the sequential rise of civil, political, and social citizenship).[6] In regard to 'social citizenship', extensive legal intervention has been required to secure the conditions for full and equal participation in the strong sense of 'participatory democracy'. This requires the redesign of organizational structures that make the institutions (corporations, semi-public associations, mass media, educational institutions) sensitive to the conditions that determine the differential capacity of differently situated groups and individuals to participate. But the expansion of such intervention, particularly in the period since 1945, while securing major changes in the circumstances of large sections of the populations of industrialized societies, also resulted in a rampant bureaucratization and an increasing institutional complexity as waves of substantive legislation strained the capacity of the legal system to deliver effectively the promise of 'social citizenship'.

Yet at the same time, the 'new social movements', such as the women's movements and environmental movements, which sought to advance sectional interests, not only came to rely increasingly on strategies of legal intervention, but as they became enmeshed in legal processes they also experienced increasing frustration with the limitations of these processes. The general form of these demands for the expansion of law constituted what has been called 'the rights revolution'. From the 1960s on the range of legal rights expanded considerably for wide sections of the population who had previously not been significantly touched by legal rights. Two examples can serve to identify these processes. Employees became protected by a broad spectrum of rights about their working conditions and their entitlement to benefits, to membership in trade unions, to redundancy payments, and many others. Second, divorce and other forms of matrimonial legal remedies came within the reach of virtually all married persons. The expansion of social law and the rights revolution combined to generate a preoccupation with the capacity of law to secure 'social justice'.

Theorists drawn from a diverse range of intellectual and political positions began to problematize the link between expanding legislative strategies and what came to be described as the 'crisis of the welfare state'. Some, like F.A. Hayek, rejected the very project of 'social justice' through legal interventionism; not only did he deny the validity of the goal of social justice, but he argued that planned legal change, what he called 'constructivism', was impossible because the legal system and those within it could never acquire sufficient knowledge of the choices of dispersed subjects to allow for legal intervention with predictable effects.[7] Hayek played a decisive role in providing an intellectual rationale for populist political projects designed to 'roll back regulation'. Other theorists, more sympathetic to the goal of social justice, began to detect significant flaws in legal interventionism. From both the political left and right came criticisms of what seemed the inescapable bureaucracy that welfarism gave rise to; rather than empowering its intended beneficiaries, welfare programs appeared to render them passive, requiring the so-called beneficiaries to stand in line dependent on the problematic discretion of officials. From the left came the further criticism that dependence on welfare amounted, albeit unintentionally, to an expansion of social control.[8] The whole welfare complex came under mounting political attack from the late 1970s on.[9]

Beyond Welfarism

Beginning in the late 1970s a sharper political divide emerged over the assessment of the welfare complex. From the political right there came an offensive to dismantle the welfare apparatuses and to abandon or displace projects of social justice. The valorization of the market was articulated through the discourses of 'deregulation'. On the political left the responses were more diverse. It is significant to note that certain features of the welfare state have exhibited a remarkable tenacity, most clearly visible in the continued widespread popularity of the public health system. This support has proved to be a considerable impediment to the wider deregulation projects of the right. It is against this background that we can track the divergent academic and political responses to the 'crisis of law' that has gone hand in hand with the 'crisis of welfare'.

However, it is important to insist that these crises are by no means symmetrical. Such an analysis must be situated against the background of two contradictory responses. On the one hand, a generalized positive commitment to 'the rights revolution' is exemplified in the popular acceptance of the Charter of Rights and Freedoms; on the other hand, as noted above, there has been an intensifying dissatisfaction with the criminal justice system. The more important theoretical attempts to come to grips with the crisis of law are 'legal pluralism' and 'reflexive law'.

Legal Pluralism

Legal pluralism proceeds from an interrogation of one of the persistent features of professional legal discourses, namely, the presupposition that law is a unitary phenomenon captured in the ubiquitous phrase 'the Law'.[10] The decisive step is the suggestion that a plurality of coexisting and competing legal forms exists.[11] The formal and professionalized law of the highly visible litigation courts is one significant legal form, while the 'bureaucratic justice' of courts routinely administering non-adversarial legal process to motoring offenders or debtors involves quite distinctive legal forms. These in turn can be distinguished from the policy-focused and non-litigious routines of policy interpretation found in diverse fields of regulatory law. Legal forms also exist outside the formal setting of courts, as is evidenced in the diversity of informal, albeit bureaucratized, processes administered within the welfare system.

The most significant manifestation of legal pluralism has been the 'alternatives to law' movement. As a reaction against the defects of legal formalism, interest has been stimulated in mediation, arbitration, and other quasi-legal techniques. This quest has broadened into an influential 'alternative dispute resolution' (ADR) movement. While there is much to be learned from this experience, it is problematic insofar as it 'turns its back on law' and assumes that alternatives, by virtue of being 'not-law', overcome the deficiencies seen as being inherently associated with the legal form. For example, one criticism of the legal form is directed against its professionalism. Lawyers are viewed with suspicion because they secure their professional and economic interests through techniques promoting a technicality of procedure and language that distance litigants and other participants from active participation. However, many forms of ADR create a 'new professionalization' in which arbitrators, mediators, and counsellors function in much the same exclusionary manner; in many cases these 'new' roles come to be colonized by professional lawyers. In short, 'informalism' tends towards an expansion of regulation and further professionalization.

Reflexive Law

The second response to the crisis of law has been a group of projects encompassed by the concept of 'reflexive' or 'responsive law', which denotes a new phase of legal development following that of 'formal' or 'autonomous law'.[12] This shift results from the emergence of an increasingly critical attitude towards authority. It is also associated with the rise of an instrumental demand that law be used to respond to social problems. When a social problem becomes recognized there is an increasing tendency for it to be followed by a demand that there should be a law purposely designed to address the problem. There is also a tendency towards an increasing politicization of legal demands by social movements.[13] Thus law increasingly comes to be perceived as a major means of societal response to social problems and is required to be purposive and remedial. The result is that law is expected to meet increasingly high aspirations that inevitably it tends to be less and less able to fulfill. Reflexive law seeks to break this circle of demand followed by frustration.

The central idea pursued by reflexive law theory is to enhance those aspects of legal mechanisms oriented towards the interests of non-professional users of the legal system and thereby to increase user satisfaction. In brief, the goal is to customize law, to make it more user-friendly, such that agents/subjects are able to reflect on their circumstances and find help from the legal system to change their circumstances in desired ways.[14] The reflexive critique of law involves elements already encountered: that law is tainted with the defect that it reproduces inequality and domination, and that it denies the possibility that justice can be secured through an impartial autonomous system of rules and procedures.[15] Reflexive law comes increasingly to make the political judgment that it is necessary to abandon the goal of comprehensive social legislation as being utopian and counterproductive, and to concentrate instead on strategic intervention where specific circumstances require legal rather than other forms of intervention.

Responsive law encourages an attitude of 'legal restraint'; this restraint is to be secured either by an abstention that queries whether further law-making is necessary or by various kinds of 'proceduralization'. Legal regulation should refrain from attempting substantive codification and instead prescribe procedures that leave the content of rules or policies to be arrived at by the participants themselves. The objective is to increase the self-regulatory capacity of those concerned.[16] Reflexivity seeks to stimulate broad participation and institutional redesign to ensure the adequate representation of varied interests and an ongoing concern with 'substantive justice'. The assessment of such a reflexive strategy requires consideration as to whether the likely result is a diminution of legalism or whether it might not add to existing legal regulation without replacing it.[17] The danger is that what advocates of reflexivity have hailed as a breakthrough in legal development may turn out to involve both an expansion and an intensification of legalization.

The concern to promote informalism and responsiveness in law needs to be weighed against the possibility that the core of law and legality is invested in its formality. Stanley Fish argues that law, in order to be law, has to be insulated and unresponsive.[18] The rhetorical practices of law are self-generating and self-perpetuating; law is largely impervious to theory and external criticism. Thus critics are wrong to think that law could be any other way. Law is resilient in its formalism, incorporating extralegal concerns in such a way as to sustain its own autonomy. While it is not necessary to assume that law has a necessary and fixed form, such considerations should temper any overly enthusiastic embrace of informalism.

Law, Governance, and Self-Regulation

The ongoing debates over the future of law have not yielded any general approach that seems capable of offering a long-term remedy for the problems besetting the legal system. There does, however, seem to be a limited consensus that brings together otherwise divergent positions in an uneasy espousal of two propositions. First, caution should be exercised before embarking on any major enlargement of the arena of legal regulation. Second, attention should be given to the potentiality of a refashioning of legal instruments to facilitate the self-regulatory capacity of citizens.

In order to explore whether these themes can be developed more fully it may prove helpful to recast the issues under discussion in the light of the idea of 'law as governance'.[19] But before this can be done some clarification of the term 'governance' is needed. The term is used with increasing frequency in political science and policy studies literature in a descriptive sense to refer to the processes and techniques of governing employed by governments and administrative agencies. I will be using it in a different sense that draws on Foucault's use of the terms 'government' and 'governmentality'. It is worth quoting his conception of government at some length.

> This word [government] must be allowed the very broad meaning which it had in the sixteenth century. 'Government' did not refer only to political structures or the management of states; rather it designated the way in

which the conduct of individuals or states might be directed: the government of children, of souls, of communities, of families, of the sick. It did not cover only the legitimately constituted forms of political or economic subjection, but also modes of action, more or less considered, which were designed to act upon the possibilities of action of other people. To govern, in this sense, is to structure the possible field of action of others.[20]

Thus, for Foucault, to govern is to engage in attempts to act upon the actions or conduct of others. Hence, law is one of the forms of governance, a distinctive set of techniques that impact on the conduct of others.

One important feature of modern liberal governance is a self-conscious concern about the proper limits of government. This generates an inclination to locate responsibility for governance not upon some state institution, but rather within civil society. Of particular importance are those situations where social agents are constructed as responsible for taking charge of some aspect of their own condition. This focus becomes elaborated in neo-Foucauldian accounts of the mechanisms of 'responsibilization' in which governance takes the form of transferring a responsibility onto an actor who accepts and internalizes the obligation. A commonplace instance is provided by contemporary health discourses that 'responsibilize' the individual by making each of us answerable for our diet, our exercise, and the substances we consume.

Governance of social life always involves the attribution of responsibility to some designated agent. Some of these allocations of responsibility we treat as entirely natural, for example, the responsibility of parents for raising their children or of doctors for the well-being of their patients. One key aspect of the allocation of responsibilities has immediate and familiar implications for law. Responsibility is assigned over many practical matters for the distribution of risk concerning who is responsible for straying animals, falling masonry, industrial pollution, etc. Some, but not all, such responsibilities take legal form; individuals have a legal responsibility to go to the aid of some categories of 'drowning persons' but not others. The allocation of legal responsibilities is subject to contestation and may change over time, for example, the shifts in negligence liability since industrialization.

Modern liberalism is characterized by a trend that has been described as 'prudentialism'.[21] Individuals are increasingly required to take responsibility for their own interests. For instance, the police less and less purport to catch housebreakers and recover property; householders today must make prudential calculations about the cost of installing security alarms against reduced premiums for household insurance. The police now fulfill a clerical role by providing reports of housebreaking that serve to back up insurance claims. Currently we are witnessing a new phase of prudential family policy, with debates over the forms of parental social and legal liability for the actions of children. During most of the twentieth century social or collective policies played a large role in the socialization of the individual. A plethora of interventions directed at the family operated to regularize habits and morals, providing family members with some measure of security across the course of their lives and actively rearing children with the technical and moral capacity for wage-labour. This mode of governance was liberal in that the governance occurred *through* the family; significantly, many of these responsibilities were devolved upon mothers.

Another new feature of modern liberal modes of governance has been the emergence of new quasi-legal mechanisms that anticipate problems by stipulating predetermined service standards and remedies for their breach. In their most banal form such mechanisms work through the medium of commercial advertising, as in 'Free pizza if not delivered within 15 minutes.' More significantly, they are found in professional codes of conduct that designate consumer rights and are increasingly penetrating relations between state administrative agencies and citizens. These developments can be understood as manifestations of struggles over the regulation of ever more complex and proliferating expertises; at stake are new ways of rendering the conduct of professionals governable in an environment of escalating levels of distrust between state functionaries and experts.

Individualizing mechanisms are present in governance practices that act upon and through the conduct of the governed by invoking the self-governing capacities of individuals. An important contribution of neo-Foucauldian scholarship has been to draw attention to the complex configurations of self-governing individuals, proliferating expert discourses, legal mechanisms, and new techniques of governmentalized non-state agencies. Not only are there increasingly calculated responsibilities in everyday life—whether to eat butter, to smoke cigarettes, to give up drinking red wine (and then to take it up again). Note here the key role played by expert knowledge and the issue of trust (who to listen to, who to believe). In its most developed form the governance of the self becomes a project of the financial management of the future. Those who are able become financial planners, often with a key role being played by an array of financial advisers through investment and insurance techniques. Instead of viewing the individual as a self-governing agent, a better alternative is to focus on how liberal rule governs through a complex mix of the family, prudentialism, and, increasingly, the market.

It is important to stress that many of the emergent liberal techniques of governance are far from 'liberal' in their consequences. The emphasis on governance through the family, on self-governance and 'prudential' governance, requires individuals to have access to cultural and economic resources. Those who lack such means are separated from others by what Foucault calls 'dividing practices'.[22] The result is that modern liberalism creates a large pool of people, sometimes referred to as an 'underclass,' who fall outside the scope of liberal governance and then become subject to distinctively illiberal forms of governance, increasingly subject to surveillance and disciplinary practices and also prone to moralization through such labels as 'welfare scroungers' and 'single mothers'.

The major focus of governmentality studies has been on the practices and discourses of 'rulers', attending to the techniques and tactics of rule and the rationalities and logics within which these practices are located. I will turn the focus around by asking a series of linked questions:

- What does governmentality have to contribute to understanding the experience of being 'governed'?
- Is it possible that legal governance can be experienced, not as the imposition of constraints and penalties, but as facilitating the self-governance of legal subjects?
- Can legal rights be a means for empowering legal subjects?
- Can such a strategy empower different categories of legal subjects, including the less advantaged, or will such a strategy tend to reinforce existing inequalities in the access to social and economic resources?

Conventional treatments of the relation between law and its subjects, in the traditions of both legal theory and the sociology of law, have relied on some version of 'legalism', the view that the essence of law is the external imposition of rules upon 'legal subjects', and give effect to some external standard of authority, morality, or policy.[23] Legalism involves a commitment to legal rules as a stabilizing force designed to promote the security and predictability of social relations.[24] This vision increasingly conflicts with two important features of contemporary conditions. One distinctive feature of modern law is the part played by democratically elected legislatures in providing powerful legitimations by democratizing both political sovereignty and substantive law. Yet such accounts of the place of the individual citizen or legal subject are problematic. They view subjects as the objects of law's regulation; yet at the same time they are subjects who possess rights bestowed by law. The second feature unaccounted for by orthodox accounts of the legal subject is that law is only one among many mechanisms of governance that impact on legal subjects. Not only do we need to account for the extraterritorial mechanisms associated with globalization (NAFTA, IMF, etc.), but also, more locally, the impersonal force of market regulation, of the regulatory regimes of the workplace, of social institutions and public spaces.

In general, law functions in conjunction with other regulatory regimes; sometimes they reinforce one another, but at other times they are in tension and may conflict. Legal techniques similarly combine with extra-legal processes. Hence we should avoid posing questions in terms of the 'efficiency' or 'effectiveness' of legal rules and procedures because to do so reinforces the assumption that law is autonomous and can be evaluated without examining the interconnected network of governance within which it is located.

In order to unpack the ambiguity of the legal 'subject' we can draw again from Foucault. He focuses attention on the need to distinguish between two closely connected processes—'subjection' (to be made subject to control or domination) and 'subjectification' (to be formed as a self, to be made into an ethical subject).[25] To be a legal subject shares precisely these two dimensions; it is to be made subject to law, a subject of obedience to law's rule, and to be constructed as a subject of rights and responsibilities. It is important to note that what may be called 'legal subjectification', that is, to be made into a legal subject, does not imply only the capacity to activate rights as an exercise of choice, but rather that such self-forming actions involve recognition of legal authority. Rights are significant mechanisms of 'social recognition'; they allow individuals and groups to be recognized by the community that grants them rights and imposes responsibilities.

The processes of subjection-subjectification have a historical dimension. In the formation of modern legal regimes there was a close link between the development of legal systems and the processes of state formation that expressed themselves in the extension of sovereignty, first of monarchical power and later of the sovereignty of legislative assemblies or parliaments. With the long march towards the democratization of political power through the expansion of political and civil rights, the scope and inclusiveness of legal rights expanded. But we should avoid falling into the trap of positing some general transition from 'sovereignty' to 'rights' as the legal manifestation of a shift from subjection to subjectification. It must not be forgotten that the consolidation of constitutional legitimacy of the great monarchies involved forms of political alliances through the extension of bundles of rights to various forms of landed property interests and later commercial interests. In other words, we should not make any simple equation between democratic constitutionalism and legal rights; in particular, it should be borne in mind that rights are not simply accumulative—the granting of rights to some category of legal subjects may restrict or impede the rights of others, as when property rights over land result in the denial of access to others.

The important question is: Can legal techniques significantly contribute to the capacity of legal subjects to order their own lives and enhance their relations with others? Here it is necessary to consider law's capacity as a means of empowerment of citizens as active agents in their own governance.

Empowering Agents

In recent years there has been much discussion of the concepts 'empowerment' and 'agency'.[26] Before we consider the ways in which law may provide a mechanism of the empowerment of agents it is necessary to place in context this interest in empowerment. The concepts 'agency', 'empowerment', and 'identity' have emerged as important in current debates within a strongly individualistic frame of reference. Issues have been articulated in the form: How may individuals be accorded greater capacity to assert and promote their interests? And this question is part of the distinctive political rationality of contemporary liberalism, for it revolves around a concern with 'individual rights'. The question might be reformulated as: How may individual rights be both extended and made effective?

It is important to problematize this presentation of the issue. The concepts 'agency' and 'empowerment' are conceived in much the same way as the pre-Foucauldian conception of power, that is, as something that one can have more or less of. It leads to posing questions in the form: How can law increase the empowerment of agents? Such an approach oversimplifies the issues at hand. If 'agency' is to be preserved as a useful concept we need some additional concept such as 'modes of agency' to identify different forms of the 'capacity of agents'.[27] Where an agent has ready access to the relevant resources, she or he can act directly. For example, a phone call can establish how to register the birth of a child; but to challenge a property tax assessment requires the resources and capacities needed to submit an objection, attend a hearing, and possibly follow through an appeal.[28] It is also necessary to encompass the many ways in which agents act upon themselves and others and are acted upon by others. This makes it possible to grasp how the individual agent is also a product of administrative and disciplinary action, is a 'normalized' rather than a 'free-willing' agent.

Contemporary political rhetoric draws heavily on themes of empowerment, family values, community, and personal responsibility. These slogans are promulgated while the institutions of family, marriage, and parenthood confront forces that systematically undermine them. Care needs to be taken in exploring the theme of agency to insist that to act always requires a complex of material and cultural capacities and resources. This does not imply that the problematic of 'individual rights' should be rejected, but rather that these rights must be seen as only part of a wider and more complex problem. Glimmerings of this are already present; for example, in the context of the contemporary political concern with 'human rights' there is widespread recognition that many of the most pressing human rights (education, security, water, etc.) are 'social' or 'collective' in character. So it follows that we must also reinterrogate 'agency' in such a way as to take account of a range of different 'forms of agency' so as to embrace these other forms of social relations.

We may now restate the problem. What are the implications of techniques that employ an expansion of rights to enhance the capacity of legal subjects to take advantage of an enlarged participation in legal resources? This question is especially important with reference to those with limited economic, social, and cultural capital. Joel Handler asks how 'dependent people', the disadvantaged, can be sufficiently empowered to provide them with effective participation. He argues for 'informalism', decentralization, and alternative dispute resolution.[29] Another solution is advanced by Roberto Unger, who urges the explicit introduction of what he terms 'destabilization rights' to empower the disadvantaged in law.[30] The promotion of destabilization rights provides the capacity to challenge official actions and taken-for-granted practices. They are a new type of entitlement that can serve as weapons for a legal 'undermining' of the status quo in a process directed towards social transformation. Such destabilization rights define a new type of entitlement that operates, intentionally, to undermine the status quo and may thereby advance processes of social change. Examples of such rights are access to information and the protection of 'whistle-blowers'.

The difficulty confronting strategies intended to make law more responsive is that such interventions tend to expand what Habermas has described as 'juridification tendencies'.[31] His solution is not that of legal abstentionism, and he resists the temptation to create further regulation of substantive issues. Rather, he advocates a 'proceduralization' of law that would put criteria in place to allow one to distinguish, at least analytically, between law's capacity to guarantee freedom and its capacity to destroy it. He uses the example of 'school law', where legislation can protect basic rights of parents and children against the school administration insofar as it frees the education process from bureaucratic constraints. When such an 'external constitution' or procedural framework is provided the education system can function to define its own goals, criteria, and mode of operation. It should be noted that Habermas is at pains to stress the paradoxical implications of the juridification tendencies, in that while law may enhance freedom, it may be also serve as a prime agent of the colonization of the lifeworld. Teubner describes this strategy in the following terms: 'The task of the law then is still to control power abuses, but the central problem becomes rather to design institutional mechanisms that mutually increase the power of members and leadership in private institutions.'[32] In general, this proceduralization of law involves attempts to avoid the imposition of prescriptive or substantive rules, seeking to identify procedures for the self-governance by the parties to specific forms of social relations.[33] Such procedures are premised upon the view that for them to be effective they must facilitate dialogue, understood as attempts to secure mutual understanding, between the participants.

The key question is the extent to which legal techniques are able to stimulate the self-activation of legal subjects. In particular, we need to be concerned with the extent to which law is able to facilitate co-operative self-regulation. It is relatively unproblematic to empower individual legal subjects by conferring new or expanded rights, remembering to ensure that appropriate, accessible, and cost-effective means of making use of such rights are in place. It is more problematic when we seek to empower a plurality or collectivity of potential legal actors. Where such actors share a common interest the legal techniques required have been enhanced over recent years; test cases, representative actions, and class actions provide effective means of joining dispersed legal subjects who might not in their individual capacities have the economic and other resources to pursue their interests. These developments, for example, are closely related to the growing interdependence between legal remedies and the insurance business. We are increasingly familiar with medical malpractice suits, environmental damage actions, and transportation accident claims.

However, two considerations should check our enthusiasm for such developments. The first is a general apprehension about whether such developments mark a move towards a 'litigation society' of the form experienced in the United States. There is a tendency for an expansion of liability to draw into litigation a widening pool of less and less directly affected persons, the implication being that legal remedies increasingly take the form of a lottery in which individuals are induced by litigation-minded lawyers to 'cash in' on someone else's loss or injury. Perhaps more significantly, such an expansion, largely of negligence actions, tends to displace other (non-legal) methods of handling social problems. It may be, for example, that the capacity for self-regulation of the professions has been undermined by the increase in negligence litigation. Rather than focusing attention on internal mechanisms of surveillance of incompetent doctors, lawyers, and accountants, such professional bodies become increasingly engaged in negotiating liability insurance contracts. It would seem preferable for professions themselves to protect the public from malpractice than for the courts to remedy such defects after the harm has been done. The paradox of such forms of action is well illustrated in the rash of current actions on behalf of tobacco smokers, or those that bear their health costs, against cigarette manufacturers. While such actions can be viewed as a means of redistributing health-care costs, they can also be regarded as derogating from the responsibility that individuals should properly bear for the risks they voluntarily assume. Such a paradox can coherently be resolved by approving of actions brought by agencies that bear the costs of treating smokers, while disapproving of recovery by individual smokers.

The second reason for caution about burgeoning group litigation is that it does not necessarily serve to empower litigants. It places the key decision-making powers in the hands of lawyers. Since many of such actions result in negotiated compensation agreements, the parties are often left in a take-it-or-leave-it situation, either to accept the compensation offered or to face the unrealistic prospect of soldiering on alone. It is by no means evident that such actions should be viewed as enhancing the competencies and capacities of legal subjects. The inflation of litigation costs associated with such actions may in practice escalate the ordinary litigation costs of citizens seeking to pursue more traditional litigation to protect their interests.

Further difficulties arise with respect to projects designed to employ law in projects of empowerment where the participants do not share a common interest. Here the question to consider is the extent to which legal proceduralization facilitates complex social aggregates in the self-determination of their mutual affairs. An important contemporary example of such proceduralization is provided by 'pay equity' legislation. The important feature of such legislation is that it does not mandate substantive outcomes, but rather requires the parties to engage in processes of arriving at consensual procedures for decision. In this respect such legislation has been greeted by some as an alternative to the expansionary tendencies of traditional welfare legislation. However, the experience of such projects has given rise to two problem areas. First, such proceduralization, although designed to avoid complex substantive legislation, results in its own complex form of juridification; it produces its own necessity to rely on arcane expertise, breeds a new brand of 'experts', and renders the process inaccessible to those affected by its processes. Second, it does not guarantee that internal relativities are handled in a manner satisfactory to the different interest groups, and may thus lead to resentments and the storing up of discontents. It may well be that such proceduralization of the wage relation amounts to a new form of discipline in which unions and their officials 'police' income relativities.

Thus, legal intervention can be viewed as disrupting, rather than contributing to, dialogic or participatory processes. To take another example, recent concern over cruelty to children has generated increasing legal intervention in the relations between parents and children. The prohibition of corporal punishment of children, while promoted in terms of children's rights, has the 'unanticipated' consequence of further intensifying the surveillance of 'problem families'.[34] This points to the need to recognize that 'participation' can lead to the advance of formal representation of plural interests through institutional advocacy movements, so that social movements come to substitute themselves for the direct participation of those directly affected. Such forms of representation have problematic participatory implications in that while serving to overcome unequal power-knowledge relations, they may themselves become bureaucratized and, as a result, replace meaningful representation of the actual participants by pursuing specific political agendas. Such dilemmas beset well-intentioned policies such as the mandatory prosecution in domestic violence complaints; while securing symbolically significant increased rates of conviction, such policies restrict the options open to the participants themselves.

The Limits of Legalism

The recent history of law reform exhibits a perplexing paradox. Determined efforts have been made to grapple with the mounting criticism of the legal order that has become more emphatic since the 1960s. Law stood charged with being tainted with the endemic defects of a liberal social order, namely, that it was itself a major participant in the reproduction of social inequality. Its very formality, preoccupation with due process, and rationality laid it open to the charge of being a mechanism of social domination. These criticisms generated conscious attention to the instrumental use of law as a vehicle of social justice. Yet these problems appear intractable. While many advances have been secured as a result of pragmatic legislative reform, the deep layers of social disadvantage seem to remain impervious to attempts to use legislation to deliver social justice. The accumulative result has been mounting doubts about the capacity of a legal system founded on an autonomous system of rules and procedures to deliver justice. From the 1980s on, attempts to engage more systematically with gender and racial discrimination have further exacerbated these problems. The inflation of expectations concerning legal redress needs to be recognized as paradoxical. While it does much to stimulate resistance to many forms of social injustice, at the same time it widens the gap between legal promise and social reality.[35]

The uncomfortable conclusion is that Canada, along with other advanced liberal democracies, has reached 'the limits of legalism'. A telling exemplification of this thesis is provided by Jeff Spinner's discussion of civil rights in the United States.[36] He argues that while formal rule-making in the public sphere (for example, in the form of civil rights legislation) has been significant, it cannot alone ensure the transformation of race relations. Further law-making will be unlikely to advance the cause of equal rights of African Americans and other minorities. Real change, he argues, depends on changes in civil society, at the level of mundane everyday interactions. Little will change until minority citizens can get the same smile of welcome on entering a restaurant as a white customer.

One implication of this line of argument is that law has little to contribute to securing the elementary forms of civility.[37] However, some caution may be in order before reaching this conclusion. The advocates of civility commonly suggest that law, even if it cannot mandate civil face-to-face conduct, nevertheless contributes by stipulating principles and procedures conducive to treating others with the respect that is the core requirement of civility. This is not the place to explore the relation between law and civility in more detail, but the question of law's potential contribution to the conditions of civil relations deserves further attention.

Additional problems arise once it is recognized that issues of difference and their recognition cannot be separated from questions of inequality. Any effective democracy must solve both the affirmation of difference and the reduction of inequality; it is immediately evident that this raises central and deeply controversial issues of economic redistribution and multicultural recognition. It may be noted in passing that while the current preoccupation with identity is significant, if pursued alone or privileged over issues of equality, there is a danger that identity politics will forget that multiculturalism can be meaningfully sustained only on the basis of social equality.

These arguments are intentionally discomforting, but should not be taken as conclusive. For present purposes it is sufficient to stress the interactive links between law and other modes of governance.[38] As emphasized above, one of the merits of a focus on law as governance is that it draws attention to the fact that law always operates in conjunction with other regulatory mechanisms and techniques. One of the most interesting (and also perplexing) features of Foucault's discussion of the role of law in late modernity is captured in his attempt to encapsulate the contemporary role of law.

> I do not mean to say that law fades into the background or that institutions of justice tend to disappear, but rather that the law operates more and more as a norm, and the juridical institution is increasingly incorporated into a continuum of apparatuses (medical, administrative, and so on) whose functions are for the most part regulatory.[39]

The puzzle is to decide what he intended by the formula that law comes to operate as a 'norm'.[40] My suggestion is that his point is similar to that made by the theorists of reflexive law, namely, that modern law becomes less focused upon laying down prescriptive or external rules of conduct, but rather functions as a mechanism for articulating general policy prescriptions. The 'norm' is embodied in the policy that frames the specific legal or administrative intervention. This line of thought serves to focus attention on how law provides a means of articulating the symbolic or legitimating pronouncements of normative values.

Returning to the instance of civil rights legislation, we should read such law as providing an authoritative pronouncement of an ethics of anti-racism. The important point is that we should not evaluate such legislation simply by the extent to which its implementation and enforcement secure the diminution of racist conduct. Rather, the strategy of law and governance should point the inquiry towards the consideration of how legislation, in conjunction with other agencies, policies, and practices, serves to advance the normative goal. One example may serve to illustrate this point. An important arena of race relations revolves around police practices. In addition to policy measures, recruitment and training, it is increasingly evident that these need to be supplemented by mechanisms of inquiry into incidents of alleged police racism conducted by an agency independent of the force against which the complaint has been made. It is in this context that the declaration of general anti-racist legislative policy is significant. Not only does it serve to endow anti-racist policies with legitimacy, but it also encourages those adversely affected by police racism to pursue complaints and, in addition, stipulates the inquiry procedure. This combination of norm and procedure—Is the procedure appropriate to the norm?—should be the focus of evaluation exercises.

The general implications of my argument are twofold. First, all demands to solve social problems by legislative action should be subjected to a cautionary presumption that legislation should only be embarked on where there is unambiguous evidence of potentially positive effects. The tendency to legislate in order to meet a demand that 'something must be done' should be consciously and explicitly resisted. This somewhat severe restriction may be made less rigid by recognizing a significant role for symbolic legislation. Legislation that plays a declaratory role by conferring legitimacy on fundamental social values already plays a significant role in the constitutional arena, for example, the equality provision (section 15) in the Charter of Rights and Freedoms and the human rights recognized in the Universal Declaration of Human Rights.[41] The important implication of this argument is to stress that the symbolic dimensions of legal enactments are not restricted to their immediate effectiveness—the securing of formal rights can itself be a precondition of the forms of legal and political action to render them ineffective.[42] The second general implication of my argument is that all projects for legal change should be considered within a framework that explicitly interrogates the interrelationship between legal and other forms of governance.

This exploration of the relationship between law and the agency of citizenship suggests some further directions for research activity. First, the general case has been made for the significance of the interaction between legal and other modes of governance, what may be called 'hybrid governance'. Studies exploring specific patterns of interaction between legal and other forms of governance could be informative. Of particular significance are fields of social life where legal policy seeks to encourage citizens to pursue consensual settlements in conflicts of interest and values, but which have in recent decades tended to generate increasingly intense legal regulation, as illustrated above with respect to family and divorce issues. Such projects might open up policy implications as to which other regulatory techniques and forms of governance might usefully be associated with specific legal reforms under consideration.

Second, a law-and-governance approach can help to identify a general framework of inquiry for empirical projects. It suggests the need to address the following elements. What are the characteristics of the specific sites or arenas of governance? What types of social relations are involved and what potential techniques of governance are available? Within this context, inquiry into the place of law in this specific instance of governance can be explored. What are the distinctive features of the specific project of legal intervention that are most significant for those who are subject to governance? What resources, conditions, or capacities can be provided or be removed from those directly affected? More generally, this question suggests the need to explore potential legal governance from the standpoint of the governed. What forms of problems and organizational arrangements does the project under consideration present when considered from the standpoint of those institutions and forms of expert knowledge that will be implicated in the administration of a projected legal initiative? Finally, what will be the implications of these three aspects for the general legitimacy of the governance project and its capacity to realize the declared governance values, which may variously be articulated in terms of democratic, accessibility, or participatory goals?

Notes

1. This chapter was originally prepared for the Law Commission of Canada and first presented at a seminar organized by the Commission on 'Law and Governance Relations' at the Centre of Criminology, University of Toronto, 8 June 1999. My thanks to the participants for helpful comments and suggestions; they are absolved of any responsibility for the content, which remains, as ever, with the author.
2. It should not be thought that the attempt to ground truth and value claims has been abandoned; new projects are always on offer. The neo-liberal attempt to revitalize faith in the market is a paradoxical amoral quest for certainty in the very uncertainty of the impersonal forces of the market. More interesting in the present context is the project of 'human rights' as not only a grounding for international relations, but a new universalistic framework for social values.
3. R. Pound, 'Causes of Popular Dissatisfaction with the Administration of Justice', *American Law Review* 40 (1906): 729.
4. Forms of law exhibit significant variations in the manner in which law is related to the social relations it seeks to regulate. Thus 'legislation' and 'case law' differ in terms of the source of their legitimacy; the former presupposes some process of representative deliberation while the latter is the result of autonomous deliberation within the judiciary. Feudal and bourgeois, private and public, criminal and civil are other significant variant forms of law.
5. P. Nonet and P. Selznick, *Law and Society in Transition: Toward a Responsive Law* (New York: Harper & Row, 1978).
6. T.H. Marshall, 'Citizenship and Social Class', in Marshall, *Sociology at the Crossroads* (London: Heinemann, 1963), 67–127.
7. F.A. Hayek, *Law, Legislation and Liberty*, 3 vols (London: Routledge & Kegan Paul, 1973–9).
8. This line of analysis was given rich theoretical grounding through Foucault's *Discipline and Punish: The Birth of the Prison* (New York: Pantheon Books, 1977).
9. It is significant that those who came to the defence of welfarism were predominantly either beneficiaries or employees of welfare bureaucracies with a vested interest in its maintenance, and this further politicized the anti-welfare offensive.
10. R. Abel, *The Politics of Informal Justice*, 2 vols (New York: Academic Press, 1982); H.W. Arthurs, *'Without the Law': Administrative Justice and Legal Pluralism in Nineteenth Century England* (Toronto: University of Toronto Press, 1985); M. Galanter, 'Justice in Many Rooms: Courts, Private Ordering and Indigenous Law', *Journal of Legal Pluralism* 19 (1981): 1–47.

11. For present purposes I refer to the general idea of a plurality of 'legal forms'; a more extended account would need to pay attention to different dimensions of legal phenomena, for example, by distinguishing between institutional variations and the long-standing importance of different types of 'legal reasoning', an approach that finds its classical articulation in Max Weber's sociological theory of law. M. Weber, *Law in Economy and Society*, ed. M. Rheinstein (Cambridge, Mass.: Harvard University Press, 1966).
12. The theme of 'reflexive' or 'responsive law' was first introduced by Nonet and Selznick in *Law and Society in Transition* (1978). It was later taken up and extended by G. Teubner: 'Substantive and Reflexive Elements in Modern Law', *Law and Society Review* 17 (1983): 239–85, and Teubner, 'After Legal Instrumentalism? Strategic Models of Post-Regulatory Law', *International Journal of the Sociology of Law* 12 (1984): 375–400.
13. For example, feminist demands to ease the burden on rape victims by restricting the rights of cross-examination by the accused raise complex issues around 'due process' considerations.
14. Some proposals amount to little more than partial responses; for example, the 'plain language movement', while having undoubted merit, cannot be regarded as more than a minor palliative.
15. One of the paradoxical implications of reflexive law theory is that while critical of procedural formalism, its own solutions increasingly rely on the deployment of procedural solutions; this is particularly evident in the writings of both Habermas and Teubner.
16. It may be noted in passing that legal restraint bears close similarities with Nikolas Rose's account of modern liberalism's concern to avoid 'governing too much'. See N. Rose, 'Government, Authority and Expertise in Advanced Liberalism', *Economy and Society* 22, 3 (1993): 283–99.
17. Blankenburg argues that any increase in the use of reflexive law indicates increasing regulation of formerly unregulated social arenas rather than attempts at deregulation. See E. Blankenburg, 'The Poverty of Evolutionism: A Critique of Teubner's Case for "Reflexive Law" ', *Law and Society Review* 18 (1984): 275.
18. S. Fish, 'The Law Wishes to Have a Formal Existence', in A. Sarat and T.R. Kearns, eds, *The Fate of Law* (Ann Arbor: University of Michigan Press, 1991), 159–208.
19. A. Hunt and G. Wickham, *Foucault and the Law: Towards a New Sociology of Law as Governance* (London: Pluto, 1994).
20. M. Foucault, 'The Subject and Power', in H. Dreyfus and P. Rabinow, *Michel Foucault: Beyond Structuralism and Hermeneutics* (Chicago: University of Chicago Press, 1982), 221.
21. P. O'Malley, 'Risk, Power and Crime Prevention', *Economy & Society* 21, 3 (1992): 252–75.
22. Foucault, 'The Subject and Power', 208–9.
23. J. Shklar, *Legalism: An Essay on Law, Morals and Politics* (Cambridge, Mass.: Harvard University Press, 1964).
24. Weber, *Law in Economy and Society*.
25. These two processes Foucault designates by the term 'assujettisement'. See 'On the Genealogy of Ethics: An Overview of Work in Progress', in Paul Rabinow, ed., *The Foucault Reader* (New York: Pantheon, 1984), 353.

26. B. Cruikshank 'The Will to Empower: Technologies of Citizenship and the War on Poverty', *Socialist Review* 23, 4 (1994): 29-55; J. Handler, *Down From Bureaucracy: The Ambiguity of Privatization and Empowerment* (Berkeley: University of California Press, 1996).
27. The concept 'agent' and 'agency' as used here is concerned with the degree of autonomy of a legal subject. It is thus almost the reverse of the narrower legal sense of 'agent' as one who acts with limited powers on behalf of a principal.
28. It is worth noting that to be an agent does not always involve taking positive action; agency may involve resistance, a refusal to act. But in turn this may require the mobilization of resources to defend that inaction; for example, if I object to the imposition of admission charges for access to Crown lands, I may later have to defend a trespass action. My capacity as an agent is qualitatively affected if I can mobilize the resources so as to act in concert with others.
29. J. Handler, *Law and the Search for Community* (Philadelphia: University of Pennsylvania Press, 1990).
30. R. Unger, 'The Critical Legal Studies Movement', *Harvard Law Review* 96 (1982): 561-675.
31. J. Habermas, *The Theory of Communicative Action*, vol. 2: *Lifeworld and System*, trans. Thomas McCarthy (Boston: Beacon Press, 1987). The concept 'juridification' refers not so much to a quantitative expansion of law, but rather to the transformation of a wide range of social processes into ones that partake of distinctively legal form, for example, the spread of the requirement of notice of 'charges' in a wide variety of disciplinary procedures. Juridification also involves belief that most, if not all, social relations can be comprehensively shaped by legal means and thus perpetuates the illusion that there can be a legal answer to most social problems.
32. Teubner, 'After Legal Instrumentalism?', 394.
33. J. Habermas, *Between Facts and Norms: Contributions to a Discourse Theory of Law and Democracy*, trans. W. Rehg (Cambridge, Mass.: MIT Press, 1996).
34. As Donzelot observed in the context of an earlier phase of the history of the regulation of the family: 'There are no longer two authorities facing one another: the family and the [state] apparatus, but a series of concentric circles around the child: the family circle, the circle of technicians (e.g., social workers), and the circle of social guardians (e.g., magistrates) . . . the more [social] rights are proclaimed, the more the stranglehold of a tutelary authority tightens around the poor family.' J. Donzelot *The Policing of Families: Welfare Versus the State* (London: Hutchinson, 1980), 103.
35. A similar conclusion emerges from a different line of reasoning. The 'systems theory' approach associated with Niklas Luhmann reasons as follows. The current crisis of law stems from it being 'overburdened' as a result of the expanding demands for law to provide substantive responses to social problems. Luhmann argues that law cannot 'understand' the complexities of the fields it is called upon to regulate; therefore it should refuse such demands since law lacks a 'conceptual system oriented towards social policy'. Cited in Teubner, 'Substantive and Reflexive Elements', 264.
36. J. Spinner, *The Boundaries of Citizenship: Race, Ethnicity, and Nationality in the Liberal State* (Baltimore: Johns Hopkins University Press, 1994).
37. The call for a return to civility has made frequent appearances in contemporary political discourse. Both President Bill Clinton and Vice-President Al Gore appealed

for a return to civility in the wake of the unsuccessful impeachment of Clinton over the Monica Lewinsky affair process. It has also attracted considerable academic interest, which, while resonating most sympathetically with conservatism, has increasingly been articulated from a diversity of intellectual and political positions. See S.L. Carter, *Civility: Manners, Morals and the Etiquette of Democracy* (New York: Basic Books, 1998); F. Hearn, *Moral Order and Social Disorder: The American Search for Community* (New York: Aldine de Gruyter, 1997); M.G. Kingswell, *A Civil Tongue: Justice, Dialogue, and the Politics of Pluralism* (University Park: Pennsylvania State University Press, 1995); E.A. Shils, *The Virtue of Civility: Selected Essays in Liberalism, Tradition and Civil Society* (Indianapolis: Liberty Fund, 1997).

38. By 'mode of governance' I designate the connections between a range of specific 'fields of governance'; when they share common features it is possible to speak, for example, of 'patriarchal', 'liberal', etc. modes of governance.
39. M. Foucault, *The History of Sexuality*, vol. 1: *An Introduction*, trans. R. Hurley (New York: Random House, 1978), 144.
40. There are divergent interpretations of this key passage. See F. Ewald, 'Norms, Discipline and the Law', *Representations* 30 (1990): 138-61; A. Hunt, 'Foucault's Expulsion of Law: Towards a Retrieval', *Law & Social Inquiry* 17, 1 (1992): 1-38; V. Tadros, 'Between Governance and Discipline: The Law and Michel Foucault', *Oxford Journal of Legal Studies* 18, 1 (1998): 74-103; N. Rose and M. Valverde, 'Governed By Law?', *Social and Legal Studies* 7, 4 (1998): 569-79.
41. I have elsewhere sought to show the political importance of symbolic rights in the process of political mobilization. See A. Hunt, 'Rights and Social Movements: Counter-Hegemonic Strategies', *Journal of Law and Society* 17, 3 (1990): 309-28.
42. The securing of legal rights does not itself guarantee forms of action that render them effective, as is well illustrated by the controversies that have surrounded *Brown v. Board of Education* and civil rights legislation in the US.

Chapter 4

The Legal Regulation of Politics: Governance, National Security, and the Rule of Law in the Modern State

BARRY WRIGHT

Regulatory ordering is sometimes contrasted with repressive forms of control, a distinction that helps to explain the differences between administrative or civil proceedings and criminal prosecutions. In traditional or professional conceptions of law, regulation also tends to be associated specifically with administrative law. This collection of essays, among other things, seeks to challenge such narrow conceptions of regulation by exploring the wider social regulatory role of law and law's relationship to other forms of governance. A distinction nonetheless continues to be drawn between regulatory and repressive law, and directing attention 'beyond the coercive' is a major theme in this volume. This chapter offers a counterbalance or challenge to this emphasis on a repressive/regulatory dichotomy. By presenting a historical overview of uses of law against real and perceived threats to the state, my focus is on one of the most coercive dimensions of law. A rigorous examination of the legal regulation of politics cannot neglect this particular area of law, nor arguably can this area be divorced from forms of legal regulation administered at a much greater distance from the active hand of the state, including the most consensual forms of private dispute resolution. The notion of a regulatory/repressive continuum is more useful than posing a dichotomy between these legal modes. The degree to which the coercive is resorted to depends on the nature of the dispute or crisis, its urgency, and the level of public concern associated with it. Common to all legal modes or forms, however, is that state law constitutes, manages, and resolves conflicts authoritatively, and in a range of ways calculated to appear to be socially neutral, disinterested, or objective. This essential similarity links repressive and regulatory laws.

The patterns reflected in the resort to state security measures reveal a great deal about the relationship of the law, governmental authority, and power, as well as about the limits of formal claims of disinterestedness, non-partisanship, or objectivity that lie at the heart of the notion of the rule of law. State security laws, designed to protect the safety and sovereignty of existing state institutions and associated interests, also set the limits of political activity within it. As such they can be understood as establishing the boundaries of political life, regulating activities deemed to be threats, not only through classic political trials, where the state prosecutes offences such as treason and sedition, but also through legislated forms of executive control over sensitive information and emergency interventions, as well as intelligence operations involving surveillance and threat assessment.

This area of law itself is often subject to polarized understandings. Conservative scholars generally emphasize the exceptional nature of state security measures and accept claims that they protect fundamental common interests rather than partisan or elite interests. Critical scholars tend to see such laws as an instrument manipulated at the behest of, or in the political interests of, powerful groups. Yet the exceptionalism of this area of law tends also to be assumed, especially in recent years as critical scholars attempt to confront what appears to be the current withering away of the modern state (as it has developed since the early nineteenth century), eroded from above by forces of globalization, and from below as forces such as identity politics fragment a common public sphere and citizenship.

An examination of the historical track record of legislative enactments, judicial decisions, and institutional developments in state security law since the eighteenth century reveals a more complex picture than that portrayed in conservative and much of the critical literature. While the recent critical work moves away from crude instrumentalist conceptions focused on criminal laws, drawing much-needed attention to more subtle forms of legal regulation and non-legal modes of regulation across a wider range of social relations, we should not assume that the more repressive elements of state power as exercised through the law have become obsolete or are irrelevant. While power has become reconfigured into more complex forms in what some call the late-modern state, as responses to the events of 11 September 2001 remind us, the state has not withdrawn its most coercive laws related to public order and sovereignty.

The chapter is divided into four sections. The first surveys existing interpretations of the relationship between law and politics and the various ways in which law regulates power as expressed in political forms. State security laws are, of course, just one dimension of this relationship, a coercive mode of regulation that accompanies administrative and constitutional modes. The second section provides a very brief overview of political trials and state security measures from the late seventeenth to the early nineteenth centuries. While the overview may resemble elements of the old-fashioned scholarly genre of constitutional history, it is approached with a decidedly critical spin. Struggles against national security initiatives were not only central to pushing the boundaries of political opposition and free expression, they also highlighted contradictions between the formal claims of the rule of law and the repressive practices of state authority, contradictions that persist to this day and say something about the complexities of law in the exercise of power. The third section explores developments in security laws and bureaucracies with the rise of the modern liberal state and the emergence of more subtle means of regulating political conflict. The final section seeks to address the current situation where the state is seen by some as an institutional formation of the past, raising the question: In crisis, will such repressive measures be revealed as moribund anachronisms or will they be revealed as the state's final resource?

Perspectives on Law and Politics

There is a wide range of understandings of the relationship between law and politics and of law's role in the regulation of politics, much of it shaped by theoretical and scholarly assumptions and research purposes. Put simply, as suggested above, conservative perspectives tend to portray the relationship between law and politics as unproblematic, at least so long as established constitutional protections are respected. Critical perspectives tend to understand law as a formalized authoritative version of politics, or hold that politics to some degree directly or indirectly informs the definition of all laws and their administration.

Conservative perspectives tend to prevail in traditional legal scholarship, which takes place within the context of professional legal education. Within this vocation-training milieu there is a preoccupation with technical doctrine and expositional legal reasoning (determining law and applying it to different fact situations), practical skills thought to be central to legal practice, which, in turn, is largely oriented to processes of dispute negotiation and resolution.

This view proceeds from the assumption of a broad social consensus around the existing order and in the production of laws. The rule of law generates a separation between law and politics by formally protecting the autonomy of legal institutions and processes from partisan political pressure, which, in turn, allows for the impartial and authoritative resolution of disputes according to established and discernible legal rules. So long as these rules or laws have been produced by legitimately constituted authority and the institutional guarantees of the rule of law are respected, scholarly critique is confined to largely technical questions such as procedural lapses, the interpretation of legal rules, and the unauthorized adoption of discretionary decision-making.

It follows from this perspective, with its assumptions of social consensus, that the regulation of politics is essentially non-coercive. The relationship of law and politics is largely encompassed by what is understood as public law. This entails judicial review of jurisdiction, the procedures of public tribunals and decision-making bodies, or other areas where the implementation of public policies generates disputes that are directed to the courts for resolution.

The Canadian Charter of Rights and Freedoms adds a new dimension to this area where public measures can be challenged by private interests on the basis of alleged rights violations. Moreover, some legal scholars have interests that range well beyond this 'regulation through judicial review' focus. A wider conception of the political regulatory role of law includes examination of election requirements (constituency arrangements, official party status, political advertising, disclosure of poll results, etc.) and legislative functions (including related matters such as parliamentary privileges, Crown prerogatives, executive Orders-in-Council, and the delegation of regulatory powers).[1]

Critical perspectives, more prevalent in non-professional interdisciplinary legal studies, tend to see all law as in some degree political and its administration influenced by a range of social pressures, particularly if groups within society are understood to be unequal and in conflict. Politics directly or indirectly informs the definition of all laws and their administration, affecting the routine, day-to-day business of the courts and other legal institutions. This includes the impact of social values, which affect the decision-making, and wide but hidden discretion exercised by legal actors such as police, prosecutors, judges, and other state officials.

From this perspective the law regulates conflicts in favour of particular social interests, or authoritatively constitutes particular subjects and distinctions in accordance with such interests. The rule of law and its associated guarantees are mere formal claims that mean little in practice. The primary importance of the rule of law is found in its role as an ideological construct. It generates the perception that laws and their administration are neutral and reflect the common good. The law is always in some degree political in its ultimate impact, although this may not be readily apparent. In short, the technical preoccupations of traditional approaches to law fail to address these issues, which become more obvious when interdiscplinary methods and theoretical perspectives are used.

This portrayal of positions on law and politics is somewhat of a caricature, set out in broad brushstrokes, and there are exceptions. Critical movements within professional legal scholarship, for instance, 'deconstruct' doctrine and other technical aspects of the law to arrive at similar conclusions to many who work within interdisciplinary legal studies. Such critical movements within professional education reach back to the American legal realist movement of the 1920s and are reflected today in the critical legal studies movement.[2] Canadians who study constitutional law are familiar with critiques of the role of the courts in federalism, in particular the issue of the largely decentralizing impact of judicial review of jurisdiction under the British North America Act. The pattern of judgments in appeals heard by the Judicial Committee of the Privy Council has been explained in a variety of ways, the classic critical position being that formalism disguised a covert imperial policy of 'divide and rule'. Similar controversies in constitutional law rage around the Charter of Rights, where judges have been further empowered to balance public interests with rights, although such recent concerns echo long-standing critiques of judicial interventions in the administrative law sphere (e.g., use of procedural or natural justice principles to influence public policy). On the other side of the divide sketched out here, many scholars engaged in interdisciplinary legal studies accept the premises of social consensus and the rule of law. These include numerous scholars in the law and economics movement, as well as others, engaged in empirical studies of the administration of law, public policy formation, and law reform.

Despite the exceptions, however, most approaches to the relationship between law and politics fall broadly within the traditional narrowly conceived or critical 'law is politics' perspectives outlined above. The problem with both camps is that the complexities of the law/politics relationship tend to be neglected.[3] The next sections attempt to draw out some of this complexity by examining some historical experiences with state security laws as well as current debates about the late-modern state.[4]

State Trials and the Rule of Law: The Study of the Repressive Regulation of Politics to the Early Nineteenth Century

The British constitutional system is the main foundation for most of Canada's legal and political institutional arrangements, and the experience of political trials in both Britain and Canada reflects the importance of the courts as an arena of political struggle, especially before the mid-nineteenth century. The prosecution of offences such as treason and sedition was at the centre of constitutional battles associated with developments such as parliamentary supremacy and responsible cabinet government, limits on Crown prerogatives and discretionary executive powers, and protections associated with the rule of law, including the right to a jury trial and judicial independence. The state's resort to political prosecutions against perceived threats such as organized political opposition and the free press tested the claims associated with these British constitutional developments, the benefits of which its colonies and dominions were claimed to enjoy as well.

The study of political trials tends to be associated with the eighteenth- and nineteenth-century collections of 'state trials'. The focus was on dramatic treason trials, treason representing the most serious offence in English law as symbolized by the horrific traditional punishment of public drawing, hanging, beheading, and quartering for those convicted of it. Much of the commentary accompanying the collections of legal reports of these political trials was uncritical but the better collections were libertarian in orientation and the very best, *Howell's English State Trials*, initiated by the radical William Cobbett, did engage with some of the complexities of the emerging modern relationship of law, politics, and power.

The English state trials collections developed after the seventeenth-century conflicts and the return to stability. It was hoped by reformers that government repression and abuses of the laws would be checked by developments such as the Habeas Corpus Act, 1679 (the right not to be detained without charges and to be tried within a reasonable period of time), the 1696 Treason Act (right to defence counsel, advance look at the Crown's case and jury panel, the need for two witnesses to the act of treason, etc.), and other more general constitutional developments such as the Glorious Revolution of 1689 and the Act of Settlement, 1701, which established, among other things, the right to be tried in regular proceedings by a jury of peers, and limits on executive manipulation of the bench (tenure based on good behaviour rather than royal pleasure) and Crown prerogatives around public prosecutions. Huge differences remained between governments and their critics over the practical implications of these developments. Matters such as toleration of organized political opposition parties, limits on judicial participation in government councils, and freedom of the press remained as objectives to be struggled for in legal and political arenas.

The first collection of political trials was published in 1719 and was extended by editions in 1730, 1775, and 1781.[5] In the wake of the repression in reponse to government fears of the French Revolution and rising anxieties about urban unrest and labour organization, William Cobbett initiated a new series edited by barristers Thomas Bayly Howell and his son Thomas Jones Howell, with commentary informed by politicians John Hawles, Charles James Fox, John Horne Tooke, radical lawyers Thomas Erskine and John Philpot Curran, the historian Alexander Luders, and philosopher William Godwin. Although there were successors to this series, none approached the standards of this series.[6]

The late nineteenth-century decline of constitutional history as a discipline was matched by the rise of legal scholarship that avoided politics and historical and political studies that lost engagement with law. The narrowing of the study of law into specialized professional training, based on a positivist paradigm that focused on technical doctrine and expositional legal reasoning and assumed the validity of formal rule of law claims, as outlined earlier, had negative implications for the study of this area. The rising influence of the social sciences on the study of history and politics had a decidedly more positive impact, not only in terms of empirical and theoretical sophistication, but also because, as social historian Edward Thompson put it, popular protest and radicalism were taken more seriously and no longer met the enormous condescension of history. However, it has only been in the past 30 years that a sustained interest in law has emerged in this area. The work of social historians has been particularly important, and those concerned about 'history from below' have turned to court records for insight into the lives of those who left few records of their own. This has also generated sophisticated new understandings of law, power, and the courts as a site of control and contestation that allows for considerable refinement on the generalizations about law, power, and the state offered by social theorists. And while the Cobbett/Howell state trials series identified the contradictions between rule-of-law formal claims and repressive practices, social historians offer much deeper insights into these contradictions, notably in terms of their ideological significance.

This has opened the way for a departure from the 'reductionism' that tends to prevail in the study of state security measures. Reflecting the two camps surveyed in the previous section, professional, or what historians would term 'whiggish' reductionist interpretations, suggest that national security measures become much less important and relevant with the achievement of limits on executive authority and institutional safeguards to preserve integrity of the rule of law. Egregious cases since the nineteenth century are dismissed as anomalies and measures are limited to extreme threats to the prevailing order where decisive action is necessary. On the other hand, critical reductionist interpretations tend to portray national security measures as instruments manipulated by political elites for socially interested ends, functions displaced largely by the development of more subtle means of managing conflict.

It is perhaps axiomatic that states seek to preserve themselves; the ability to maintain institutional integrity against perceived fundamental threats is an important element of power. The means used to deal with such threats differ in important ways. Organized, politically plural modern states claim a monopoly over the legitimate use of violence and a formal adherence to the rule of law. The use of military force or unregulated state violence against security threats is avoided. This is not simply because harshness is a sign of weakness and forbearance to the point of leaving matters to the legal process is seen as a strength, but that resort to the law and the characterization of activity as criminal lends greater credibility to the repression. Part of the calculated advantage of resorting to the law rather than less regulated forms of state violence is the enhanced opportunity to shape public perceptions and lend greater legitimacy to official actions. The success of such a course depends crucially on the appearance of disinterested and impartially administered law. As Thompson explained, the advantage of greater legitimacy gained by proceeding through legal processes carries a cost; the law can only be stretched so far and its partisan biases or manipulation cannot be apparent, precisely because of the constraints of popular expectations concerning its formal claims. The formal claims of the rule of law can actually limit repression and with this insight Thompson argued that the rule of law could be shown, by his particular historical examples, to be an unqualified good: the formal claims around the law not only constrain the exercise of power, they also create opportunities to contest repression.[7]

Informed by these insights, recent historical work on the eighteenth and early nineteenth centuries reveals that the language of constitutional liberties and the rule of law were used by governments in justification of measures to protect constitutions, by opposition radicals, protest leaders, and defendants to lend weight to their criticism of government measures, to inform their counter-hegemonic strategies, and to appeal for broader support. While the level of popular engagement with this discourse is difficult to measure, it is apparent that the challenge of meeting popular expectations was clearly a prominent consideration for governments facing a choice of repressive strategies. Careful historical study also reveals that governments would usually have a range of options between the extremes of military force and regular legal procedures, demonstrating the rich complexity of law in this area. When the government resolved to pursue a legal route, intermediate procedural expedients could be deployed tactically depending on perceived public toleration. In considering these options governments were mindful of levels of public toleration and that too obvious a manipulation of process would compromise the ideological advantages of proceeding through the law. Legal forms of repression were sometimes successfully contested using rule-of-law claims, and acquittals evidently embarrassed and frustrated governments, suggesting such counter-hegemonic victories were more than calculated gestures to preserve the illusion of even-handed justice.[8]

National Security and the Rise of the Modern State

As we move beyond the early nineteenth century and consider the rise of the modern state, the role of this repressive form of the legal regulation of politics undergoes changes. Recently, some postmodernist-influenced historians have suggested that as the state develops, power is increasingly exercised at a distance from the state's direct control and repressive expressions of power become displaced by more subtle forms of social ordering and moral regulation. Attention is drawn away from what Michel Foucault calls juridical expressions of power to multiple sites of power.[9] Has the emergence of new, more subtle forms of governance rendered state security measures increasingly irrelevant to questions of power in modernity? The evidence suggests not the abandonment of national security laws, but rather their persistence and renewal within a more complex deployment of legal and social regulatory practices.

State security measures have been elaborated in the area of executive enabling legislation, largely insulated from the scrutiny of the courts or Parliament, and in crisis these have proved to have more repressive utility than high-profile criminal prosecutions. Outside times of crisis, pre-emptive, micro-managed means of handling perceived security threats have accompanied the development of new institutions of surveillance and intelligence assessment (professional policing and the development of security branches and secret service bureaucracies mandated by espionage and official secrets legislation). The state appears increasingly preoccupied with accumulating information about sources of uncertainty and instability, identifying threats to state sovereignty and to the prevailing social order. While such historical developments in national security have been relatively neglected, they have paid a great deal of attention to parallel developments in the routine administration of criminal law since the early nineteenth century. These include the rise of professional policing, the introduction of public prosecutions and lawyers in the routine processes of the criminal trial, and the development of the penitentiary. These largely utilitarian-inspired reforms to the criminal law accompanied codification and abolition of capital punishment for all but the most serious offences and appeared to limit the wide discretionary authority and justice that characterized the pre-reform system of criminal law. As more critical historians point out, these changes not only responded to a system that was breaking down under the sheer pressure of numbers; the changes also saw the 'net' thrown wider, creating a system that could penetrate much further into society and process far greater numbers than was possible before. Developments in the repressive legal regulation of politics follow a similar trajectory.

Although remaining part of the law to this day, the offences of treason and sedition are largely moribund. From after the 1820s in Britain and the rebellions of 1837-8 in the Canadas, we find only the occasional reversion to the prosecution of the classic political offences during wartime and serious insurrection. Before this time, the procedural obstacles presented by legislation such as the Treason Act and Fox's Libel Act were at times bypassed by constructive judicial interpretations, emergency legislation to override regular procedural protections, resort to residual Crown prerogatives, and, following the 1837-8 rebellions and patriot invasions, the trial of civilians by courts martial. Such expedients were less accessible as the nineteenth century progressed. The legislative process became more unwieldy (the Reform Act, responsible cabinet government), judges became more remote from the centre of government (removal from executive councils), and greater toleration of political pluralism and organized opposition made such political prosecutions controversial except in extreme security crises. Perhaps most importantly, the courtroom became an increasingly awkward site of repression as potential defendants had access to a radical bar ready to exploit procedural rights and an established track record of some counter-hegemonic success.

Sedition, for example, was largely displaced during the second quarter of the nineteenth century by the development of laws of unlawful assembly and breaches of the peace, a change in prosecutorial focus examined in detail by Michael Lobban.[10] The proactive presence and surveillance of the newly established professional police made it possible to cast the net much wider and make effective use of these measures.

As the nineteenth century progressed there was great elaboration of intelligence gathering about *potential* security threats. Instead of dealing with the commission of activity falling within the definition of political offences, lawful activity came to be subject to systematic and ongoing surveillance on the basis that it could have unlawful consequences.[11] Such surveillance and security risk assessment were first manifested in operations of special branches of the police and military, and over the course of the nineteenth and early twentieth centuries these practices developed into separate intelligence bureaucracies.

During this period, as well, the management and secrecy of information, particularly about technology, becomes a national security preoccupation, and is married to espionage and counter-espionage activities of the new intelligence bureaucracies. The development of breach of official trust legislation in the late nineteenth century was followed by the passage of the Official Secrets Act before World War I. The Official Secrets Act was further elaborated in Britain and Canada after the war, expanding the prerogatives around government secrecy into the realm of political discourse, justifying extended surveillance on the basis of security vetting, and creating a range of special procedural expedients should matters be taken to trial. US espionage legislation, first passed in 1917, achieved similar objectives.

The most notable development at the harder end of national security measures is the passage of what is best described as 'executive-enabling' measures, which effectively bypass scrutiny by means of regular legislative and judicial review on the premise that governments need a free hand to deal effectively with designated emergencies or security threats, unencumbered by the need to be accountable during the period of crisis. Such measures have their roots in Crown prerogative powers, such as the power over 'aliens', manifested typically through indefinite detention or summary deportations without recourse to the courts. Modern executive enabling legislation applies to citizens and aliens alike during defined emergencies, authorizing wide-ranging cabinet regulations (Orders-in-Council) that expand police powers, suspend habeas corpus, and allow indefinite detention and other measures such as censorship and the regulation or expropriation of property. Canada's version of such legislation is the permanently enacted War Measures Act, passed during World War I but explicitly extended to have peacetime application in situations of 'apprehended insurrection' through amendments in 1927, and recently revised and retitled the Emergencies Act, 1988. The British Defence of the Realm Act, 1914, Emergency Powers (Defence) Act, 1939, as well as the peacetime Incitement to Disaffection Act, 1934, the Northern Ireland Civil Authorities (Special Powers) Act (first enacted in 1922 and made permanent in 1933), and most recently the Northern Ireland Emergencies Provisions Act, 1975, and the Prevention of Terrorism (Temporary Provisions) Act, 1974, are similar, ostensibly temporary emergency acts that have become permanent executive-enabling measures.[12] The omnibus US Espionage Act, passed in the later years of World War I, the Smith (Alien Registration) Act, 1940, and the National Security Act, 1947, create similar executive powers and new agencies to enforce them in the United States. It remains to be seen what impact new anti-terrorism, public safety, and military justice laws will have in these jurisdictions.

Canada's experience with these state security laws is extensive. Such measures have been frequently deployed not only in situations of war, invasion, and insurrection but also against manifestations of political opposition, as well as labour organizers and minority ethnic groups ranging from Aboriginal populations and non-loyalist American and Irish immigrants in the nineteenth century to Asian and East European immigrants in the twentieth century. Quebec nationalism has presented continuing security anxieties to British and later Canadian governments, from the period of the French Revolution to the use of the War Measures Act during the October Crisis of 1970 and the 'dirty tricks' that followed. Full use was made of treason and sedition laws and deportation in the early period. British North American governments that remained after the American Revolution readily apprehended potential sources of revolution in the late eighteenth and early nineteenth centuries. After the crisis of the Canadian rebellions, the ascendency of reform politicians and the policies of Lords Durham and Sydenham defused the deepest political conflicts but introduced new interventions to manage social and political order. Sir John A. Macdonald's 'National Policy' introduced new state-building imperatives. Sometimes old measures were reverted to (the treason-like offence of lawless aggression applied to the Fenian invaders, treason prosecutions in response to the 1885 Northwest Rebellion, sedition prosecutions during the 1919 Winnipeg General Strike), but the new interventions were fully exploited to 'better manage' perceived security threats. Examples include police harassment and surveillance (e.g., creation of the RCMP Security Branch, s. 98 of the Criminal Code), preventative internments and property confiscations (e.g., the treatment of Japanese Canadians during World War II under the War Measures Act), repatriation (labour and ethnic leaders, as well as Japanese Canadians, under sections 41 and 42 of the Immigration Act), and the administration of policies promoting segregation and assimilation and prohibiting cultural customs (e.g., amendments to the Indian Act banning the potlatch and sun dance).

Recently, many critical scholars, despite rejecting claims about the democratic achievements of the modern state, have tended to view repressive laws as decreasingly relevant to understanding power in modernity. Michel Foucault has argued that we must eschew the model of Leviathan in the study of power, alluding to Thomas Hobbes's anticipation of the omnipotent power of the modern state in forming the will of its subjects.[13] The postmodern interest in shifting attention to a multiplicity of sites of power, away from the practices of the state and its repressive actions in particular, has been a mixed advance. On one hand, it moves beyond whiggish/professional and critical reductionist accounts by examining the complexities of the exercise of power within and beyond the modern state. On the other, it has diverted attention away from repressive state-centred measures when their evident persistence and renewal demand further attention.

For instance, the recent historical work on 'state formation' and 'moral regulation' is concerned with the nineteenth-century transformations that led to the institutional practices and social relations of modernity.[14] Drawing upon problematizing postmodernist practices, this literature moves beyond whiggish reform narratives (stressing liberalism, utilitarian rational governance, technology, and production) and more critical narratives (stressing contradictions, conflict, and their accommodation by new state institutions). The result is more complex accounts—of the rise of police, penitentiaries, schools, hospitals, and other domains of social regulation—that explore the relations among institutional development, administrative practices, and the processes of moral regulation and self-formation. Universalized identities are promoted, which arch over and diminish cultural differences such as class, gender, age, race, and ethnicity and are effectively regulated by an array of new institutions and specialists.[15] Some scholars, such as Mitchell Dean, drawing on Foucault's concept of 'governmentality', suggest even this new literature remains too state-centred. It assumes an overly rational coherence of state practices and is preoccupied with 'juridically discursive' conceptions of power. More attention needs to be directed to domains of moral regulation that operate at a distance from the state and to forms of ethical self-regulation linked to the formation of identity.[16]

If the practices of moral regulation or governmentality operating at some undefined distance from the state minimize the need for repressive measures, why is it that repressive forms of regulation continue to be resorted to? Canada's extensive historical track record here, as outlined above, cannot be cavalierly ignored. And as we confront current emergencies and potential future political crises and reflect upon the potential response of the state, should we not have a rigorous understanding of these historical experiences?

National Security in Late Modernity

Has the emergence of new, more subtle forms of governance rendered national security measures increasingly irrelevant to questions of power? Are they an anachronistic holdover, residual pre-modern forms of state power? These questions are pressing in the current day when the nation-state is increasingly seen as an institutional formation of the past, eroded from above by globalization and from below by forces such as increasingly diverse identity politics. In an era of a fragmenting public sphere and new discourses of political rationality, will state security measures be revealed as moribund anachronisms or will they be revealed in crisis as the state's final resource?

While in some circles postmodern analyses engender pessimism about the possibilities of resisting power in the face of innumerable sites of power, others embrace more positive predictions about the withering away of the modern state and national institutions, which are seen to create more possibilities for liberation. Such 'deconstruction' promises a postmodern constitutional formation where diverse identities will no longer be subsumed under monolithic national citizenship and the range of state-enforced demands that flow from it.[17]

Carl Stychin, for example, has suggested that Canada is a prime candidate for such a development. Indeed, we may become the first 'postmodern state', based on what he describes as Canada's historical failure to form a coherent national state identity and the Charter of Rights, which opens up an alternative arena through the courts for the resolution of conflicts between identity groups.[18] So long as identities do not coalesce in the form of a demanding nationalism (or completely fragment and destroy common links with other communities), the need for repressive national security institutions will, under such a scenario, presumably disappear.[19]

Such optimism does not seem adequately grounded historically or empirically, as is evident from the political repression surveyed in the previous section. As I have suggested elsewhere, Stychin not only offers a mostly whiggish interpretation of Canada's political and legal history but a most uncritical evaluation of the current potential of the courts as a socially neutral arena of struggle.[20] His analysis conveniently neglects basic issues such as who has access to the courts, the social location and political allegiance of judges, and the ideological work of legal discourse. The optimistic scenario of Canada as the first postmodern state echoes liberal narratives that assume the benign nature of the modern democratic state and endless possibilities for contestation within it.

State institutions and laws are indeed being reconstituted under new discourses of political rationality. However, as Chapters 2 and 3 in this collection suggest, the appearance of the state's withdrawal is deceptive. Hunt, in Chapter 3, points out how replacing direct legal regulation with more informal methods actually increases 'legalization' of relations. In Chapter 2, Purvis argues that while the state is undergoing significant reconfiguration, power remains orchestrated by the state. A good example is the debate around the postmodern fragmentation of that apparently paradigmatic institution of modernity, professional policing. Decentralizing forces and internationalizing pressures are said by some to be leading to the dispersal of law enforcement, as the state struggles with complexity and cost, including the move to community-based policing, the franchising of crime prevention initiatives, and the rise of private policing. While these developments need to be accounted for, they can be equally read as reflecting the ever more pervasive presence of policing in peoples' lives under new discourses of political rationality orchestrated by the state, shifting citzens from passive recipients of services to active citizen participants and suppliers.[21] Law enforcement initiatives take ever-increasing proportions of government resources, relative to other services, even among the most strident deficit-cutting governments.

Another example is citizenship. Citizenship may be understood as the legal status under which individual rights and entitlements are allocated and the social conditions under which integrative or self-disciplined civic-mindedness is encouraged.[22] Some argue that a coherent national citizenship is increasingly difficult for states to maintain. Yet concerns about developing and defining an inclusive national citizenship and constitutional change as a means of fostering national identity and reconstituting politics preoccupy politicians in most late-modern states.[23] All are attempting to encourage citizenship's 'active attributes' of civility, economic self-reliance, and participation beyond the 'passive qualities' of rights, justice outcomes, and entitlements.[24] Anxious reflection by governments about the attitudes of citizens comes in the context of stresses on the public sphere, struggles to accommodate increasingly mobile multicultural populations, and attempts to reconcile legal recognition of minority rights and differentiated cultures.[25] Such efforts are relevant to the processes of governmentality and are not disconnected from the continuing repressive operations of the late-modern state.

The state remains challenged and preoccupied with accumulating information, assessing threats to state sovereignty, identifying sources of contingency, and evaluating conduct. The promotion of active citizenship within the context of decentralizing and internationalizing pressures suggests enormous potential for conflict and instability, not only in terms of narrow conflicts between group rights and individual rights, but also in terms of broader perceived threats to the state, such as secessionist pressures and migration. Michael Dillon points out that in drawing attention to how power operates beyond juridical discourse and the activities of the sovereign state, Foucault did not claim that state power disappears. A continuing intersection between sovereignty and governmentality is manifested in national security:

> For the practices of statecraft, especially in the core areas of foreign and defense policymaking and of 'national security,' for example, are particularly preoccupied by and continually challenged with accumulating the knowledge and expertise concerned to specify the norm around which the distribution of friends, enemies, allies, terrorists, subversives, and so on may be produced.[26]

Even with the disappearance of the crude narratives of the Cold War, late-modern states continue to devote enormous resources to information acquisition and security assessment in the attempt to identify and respond to sources of domestic uncertainty and foreign threats, whether manifested in the form of migration, economic entities, terrorism, or the actions of other states.[27] Recent events will result in even more resources devoted to this area.

Two examples, technology and international trade, help to demonstrate the point further. The advent of electronic and video surveillance, both in state and in ostensibly private forms, and information control laws are late-modern extensions of continuing state security preoccupations.[28] While many hail the liberating effect of the Internet and emphasize the real difficulties facing nation-states in policing information flows and enforcing laws around them, affluent states fully exploit new technologies to monitor and assess security threats and preserve the confidentiality of sensitive information. Many of the issues examined by Michael Mac Neil in Chapter 13 are directly relevant: What role does technology play to facilitate an even more intrusive state and what potential exists for resisting state-centred monitoring and control of citizens' activities?

In a similar vein, developments in international trade raise a number of questions about the persistence of national interests, often hidden, within globalization trends. The US government's resort to the national security exception (Article XXI of GATT) to justify violation of international trade agreements by the Helms-Burton Act (which extends punitive sanctions to non-American citizens who trade with Cuba) is but one of the more transparent examples of national security prevailing over economic liberalization. International free trade policies do not dissolve national interests but redeploy them in complex and covert ways. State security resources are increasingly deployed to deal with transnational economic activity and competitiveness. And as the responses to anti-globalization protesters at the recent Asia-Pacific and Americas summits in Vancouver and Quebec City suggest, the old patterns of confusing dissent with state security and of the political deployment of the RCMP and other security services continue.

In addition to the challenges of technology and trade there are ongoing concerns about the management of refugees and migration. While there are certainly new discourses of political rationality around state security, international co-operation, and national interests, the nation-state continues to orchestrate a range of responses that include the law and security institutions. This has become even more apparent in the responses to the terrorist attacks of 11 September 2001 in the United States. The sweeping Canadian anti-terrorism and public safety measures (currently Bills C-36 and C-42), which supplement the existing laws and institutions described earlier, raise the spectre, much experienced in Canadian history, of loose definitions of state security, preventive and indefinite detention, martial law, and enforcement priorities that compromise judicial independence and freedom of information.

Within the specific Canadian context one could also speculate that the aspirations of francophone Quebecers, Aboriginal peoples, and many other groups will be accommodated within new pluralist constitutional structures. It is equally plausible that they will be suppressed by national security measures by a central state claiming to act according to national interests. The Canadian state most likely will continue to follow the historical pattern of marginalizing identity groups, playing them off against each other, a process that has broken down on occasion. If a crisis does occur, the prospects are not reassuring if indeed we are informed by historical examples.[29] For this reason alone, it is imperative that we have a rigorous grasp of the repressive legal forms that continue within the legal regulation of politics.

Conclusion

Recognition of the complexities of power and the emergence of new forms of political rationality should not lead to our neglecting the very real remnants of repressive state-centred power manifested in such areas as state security laws and institutions. This demands some skepticism be directed at the more sweeping postmodern claims about the dispersal of power in late modernity. Canada still has repressive laws whose origin predates the rise of the modern state (treason and sedition and a range of other political offences are still part of the Criminal Code), and state security institutions and measures developed in the nineteenth and early twentieth centuries and currently under yet further elaboration. To be sure, changes are taking place in the regulation of political life, but a state-centred and repressive core to power continues. State security measures may be understood as the state's final resource, short of war, in dealing with crisis.

While the state's presence is obviously much more remote when it comes to the regulatory modes of law, the hand of the state remains, proclaimed and justified in terms of public or national interests. All legal interventions are rationalized in these terms and are presented, through the ideological discourse of the rule of law, as autonomous from partisan or powerful interests. The law manifests a wide array of forms within a repressive/regulatory continuum. If historical patterns are taken seriously, skepticism is warranted in response to optimism about the withering away of state interests and the consensual potential of law in the new millennium.

Notes

1. See, for example, G. Tardi, *The Legal Framework of Government: A Canadian Guide* (Aurora, Ont.: Canada Law Book, 1992). Tardi describes this area as 'political law', which not only encompasses the interactions of courts and government understood as 'public law' but also analysis of instruments of governance and legal rules of public affairs of interest to political scientists and public policy specialists as well as lawyers.
2. For an examination of the relationship between law and politics from this perspective, see David Kairys, ed., *The Politics of Law: A Progressivist Critique*, 3rd edn (New York: Basic Books, 1998).

3. Alan Hunt identifies this artificial law is/law is not above politics polarization in 'The Politics of Law and the Law of Politics', in K. Touri, Z. Bankowski, and J. Uusitalo, eds, *Law and Power: Critical and Socio-Legal Essays* (Liverpool: Deborah Charles, 1997), 51.
4. The material that follows summarizes some of my research and ideas published elsewhere. See F.M. Greenwood and B. Wright, eds, *Canadian State Trials*, vol. 1: *Law, Politics and Security Measures, 1608–1837* (Toronto: University of Toronto Press, 1996); Greenwood and Wright, eds, *Canadian State Trials*, vol. 2: *Rebellion and Invasion in the Canadas, 1837–9* (Toronto: University of Toronto Press, 2002); B. Wright, 'Quiescent Leviathan? Citizenship and National Security Measures in Late Modernity', *Journal of Law and Society* 25 (1998): 213–36.
5. Thomas Salmon, *A Compleat Collection of State-Tryals, and Proceedings upon Impeachments for High Treason, and other Crimes and Misdemeanours, from the Reign of King Henry the Fourth, to the End of the Reign of Queen Anne* (1719). New editions were edited by Sollom Emlyn and Francis Hargraves.
6. None of the editions attempted definition and the term 'state trial' derived from the use of public prosecutorial authority during a time when most criminal trials were prosecuted privately. A modern definition is offered in Greenwood and Wright, *Canadian State Trials*, derived in part from George Rude, *Protest and Punishment: The Story of the Social and Political Protesters Transported to Australia 1788–1868* (Oxford: Clarendon Press, 1978). Another useful approach is Edward Thompson's distinction between crimes for private purposes and those involving acts of natural justice based on common rights. Others draw on Robert Merton's distinction between non-conforming and aberrant behaviour, e.g., A. Turk, *Political Criminality: The Defiance and Defence of Authority* (Beverly Hills, Calif.: Sage, 1982). See also Otto Kircheimer, *Political Justice: The Use of Legal Procedure for Political Ends* (Princeton, NJ: Princeton University Press, 1961).
7. See E.P. Thompson, *Whigs and Hunters: Origins of the Black Act* (London: Allan Lane, 1975), 462–3. Whether the rule of law is indeed an 'unqualified good' has fuelled much theoretical debate over the limits and possibilities of rights struggles. See, e.g., P. Beirne and R. Quinney, eds, *Marxism and the Law* (London: Hutchinson, 1982). Recognition of the repressive limitations and the 'counter-hegemonic' possibilities of the law does not entail positive judgment of character of the laws applied nor does it deny the possibility of their covert manipulated administration.
8. The insight into law's utility as an ideological force opens up a new dimension in the study of the repressive uses of the law, which historians have begun to explore in Britain, Canada, Australia, and, within a slightly different constitutional tradition, the United States. See, e.g., Thompson's classic 'The Moral Economy of the English Crowd in the Eighteenth Century', *Past and Present* 50 (1971): 76–136; D. Hay et al., eds, *Albion's Fatal Tree: Crime and Society in Eighteenth Century England* (New York: Pantheon, 1975); S.H. Palmer, *Police and Protest in England and Ireland, 1780–1850* (Cambridge: Cambridge University Press, 1988); J. Barrell, ed., *The Birth of Pandora and the Division of Knowledge* (Philadelphia: University of Pennsylvania Press, 1992), 119–43.
9. M. Foucault, 'Two Lectures', in C. Gordon, ed., *Power/Knowledge: Selected Interviews and Other Writings, 1972–77* (Brighton: Harvester, 1980), 102. Foucault

describes this diffused exercise of power 'governmentality'.
10. M. Lobban, 'From Seditious Libel to Unlawful Assembly', *Oxford Journal of Legal Studies* 10 (1990): 307-52.
11. See, e.g., E. Grace and C. Leys, 'The Concept of Subversion and its Implications', in C.E.S. Franks, ed., *Dissent and the State* (Toronto: Oxford University Press, 1989), 62.
12. Although the Emergency Provisions Act was introduced in 1975 as a temporary measure, routinized renewal debates have effectively made it permanent, as is the case with the Prevention of Terrorism (Temporary Provisions) Act, 1974. See R. Devlin, 'The Rule of Law and the Politics of Fear: Reflections on Northern Ireland', *Law and Critique* 4 (1993): 174; also J.E. Finn, *Constitutions in Crisis: Political Violence and the Rule of Law* (New York: Oxford University Press, 1991); K.D. Ewing and C.A. Gerty, *Freedom Under Thatcher: Civil Liberties in Modern Britain* (Oxford: Clarendon Press, 1990).
13. Foucault, 'Two Lectures'.
14. For Britain, see P. Corrigan and D. Sayer, *The Great Arch: English State Formation as Cultural Revolution* (Oxford: Blackwell, 1985). For the Canadian and Australian contexts, see A. Greer and I. Radforth, eds, *Colonial Leviathan: State Formation in Mid-Nineteenth Century Canada* (Toronto: University of Toronto Press, 1992); A. Davidson, *The Invisible State: The Formation of the Australian State, 1788-1901* (Melbourne: Cambridge University Press, 1991).
15. See especially Corrigan and Sayer, *The Great Arch*. M. Dean, *Critical and Effective Histories: Foucault's Methods and Historical Sociology* (London: Routledge, 1994), offers a detailed critical overview of this literature.
16. Dean, *Critical and Effective Histories*, 151-2, 155-6. See also A. Hunt and G. Wickham, *Foucault and the Law: Towards Sociology of Law as Governance* (London: Pluto Press, 1994).
17. See work of Boaventura de Sousa Santos. Also, P. Evans, D. Rueschemeyer, and T. Skocpol, eds, *Bringing the State Back In* (Cambridge: Cambridge University Press, 1985).
18. C.F. Stychin, 'A Postmodern Constitutionalism: Equality Rights, Identity Politics and the Canadian National Imagination', *Dalhousie Law Journal* 17 (1994): 61-82; Stychin, *Law's Desire: Sexuality and the Limits of Justice* (London: Routledge, 1995), 103-7.
19. Stychin, *Law's Desire*, 103, 114-15. Stychin makes no direct reference to repressive state action. He suggests that modernist tactics, such as the sentimental deployment of nationalism, will be futile, although 'some conception of loyalty might still prove necessary to prevent rupture of the nation into its component parts. . . . It may well be that loyalty now only can be expected and only will be forthcoming in those circumstances where the individual is convinced that through membership in the postmodern state, the right to identify variously and to express her identities will best be secured. Loyalty thus stems from the values which underpin the equality guarantees of the Canadian Constitution' (p. 115).
20. Wright, 'Quiescent Leviathan?'.
21. See P. O'Malley, 'Post-Keynesian Policing', *Economy and Society* 25 (1996): 137-64.
22. Citizenship debates have tended to focus on the distinction between classical active citizenship and liberal passive citizenship, and whether social democratic econom-

ic and welfare rights were compatible with liberal rights. See J.G.A. Pocock, *The Machiavellian Moment: Florentine Political Thought and the Atlantic Republican Tradition* (Princeton, NJ: Princeton University Press, 1975); Bryan J. Turner, *Citizenship and Capitalism: The Debate over Reformism* (London: Allen and Unwin, 1986); W. Kymlicka and W. Norman, 'Return of the Citizen: A Survey of Recent Work on Citizenship Theory', in R. Beiner, ed., *Theorizing Citizenship* (Albany: State University of New York Press, 1995), 307–22; D. Burchell, 'The Attributes of Citizens: Virtue, Manners and the Activity of Citizenship', *Economy and Society* 24 (1995): 540–58; C. Tilly, ed., 'Citizenship, Identity and Social History', *International Review of Social History* 40, suppl. 3 (1995): 1–17.

23. See, e.g., Commission on Citizenship, Great Britain, *Encouraging Citizenship* (1990); Senate of Canada, *Canadian Citizenship: Sharing the Responsibility* (1993); Senate of Australia, *Active Citizenship Revisited* (1991).

24. Although the formal rights of citizenship are legally conferred in modern states, their social attributes and political implications are formed in practice. States now seek to promote self-disciplined civic-mindedness. See Burchell, 'The Attributes of Citizens'.

25. See Kymlicka and Norman, 'Return of the Citizen'.

26. Michael Dillon, 'Sovereignty and Governmentality: From the Problematics of the "New World Order" to the Ethical Problematic of the World Order', *Alternatives* 20 (1995): 337–8.

27. See, e.g., B. Castel, ed., 'The Future of Espionage: New Players, Old Game', special issue, *Queen's Quarterly* 100 (1993).

28. See M.J. Shapiro, 'Stategic Discourse/Discursive Strategy: The Representation of "Security Policy" in the Video Age', *International Studies Quarterly* 34 (1990): 327–40.

29. In October 1995 the sovereignty referendum in Quebec proposing to open negotiations with Ottawa on Quebec's independence was narrowly defeated by 50,000 votes (50.6 to 49.3 per cent). The recent Supreme Court reference and the proposed federal 'clarity legislation' constitutionally acknowledge the possibility of secession and attempt to set out ground rules for negotiation. Even if the Canadian government recognized a subsequent verdict in favour of sovereignty, federalist francophones, anglophones, and Native communities have declared they, too, would demand the right of self-determination, which the federal government has said must be accommodated. Would not the possible partition of Quebec create a situation similar to the partition of Ireland? With prospects of violent conflict would the response of the Canadian state be all that different from that of the British in Northern Ireland?

Chapter 5

Radical or Rational? Reflexive Law as *Res Novo* in the Canadian Environmental Regulatory Regime

DAVID J. SCHNEIDER

The Dilemma of Law in the Welfare State: A Failure of Law?

In Canadian society an important and traditional role of law is to provide a means to regulate human behaviour. Government's use of law to prescribe human activity in accordance with policy objectives is the essence of legal regulation. However, legal and social theorists have begun to question whether the complexity of socio-economic processes has pushed us to the limits of our traditional political-legal mechanisms of control.[1] The current concern, often discussed under the rubric of 'the dilemma of law in the welfare state',[2] is that law, or more precisely the customary version of legal regulation, is losing its effectiveness as an instrument of social control. The difficulty of achieving effective legal protection for the Canadian natural environment is but one manifestation of this dilemma.

Since the 1940s and 1950s, the governments of most industrialized countries, including Canada, have assumed many allocative and distributive functions in an attempt to pursue a wide range of social goals. This new style of 'social regulation'[3] cuts across economic sectors and permeates contemporary society. Social regulation pursues a multitude of goals with respect to health, safety, environmental quality, and related issues such as equality of opportunity in education, employment, and housing.[4] Increasingly, law has become an instrument to impose social and political imperatives on the economy, with these strategies of intervention both describing and defining the transformation of 'a government of laws' into the modern welfare or regulatory state.[5]

Contemporary research has generated a large body of critical literature that challenges welfare state claims to social control through law.[6] Implementation research, however, does not necessarily always implicate law as the principal reason for regulatory failure. Some studies of Canada's environmental regulatory regime, for example, draw attention to the historical, political necessity of socio-economic compromise (a failure of politics) as the source of its problematic nature.[7] Other criticisms link environmental regulatory difficulties to governmental budgetary restraints (a failure of economics), while another category of critical evaluation highlights scientific uncertainty as a root cause of environmental degradation.[8]

This chapter presents Gunther Teubner's conception of reflexive law as a theoretical solution to the problem of improving the current Canadian approach to environmental regulation.[9] Reflexive law then is linked to the notion of responsive regulation as formulated by Ian Ayres and John Braithwaite to suggest a new and more effective model for Canadian environmental regulation.[10] The focus of inquiry, therefore, is restricted to an examination of regulatory intervention in the context of law as an instrument of social control rather than an exploration of the auxiliary roles played by politics, economics, science, or other disciplines.

Prescriptive Models and the Debates

The central problem of environmental regulation is subjectively perceived by different constituencies. The public conceptualizes the problem in terms of the increasing risk to human health posed by the ubiquity of toxic chemicals, as well as the inherent dangers of hazardous waste, unsafe drinking water, and polluted air.[11] Spearheaded by pro-environmental interests, the public wants more environmental regulation, stricter enforcement of existing regulation, or some combination of the two.[12] At the other extreme, the common complaint from regulated entities characterizes the problem as 'too much law'. The targets of regulation frequently feel overwhelmed by the volume of rules, regulations, policies, and guidelines directed at them. In their view environmental regulation entails stringent constraints and increasingly high compliance costs.[13] Somewhere in between these polar positions, loosely described as 'environment versus economy', governments attempt to cope with the mass of environmental law they have generated.

Escalating fear surrounding the adequacy of command-and-control regulation[14] to deal with current environmental challenges, as well as future, potentially even more intractable problems, has become a catalyst for questioning the traditional approach.[15] It is being said that the usual attitude of business and government to the relationship between environmental regulation and the goal of maintaining growth and prosperity is now thoroughly out of date.[16] It is also suggested that traditional forms of regulation produce a fundamental misunderstanding of the potential effectiveness of any form of regulation.[17]

There is a debate between supporters of strict government regulation and those who advocate a move towards deregulation.[18] There is also debate between those who would respond to environmental harm with sanctions and penalties and those who favour harnessing the economic self-interest of potential transgressors.[19] Within the contours of these debates, critics from the left point to the gap between expectations and results of government policies while the right resents the growing intrusiveness and associated costs of government policy instruments. Caught in the middle, governments are subject to criticism, either for shirking their environmental responsibilities or for unnecessarily constraining enterprise.

The immediacy and potential magnitude of environmental dangers, together with the awareness of the failings of traditional regulation, have spurred a quest for a new conceptual basis for environmental regulatory policy.[20] Three competing, prescriptive, strategic models have emerged.[21] The first, labelled 're-formalization', is a reaction to the economic and social costs of traditional forms of regulation. Advocates of this model view state intervention as a bar to the attainment of allocative efficiency. They assert the function of law is merely to furnish a framework for a sphere of autonomous activity with fixed boundaries to protect the property rights of private actors. Accordingly, this prescription seeks a certain 'de-legalization' or ordered retreat of the law through withdrawal of its regulatory functions.

A second strategic model, exemplified by a new, market-oriented regulatory style, is known as the 'implementation approach'. Its proponents, mainly economists, seek to enhance the instrumental efficiency of law by importing economic analysis.[22] Within this model there is a further debate between those who favour a directives model of regulation (typified by the penalties and sanctions of command and control) and those who prefer appealing to the economic self-interest of private actors. The latter propose moving to an incentives-based model in which either negative incentives (taxes/charges) or positive incentives (tax concessions and 'green' marketing) are used to induce desired behaviour.

The third alternative, a systems theoretic approach developed by Gunther Teubner, defines the current failings of regulation in terms of incompatibility among the internal logics of various social systems. The legal reticence that characterizes Teubner's prescription is meant to relieve law of the onus of direct regulation by assuming a subtler, more abstract and indirect control of self-regulatory processes.[23] The role of law shifts, therefore, from the planning implicit in both the traditional command and control and the newer market-based regulatory style to one of seeking ways to influence the development of self-regulating processes within other social systems.

The Evolution of Modern Legal Rationality

Writing against a background of mounting frustration with the 'goals, structures and performance of the regulatory state',[24] Teubner builds upon previous neo-evolutionary work and observes that his predecessors all deal with the same problem, namely, the crisis of formal legal rationality.[25] Formal rationality is the hallmark of modern law and the shift away from it (or the 're-materialization of law'[26]) is evidenced by the increasing 'legalization' of various spheres of social life. Elsewhere this process is referred to as 'juridification'. In his work, Jürgen Habermas draws a distinction between two spheres of social life: system and lifeworld. The lifeworld is based on communicative rationality with a goal of resolving conflict through communication and understanding. In contrast, system imperatives are efficiency and consistency. In the system, law is used instrumentally to promote system integration. As Habermas warns, this inexorably leads both to a proliferation of law and to the extension of law's reach into areas of the lifeworld previously unregulated by law. The juridification process, described by Habermas as the 'colonization of the lifeworld', leads to the replacement of the communicative rationality of the lifeworld by system imperatives.[27]

Expressing his concern with the transition to substantive modes of legal thought and practice in the welfare state, Teubner characterizes the shift as a 'crisis' of legal and social evolution.[28] The development of this crisis is illustrated by the assertion of a threefold sequence in the structure and function of modern legal rationality. Teubner identifies each developmental stage with an ideal type of law: first formal, followed by substantive, and finally, reflexive. Each stage is distinguished by reference to the trichotomy of justification of law, its external functions, and its internal structure.[29]

Historically, Teubner's first stage of legal development is contemporaneous with the appearance and growth of the capitalist state. Law is justified by reference to the values of individualism and autonomy and the creation of spheres of activity for private actors. The focus of law is the rule-oriented resolution of private disputes. Its external (or social) function is to provide the structural premises for the development and allocation of resources in a market society. At the same time, the apparent neutrality and autonomy of law provide legitimation for the emergent legal and political systems in capitalist societies.

The second stage of legal development in Teubner's schema is identified with the appearance of the welfare state and the concomitant increase in state intervention in market structures. The justification for the transition to substantive rationality is the perceived need to regulate economic and social activities in order to compensate for market inadequacies. Substantively oriented law is used to attain 'specific goals in concrete situations' and tends to be 'more general and open-ended' yet 'more particularistic' than formal law.[30] Externally, this means that law shifts its focus from autonomy to regulation as it seeks to alter market-determined patterns of behaviour. Law therefore becomes aggressively instrumental in order to determine outcomes. Internally, the structures of law become purpose-oriented. Government-created control agencies exercising delegated authority enforce regulations and standards.

Teubner's third stage of legal development is imbued with a reflexive rationality that holds self-regulation as an ideal. In its external dimension, law is concerned with creating the structural premises required for the encouragement of self-regulative processes in other social subsystems rather than with attempting to provide content to solutions. Teubner maintains law has only 'to decide about decisions, regulate regulations and establish structural premises for future decisions in terms of organization, procedure and competencies'.[31] Internally, reflexive rationality switches legal thinking from a substantive to a procedural orientation.[32] The legal control of society under a regime of reflexive rationality becomes indirect as the legal system attempts only to set out the organizational and procedural basis of future action rather than to dictate or plan the outcome of future action. The role of law, therefore, is no longer the aggressively instrumental one typified by command-and-control regulation; rather, law is 'limited to providing forms of organization, procedures and competencies for relationships within and between organizations.'[33] Teubner describes this as communication through organization. His explanation of this process recognizes that the major subsystems of society are incapable of collective action. Formal organizations within these various subsystems, as collective actors with some power over their members, do however enjoy the ability to communicate among themselves and across their respective boundaries. This leads to a form of 'inter-system relationships' that become autonomous, and includes 'discussion groups, collective bargaining, interrogation procedures and concerted action'. The process of communication through organization provides access, although on an indirect basis, to the central mechanisms of self-regulation located within the various subsystems. Thus politics, for example, 'gains access to the central control mechanisms of firms, trade unions and interest groups.'

The Evolution of Canadian Environmental Law

The trajectory of Canadian environmental law illustrates the evolution of Teubner's ideal types of modern law. The earliest 'environmental' law was based on the concept of nuisance and was intended to protect individuals from unreasonable interference with the enjoyment of their property. Environmental protection was merely a side effect of the vindication of a collection of primarily formal, common-law rights inhering in individuals as individuals. While it can be invoked in environmental cases, nuisance law is afflicted with several serious limitations. One problem arises because of its very formality and adherence to rules. A potential plaintiff is confronted with the difficulty of demonstrating that his/her complaint is 'particular, direct and substantial, over and above the injury inflicted on the public in general'.[34] Thus, when a whaling company erected a factory at Watering Cove on Big Grady Island on the Newfoundland coast and polluted the waters adjacent to the premises of a fishing establishment, the plaintiff fishing enterprise was precluded from recovering damages in nuisance. The (formal) rationale was that the loss was merely consequential (as opposed to direct) damage resulting from the nuisance. Further, the plaintiff's loss was not peculiar to the plaintiff but was suffered in common with everyone else whose right to fish in the subject public waters was affected by the nuisance.

In addition to this type of 'formal rule difficulty', actions in nuisance are generally confined to a small number of litigants. This precludes judicial consideration of the wider implications of any particular case. The limitations of (formal) nuisance law in conjunction with a growing public environmental consciousness have been important in the development of the current (substantive) legal reaction to environmental degradation.

With the advent of the welfare state the nature of the protected interest changed. Environmental protection became a public goal and environmental law underwent a transformation in order to afford protection on a broader scale. Instead of relying on individuals to litigate their personal claims under a formal nuisance law regime, the Canadian environmental regulatory regime has proliferated, with control agencies being created to administer a plethora of new (substantively oriented) statutes and regulations.

The incremental expansion of substantive environmental regulation leads to two significant problems. First, it epitomizes the juridification phenomenon both in terms of the proliferation of law and in the sense of extending law's reach into previously unregulated areas of life. The second problem has a normative basis. The increasing volume of substantive law necessarily delegates a commensurate amount of discretion to administrative and control agencies to make and enforce the law. In effect, control agencies exercise legislative power that 'unbinds substantive law from the legitimating democratic procedures of legislation.'[35]

There is, however, an incipient awareness of adopting a new (reflexive) approach to environmental regulation. Elements of the reflexive rationality identified in Teubner's third evolutionary stage of modern law are beginning to emerge: the introduction of the European Eco-Management and Audit Scheme (EMAS)[36] and the more recent unveiling of the International Organization for Standardization (ISO)[37] 14000 series are examples of the broadening of the horizon of thought about innovative approaches to environmental protection. The EMAS is a 1993 regulation of the European Economic Community designed to promote voluntary participation by industrial-sector businesses in an eco-management and audit program. The objective is to 'promote continuous improvements in the environmental performance of industrial activities.'[38] It provides for the voluntary participation of industrial enterprises in a site-based process of environmental management. Participants are encouraged to be self-reflective and self-critical by the requirement to conduct an initial environmental review, to develop environmental management systems for registered sites, and to establish a company environmental policy.[39]

The ISO 14000 series is a set of generic standards covering environmental management systems, auditing, performance, evaluation, labelling, and life-cycle assessment. Its purpose is to provide any business, regardless of size or type, with a structure for managing environmental impacts. It is not a law or regulation, but a guideline endorsed by the Standards Council of Canada.[40] This new guideline consists of a comprehensive, proactive systems approach to improve environmental protection by using a single environmental management system across all functions of a business organization. It is crafted to allow a business to establish an environmental policy appropriate to itself, to identify the environmental aspects of its activities, to determine environmental impacts, and to identify the relevant legislative and regulatory requirements. This allows a business to establish a structure to implement its policy and to facilitate planning, monitoring, auditing, and review to ensure policy compliance, and to be capable of adapting to changing circumstances.[41]

The reflexive rationale of both the EMAS and the ISO 14000 series lies in their procedural orientation, indirect approach, and mandatory requirement for environmental self-auditing and reporting. Both programs provide participants with public recognition of their commitment to adopt a structured approach to improving their environmental performance. Rather than dictating environmental performance, both the EMAS and the ISO 14000 provide a structured process for integrating environmental issues into the overall management system of a participating business.

Differentiated Subsystems and the Closed Circle

Legal theorists employ a variety of perspectives to develop intelligible accounts of the relationship between law and society and their interaction. One such viewpoint, employed by Teubner, is systems theory. From a systems theoretic viewpoint society has undergone an internal differentiation. This means society has developed numerous subsystems in response to the complexity of its environment. Initially, differentiation took the form of segmentation, that is, the production of many essentially identical subsystems, such as villages in the Middle Ages. Later differentiation occurred through stratification, by which society split into unequal subsystems to form a hierarchy of nobility, clergy, and peasants. Today, societal subsystems are considered to be functionally differentiated, that is, they are established and identified according to the particular tasks they perform, for example, law, politics, economics, religion, science, education, and so on.[42] Traditional conceptions of these (functionally differentiated) subsystems see them as 'closed' and highlight the internal components of the system in order to identify its emerging properties. Modern notions of subsystems consider them to be 'open'. Law, for example, is seen as adapting to and at the same time shaping the social environment.[43] Both traditional command-and-control regulation and the newer, market-based proposals for environmental regulation seem to rely on a linear vision of open systems. They assume that politics can use law instrumentally to effect desired changes in the economic system.

The reflexive legal theory of Gunther Teubner offers an alternative vision of subsystems. By taking account of their increasing complexity, this vision posits that they are neither wholly open nor completely closed.[44] Teubner observes that:

> Different lines of thought about modern society converge on [an] important point. A general discourse on society is, more than ever before, confronted with 'a dissociation of its rules systems' [Lyotard], 'a multitude of language games' [Wittgenstein], and a plurality of 'semiotic groups' [Jackson]. Sociologists characterize modernity as the 'separation of spheres' [Selznick], the differentiation of the 'subsystems of society' [Parsons], the 'operational closure of autopoiesis' [Luhmann], and the plurality of the 'forms of discourse and negotiation' [Habermas].[45]

This view confounds the current notion of subsystems as completely 'open' and sharing the capacity to exert direct influence over one another since the process of differentiation includes the development of a specialized logic or discourse within each semi-autonomous unit. Communication between and among subsystems is thus rendered difficult or even impossible, and therefore 'differentiated' subsystems are less susceptible to direct intervention. This self-referential nature, or 'closed circle' of social subsystems, poses a challenge to the traditional, dominant role of law as a central integrating institution. In particular, it restricts the effectiveness of legal forms of external regulation of other subsystems. To Teubner the process of differentiation means that social and economic arrangements simply become 'too dense, complex and potentially contradictory'[46] for the current mode of interventionist control mechanisms to be effective.

Breaking Out of the Closed Circle

As a research strategy some legal theorists, most prominently Niklas Luhmann[47] and Gunther Teubner, have introduced the biological concept of autopoiesis to their systems theoretic analyses.[48] A current debate among legal theorists concerns the status of law itself as either autonomous and self-referring (a closed system) or derivative of its socio-cultural setting (an open system). Luhmann and Teubner are the chief advocates of interpreting the legal system as an autopoietic entity, that is, a system that is simultaneously open and closed. Using autopoiesis as a metaphor, Teubner explains that the legal system is normatively closed yet cognitively open (or inwardly closed yet open to its environment, which consists of the other societal subsystems). This means the acts of communication that constitute the legal system and that operate within a binary code of legal/illegal, reproduce themselves as legal acts by means of legal acts. Its self-reproductive operations are therefore normatively (or inwardly) closed.

In its operations, however, the legal system constructs a 'legal reality', which is an internal model of the external world. This construction is the basis of its cognitive (or outward) openness. The result, according to Teubner, is 'that external changes are neither ignored nor directly reflected according to a "stimulus response scheme".' Instead, they are 'selectively filtered into legal structures and adapted in accordance with a logic of normative development.'[49]

Teubner stresses an important implication of the relationship between openness and closure within a system. We must change our view of legislation as a transmitter of information to social spheres, for '[i]t is not legislation which creates order in the social subsystems. It is the subsystems themselves which deal selectively with legislation and arbitrarily use it to construct their own order.'[50] To an autopoieticist this creation of 'order from noise' is evidence that a social subsystem experiences legislation as 'noise' (or external interference). This 'noise' is interpreted through the filter of the particular subsystem's norms, values, and ideologies in such a way that only those components of the 'noise' that contribute to order in the system are selected by the system.[51] This closed circle, or self-referential quality of social subsystems, is the basis of their resistance to external forms of regulation.

In terms of contemporary environmental regulation, Teubner's insight means that neither the current form of regulation, that is, the command-and-control approach, nor proposed market-based alternatives directly affect the behaviour of their intended targets, as is posited by current legal theory. Law, or more accurately, legal regulation, is only one of a number of factors, including the cost of compliance with regulations, consumer demand and market competition, environmental group pressure, rulings by courts in environmental cases, ethics and morality, and public opinion and media attention, that influence activity in the economic sphere.

In rejecting the notion that law is able to exert a close and direct influence over other societal subsystems, Teubner poses the question: 'how are we to break out of the closed circle of the law through legislation and penetrate the closed circle of social worlds?'[52] Since the autonomy of social subsystems, rooted in their self-referential relationships, immunizes them from direct legislative intervention, Teubner maintains that only indirect intervention is feasible.[53]

As a consequence of this assertion, Teubner insists that successful regulation demands threefold compatibility among the regulating systems—law and politics—and the system targeted for regulation. Compatibility is achieved when all three systems, during their interaction, maintain their self-producing interaction of elements. Absent such compatibility, regulation is doomed to failure under one of three scenarios described by Teubner as the 'regulatory trilemma'.[54] The first, referred to as the 'incongruence of law, politics and society', maintains that regulatory action is unable to effect change in the targeted system unless it takes account of the self-reference of the regulated system. Regulatory action that fails to do so, for example by ignoring the economic costs of regulation, is simply irrelevant to the interaction of the regulated system's elements. A second reason for regulatory failure is attributable to the 'over-legalization of society', which means that law overly influences the internal interactions of elements in the regulated field to a degree that threatens its self-producing organization. Such regulatory action may follow a functional logic, but it is insensitive to its unintended or uncontemplated side effects. In environmental law this is exemplified by such phenomena as the expansion of the personal liability of corporate directors to the point at which 'directors' chill' sets in.[55] The final instance of regulatory failure explained by Teubner is the 'over-socialization of law'. This situation puts the self-producing organization of law at risk while the self-producing organization of the regulated area remains stable. Law thus becomes subject to capture, either by politics or by the regulated field. The over-socialization of law manifests itself in a variety of ways; for example, law may become merely a political instrument lacking in internal coherence because it follows only political imperatives.

Teubner further analyzes the legal system in terms of its function, performance, and reflection. The function of law is defined as its ability 'to provide congruent generalizations of expectations for the whole of society', while its performance is the resolution of conflicts originating in other social subsystems but that cannot be resolved there. The role of reflection in the legal system is described as reconciling the inherent tensions between function and performance (because they have the potential to conflict and overlap) by imposing internal restrictions on its own capacities. This is accomplished by exercising restraint over the dimension of legal performance. The conclusion is that to be an effective instrument of social control, law must reject the attempt to regulate comprehensively that is demanded by substantive legal rationale. Legal performance (as opposed to legal function) must be restricted to 'more indirect, more abstract forms of social control'.[56]

Therefore, the primary role for reflexive law is one of co-ordination.[57] In lieu of assuming responsibility for concrete social results (as in the substantively oriented approach to the role of law), law must exercise self-restraint by restricting itself to merely 'regulating regulations' through the creation of structures to promote the self-regulation of other social subsystems.[58] In Teubner's words, '[l]aw realizes its own reflexive orientation insofar as it provides the structural premises for reflexive processes in other social subsystems.'[59]

Teubner's vision of reflexive law thus has two self-referential components. First, there is an awareness of the limitations of law in dealing with the growing complexity of modern society, and second, there is the objective of encouraging other social subsystems to reflect on the tensions between their own function and performance. Internal resolution of these tensions is self-regulation.

Ayres and Braithwaite: The Idea of Responsive Regulation

The central argument of this chapter is that Canada's environmental law regime, having undergone a transformation from its early formal rationality to a new substantive orientation, must take the next evolutionary step to capitalize on the reflexive rationality identified by Teubner. The goal is to enhance regulatory effectiveness by introducing a measure of compatibility to the interactions between regulating and targeted systems. The recommended method is to forgo direct regulation in favour of proceduralization, that is, to establish only the parameters (or structural basis) for future action within the targeted system. The idea is to stimulate the system's reflective dimension to mediate between its function and performance in a manner that attaches greater awareness of and concern for the system's external effects on the natural environment. This less direct regulatory approach appears in a model of 'responsive regulation' developed by regulatory specialists Ian Ayres and John Braithwaite. Their recommendation is:

> distinguished (from other strategies of market governance) both in what triggers a regulatory response and what the response will be. We suggest that regulation be responsive to industry structure in that different structures will be conducive to different degrees and forms of regulation. Government should also be attuned to the differing motivations of regulated actors. Efficacious regulation should speak to the diverse objectives of regulated firms, industry associations and individuals within them. Regulations themselves can affect structure (e.g., the number of firms in the industry) and can affect motivations of the regulated.[60]

Ayres and Braithwaite maintain that an effective regulatory regime must be capable of more than applying sanctions for the breach of standards. Following Teubner's insistence on compatibility between regulating and targeting systems, they stress the necessity for compatibility among the regulatory style, its target, and the public policy goals that underlie the need for regulation. By implication, this means that government regulation should be integrated with existing governance mechanisms in the targeted system. Lastly, they recognize the need to alter the target's structure so that the regulatory objectives are internal and voluntary rather than external and forced.

Their innovative approach allows a regulatory target to translate underlying policy objectives into self-generated regulations that then govern how it operates. In a prototype, Ayres and Braithwaite suggest:

> Under enforced self-regulation the government would compel each company to write a set of rules tailored to the unique set of contingencies facing that firm. A regulatory agency would either approve these rules or send them back for revision if they were insufficiently stringent. . . . rather than having governmental inspectors enforce the rules, most enforcement duties and costs would be internalized by the company, which would be required to establish its own independent inspectorial group.[61]

This model of 'enforced self-regulation' explicitly rejects the simplistic stereotype of business actors as profit maximizers. By creating a structure for corporate actors to develop their own guidelines within government-defined parameters, they hope to appeal to internal mechanisms of control motivated by business responsibility, ethics, and obligations to abide by the law and to take account of non-shareholding stakeholders.[62]

Ayres and Braithwaite also emphasize the importance of effective communication between regulators and those who are regulated.[63] This stands in contrast to the sporadic and often adversarial nature of communications between regulators and those subject to regulation under traditional command-and-control regimes. Enforcement in this model takes the form of a 'regulatory enforcement pyramid'.[64] Form is more important than content in the pyramid because each regulatory arena requires an appropriate style of sanctioning.[65] State enforcement measures are proportionate to the lack of industry cooperation and range from warning letters through criminal prosecutions to revocation of operating licences. This strategy permits the control agency to escalate penalties in a fashion that responds to the degree of non-co-operation exhibited by a regulated firm or industry.

The strongest links between Teubner's account of reflexive law and the Ayres/Braithwaite notion of regulation are the shared recognition of the limits of law as a tool to regulate behaviour and the emphasis on self-regulation. The Ayres/Braithwaite model has the potential to make use of Teubner's vision of reflexive law in several ways. First, the model incorporates a legal reticence in that it recognizes the limitations of using law to set behavioural standards for other complex social systems. Instead of defining requisite behaviour, this regulatory model provides a procedure for firms or even entire industries to draft their own rules in accordance with officially established parameters. This is the reflexive point in the Ayres/Braithwaite conception of regulation. Within defined limits, firms are permitted to make their own choices with respect to the trade-offs necessary to reconcile their goals with the public policy objective. In working out the requisite trade-offs, the self-referential aspect of the firm or industry is triggered so that the reflective dimension reconciles the conflict between function and performance to take greater account of its external effects. The policy goal of the regulatory regime, for example, the need for environmental protection, is internalized by the system.

Allowing firms to write their own rules relieves law of the onus of direct regulation, which includes the high costs associated with information-gathering and the drafting of legislation. It also encourages both self-reflection and self-criticism within the decision-making processes of the firm since it compels consideration by the firm of its external effects. Acceptance of the rules and guidelines is enhanced by the fact that they are self-generated. The performance of law in this model is restricted to guiding the process of self-regulation rather than providing a detailed scheme of regulations to be rigidly followed. Including the necessity of official approval for the rules of conduct encourages communication.

The second point of convergence between the Ayres/Braithwaite model and a reflexive strategy is that requiring firms to draft rules and to be subject to 'a publicly monitored private enforcement of these rules'[66] relieves law of other burdens inherent in a traditional command-and-control regulatory regime, such as large staffing requirements and the costs of monitoring and enforcing agency-imposed rules. Some of the onus of direct regulation is transferred to the internal structures of the regulated firm or industry. Within this model, the public monitoring of the private enforcement of rules could take the form of auditing by the appropriate regulatory agency of self-reporting by the firm or industry in much the same way as tax returns are subject to review and assessment. The existence of the enforcement pyramid, while unconnected to reflexive legal strategies, promotes compliance by ensuring that the regulatory agency has an appropriate response mechanism to deal with potential transgressors.

Canadian Environmental Law Revisited

Any attempt to 'regulate regulations' in the Canadian environmental law regime should address the contextual background of environmental problems in Canada while being sensitive to the need to enlist other mechanisms of social control. The greatest source of environmental degradation occurs within the realm of business and industry.[67] Recognition of this fact inspired initiatives like the European EMAS and the ISO 14000 standards, both of which specifically target the corporate industrial sector. In Canada the corporate world is composed of over a million parts: approximately 900,000 for-profit business, including Crown corporations; about 140,000 voluntary organizations such as churches and trusts; 18,000 professional associations; 7,000 co-operatives, including credit unions, caisses populaires, and insurance co-operatives; 1,227 hospitals; 945 unions; and 249 universities, colleges, and community colleges.[68] In addition, there are 97 federal, provincial, and territorial departments and agencies and 7,524 local governments within Canada. The federal government is the largest commercial property holder in the country, owning or leasing 25 million square metres of office space. The federal inventory of buildings and facilities lists more than 50,000 items, including office buildings, laboratories, parks, and military bases.[69]

Currently, very few Canadian corporations have reporting procedures that go beyond the traditional prototype. Only about 7 per cent of large corporations (defined as having more than 200 employees) report environmental issues to their boards of directors on a regular basis. Less than 1 per cent are committed to releasing an annual environmental report for external consumption. Only 1 per cent routinely monitor and assess some aspect of progress on sustainable development practices. In medium-sized corporations (100–200 employees in the manufacturing sector, 50–200 employees otherwise) less than 0.1 per cent routinely report progress on sustainable practices, and among small businesses and self-employed individuals any such monitoring and assessment is a 'rare exception'.[70]

Annual surveys conducted by the national accounting firm KPMG indicate a lack of environmental management systems within the Canadian corporate community. For example, in the manufacturing sector only 22 per cent of respondents report having an environmental management system with all vital components.[71] The survey results also indicate the most compelling factors leading up to the corporate decision to take action on environmental issues are 'compliance with regulations' and 'board of director liability'.[72] These data suggest a defensive posture (fear of prosecution and affording protection for directors) regarding corporate compliance rather than a proactive attitude concerning the environmental impact of their operations. Defensive compliance is also evidence of the incompatibility between the internal logics of law as expressed through direct regulation and the target of such regulation because it connotes both a grudging and minimal compliance effort and that policy goals are not being fully accepted and internalized.

The contrast between the magnitude of the Canadian corporate world with the relatively small number of effective environmental management systems in place and the paucity of environmental reporting underscores the importance of cultivating a new legal attitude towards environmental regulation. Only two alternatives to the traditional approach are currently discussed: the abandonment by law of its regulatory functions (deregulation, labelled re-formalization by Teubner) or the importation of economic analysis as a rationale for regulation (which Teubner refers to as the implementation approach). The need to transcend the dichotomy of deregulation versus intensified state control calls for creative options. This entails an abandonment of the instrumental conception of regulation in favour of enlisting other, less direct control mechanisms that already exist within other social subsystems and that can be adapted to a voluntary and proactive approach towards environmental protection. Within this context the most appropriate vehicle to make the transition from direct regulation to less intrusive forms of control is a program of enforced self-regulation with an emphasis on self-auditing and reporting.

A Proposal for a Canadian Eco-Management and Audit System

A Canadian Eco-Management and Audit System should focus on industry since it is the greatest source of environmental degradation. The primary objective, of course, is to establish control of industry behaviour. The following proposal highlights four fundamental requirements: it should be a voluntary, company-based process of adopting government-approved environmental management systems that include self-auditing and self-reporting. The emphasis on self-regulation is an important feature because it is the least coercive instrument of government intervention.[73] It also enjoys the advantage of encouraging voluntary acceptance by business of government-defined parameters for environmental management systems. In addition, self-regulation contains both detector and effector capabilities. These terms, borrowed from cybernetics, refer to a government's or a control agency's ability to tell what is occurring in a given sphere of activity (the detector capability) and the further ability to act on that sphere (the effector capability).[74]

   A management system that includes an assessment of the environmental impact of all business decisions is a prerequisite for stimulating reflection structures within a corporate enterprise so that environmental concerns are considered along with balance sheets, income statements, and reports to shareholders. Self-auditing and self-reporting allow both the individual business and the control agency to assess environmental performance and also serve as the control agency's detector capability. Control agency audits of the self-generated reports provide the basis for its communication with participants and lay the groundwork for its effector capability. Unsatisfactory reports should be the trigger for regulatory enforcement action calibrated to match the seriousness of the offence.

   An essential feature of such a Canadian system is voluntary participation. At bottom, reflexive law is really an effort to (re)structure the rules of the game. Since its theoretical starting point is the increasing futility of employing direct (external) legal commands as a means of altering behaviour within other social subsystems, it must seek an alternative way of injecting desired environmental values into the (internal) structures of corporate decision-making. Defining the parameters within which businesses make decisions is only the first step. The parameters must be accepted. To encourage acceptance and thereby promote voluntary participation in the program it must be made attractive to all potential participants.

The issue of voluntarism is important because it means the desired change in attitude comes from within rather than being forced from the outside as in the command-and-control regulatory style. This means appealing to the self-referential dimensions of businesses and their decision-makers. The advantages of membership must be well conceived, well publicized, and well administered. There are many advantages to adopting an environmental management system aside from meeting legal obligations in a timely fashion: some companies that already have an environmental management system in place have found it less expensive to operate than a compliance-driven system, and a company's products will have a greater appeal to consumers seeking green products. In connection with appealing to the 'green market' there should be an explicit recognition of an organization's participation in a Canadian Eco-Management and Audit System. Participants should have an emblem to display that would signify their membership in the effort to achieve continuous environmental improvement.[75]

The Canadian Eco-Management and Audit System process should be company-based to avoid the potential problems of the site-based European Eco-Management and Audit Scheme and the fragmented approach of the ISO 14000 set of standards. A different environmental management system for each individual site, as in the European version, or a different environmental management system for different divisions of the same company, as in ISO 14000, would be too complex for efficient administration. Consideration should also be given to extending such a system to all levels of government and Crown corporations. Government in Canada is big business and as such affects environmental quality.[76] The activities of other agencies, such as Hydro-Québec and Hydro One Networks in Ontario, have tremendous potential environmental consequences and should also be subject to the proposed Canadian system.

This system should begin by requiring the top management of each participating business to define an environmental policy for the firm. The policy should be in the form of a statement of corporate intentions with respect to its environmental impact. It should take into account legislative and regulatory requirements and serve as an internal reference framework for continuous environmental improvement. A corporation's environmental policy should be unique to that company, be communicated to all employees, and be made publicly available. The policy is the starting point for setting the corporation's EMS objectives.

Participants should then be required to devise an environmental management system to ensure that the environmental policy is actively carried out.

This entails planning, implementation and operations, checking and corrective action, and management review. It is important to note that a corporation's environmental management system should not be intended to replace the requirements of existing legislation. Instead, its purpose is to provide a means of monitoring, controlling, and improving performance vis-à-vis those requirements.[77] The process of crafting both an environmental policy and a means of implementing and monitoring that policy is intended to encourage self-reflection and self-criticism within business organizations and to promote the internalization of the policy goal, namely, environmental protection. The fact that both the policy and the environmental management system are self-generated is meant to ensure its acceptability to the participant.

As part of the enforcement process an environmental management system drafted by an individual participant should then be subject to control agency review. Key tests for a self-produced environmental management are that it is comprehensive, proactive, and based on a systems approach. *Comprehensive* means the environmental policy includes all members of the participating business organization, that the environmental management system considers all stakeholders, and that all potential environmental impacts of the organization's activities are identified. *Proactive* means that the proposed EMS focuses on forward thinking and action instead of merely reacting to past command-and-control policies. A *systems approach* means the goal is to improve environmental performance by employing a single environmental management system across all functions of the business organization. Independent, authoritative scrutiny by a control agency of a proposed environmental management system should ensure its conformity to existing legal requirements and also assist in the development of industry-specific systems.[78]

A critical element of a self-generated environmental management system is the necessity for self-auditing and self-reporting. It is important to audit an EMS on a periodic basis to ensure that it is operating as planned, to provide information for management review, and to assess its capability in meeting defined objectives. Disclosure of information to control agencies through self-reporting is important because it ensures that the company has an environmental management system capable of producing the information that must be disclosed and it also defines the basis for communication between the participant and the regulatory agency.[79]

All reporting done within the context of a Canadian Eco-Management and Audit System should be comprehensive, done on a regular basis, made public, and subject to control agency verification. Environmental management and auditing would lack credibility if the results were not publicly released and subject to independent, official (government) verification. Aside from imparting credibility to the process, the public release of verified information would help address the lack of access to important environmental information that prevails under the current regulatory regime.[80] Corporate environmental reporting under a Canadian Eco-Management and Audit System, similar to the reporting under the Income Tax Act, should be subject to government scrutiny as part of the enforcement process. A regulatory enforcement pyramid containing penalties commensurate with the severity of transgressions should be available for use by the control agency to deal with instances of false reporting.

This proposed system alters the traditional role of control agencies in several significant respects. First, it creates the duty of setting parameters for the environmental management systems required of participating businesses. Second, it entails review of proposed environmental management systems to ensure both their conformity with all existing legislative and regulatory requirements and that their focus is on the improvement of environmental performance. Third, it requires the control agency to audit (randomly) the self-generated reports of participants. Cumulatively, the first three changes to the control agency role lead to the last difference, which is relief from some of the burden associated with the monitoring, inspection, and investigation duties required under a traditional regulatory program. The last change stems from the emphasis placed on self-regulation, auditing, and reporting within the economic sphere.

A proposed Canadian Eco-Management and Audit System should also include a role for the public. The rationale is simple: environmental degradation produces risk; those at risk (the public) should have input into the 'risk-bearing decision'. In considering the issue of nuclear fuel waste disposal, the Seaborn Panel[81] spent over eight years studying safety and acceptability criteria. A principal finding was the sharp distinction between scientific and social concepts of risk. The panel noted that the public 'is demanding more openness, public scrutiny and debate, and shared decision-making'[82] and concluded that standards set without broad public consultation would not reflect general public acceptance. Without public acceptance there is a likelihood of vigorous opposition. A method of involving the public, recommended by Ayres and Braithwaite, is through tripartism. This is defined as a policy to foster the participation of public interest groups in the regulatory process by granting them access to all the information available to the regulator and by giving them a seat at the negotiating table with business organizations and the control agency 'when deals are done'.[83] In terms of the proposed Canadian Eco-Management and Audit System, this means giving public interest groups input and even approval for any proposed environmental management system. An important advantage of including public representation, aside from broadening the value base that drives the regulatory process, is that it mitigates against the possibility of capture or corruption that can occur in a two-party regulatory system.[84]

The proposed Canadian Eco-Management and Audit System is an attempt to implement the reflexive legal strategy identified by Gunther Teubner within a model suggested by the Ayres/Braithwaite conception of regulation. Teubner and Ayres and Braithwaite share a vision of self-regulation as a potential solution to the current dilemma of law. Teubner's concern is to establish organizational structures designed to achieve an optimal balancing of performance and function within subsystems by having them take account of the requirements of their external environment. Enactment of the proposed Canadian Eco-Management and Audit System, by setting out the procedural requirements to be followed in drafting an environmental policy intended for implementation through an environmental management system, is meant to give statutory expression to this approach. The proposal provides for the adoption of state-approved environmental management systems as a means of encouraging reflection structures within the economic subsystem. Business corporations need to compare their actions and performance with those of their competitors. There are many available sources of information on financial performance but very little exists concerning environmental performance. Encouraging the establishment of environmental management systems presents an opportunity to develop a source of information on environmental matters for businesses and control agencies alike.

Ayres and Braithwaite, while not working within the same reflexive law paradigm, reinforce Teubner's emphasis on establishing structures for regulation within the economic sphere by recommending self-generated rules of conduct, subject to state approval, that is, environmental policy statements and management systems. Their inclusion of an enforcement pyramid provides a mechanism to ensure compliance.

In some respects, the inclusion of such a mechanism appears to be in conflict with Teubner's vision of 'self-regulation taking the form of decentralized moral self-control',[85] and equally it seems at odds with the required spirit of voluntarism. On the other hand, there must be some method of ensuring compliance and some punishment to deal with transgressors, for example, corporations that file falsified reports concerning their environmental performance. The strength of the Ayres/Braithwaite enforcement pyramid is that it is geared to respond to such transgressions in proportion to the seriousness of the breach.

Conclusion

The difficulty of providing effective environmental protection is a concrete example of the general failing of the substantively oriented, purposive law characteristic of the modern welfare state. Exclusive reliance on substantive legal solutions implies that cause and effect are directly and linearly related and that cause and effect can be isolated from their context. Resort to substantive law also implies a belief that when a given action is taken the (desired) result is inevitable and if a proposed action is not taken then the opposite (undesirable) result is similarly inevitable. The underlying message is that careful planning is a panacea for social ills and that there is a legal answer to every social problem.

In contrast, Gunther Teubner's elaboration of reflexive law, informed by autopoietic theory, recognizes the reality of complex modern social subsystems and resists the impulse to attempt direct regulation. Teubner's work not only provides insight into the reasons for the regulatory failures of substantive law but also points to a theoretical solution. The contribution of Ayres and Braithwaite complements Teubner's theoretical framework by providing a model to test the application of reflexive legal solutions to regulatory problems. Teubner and Ayres and Braithwaite do not suggest a new role for law, but rather they point to a new way for law to perform one of its traditional roles.

The key lesson from examining the specific problem of the Canadian environmental regulatory regime within the larger context of the dilemma of law in the welfare state is that regulatory approaches to policy implementation must co-evolve with systems targeted for regulation. The process of differentiation that accompanies society's evolution means that a regulatory approach appropriate to one time and setting cannot maintain its utility in another time and setting. There is growing recognition among legal theorists that changes resulting from the evolution of society call for an innovative regulatory response that is creative and adaptable to change. The implication for the legal regulation of the environment is learning to think differently about how law is used to alter established patterns of human behaviour.

Notes

1. G. Teubner, 'Substantive and Reflexive Elements in Modern Law', *Law and Society Review* 17 (1983): 268, discussing what Jürgen Habermas terms a 'rationality crisis'.
2. See G. Teubner, ed., *Dilemmas of Law in the Welfare State* (New York: Walter de Gruyter, 1986).
3. The term 'social regulation' has different meanings depending on the context in which it is used. In one sense the term describes regulation aimed at the amelioration of social conditions in contradistinction to earlier (economic) regulation, which established government oversight of entry, exit, output, and price in specific industries or economic activities, usually those affecting the public interest such as transportation and banking. Elsewhere in this volume, Trevor Purvis, in Chapter 2, writes from a Foucauldian perspective in which the term 'social regulation' implies governing all types of social conduct. In the present chapter, 'social regulation' is used in the narrower and more traditional sense of regulating economic activity only.
4. B.J. Cook, *Bureaucratic Politics and Regulatory Reform* (Westport, Conn.: Greenwood Press, 1988).
5. N. Reich, *Reflexive Law and Reflexive Legal Theory: Reflections on Postmodernism in Legal Theory* (Bremen: University of Bremen, 1988), I. In the literature this is referred to as 'politically instrumentalized law'.
6. See, for example, I. Bernier and A. Lajoie, eds, *Consumer Protection, Environmental*

Law and Corporate Power (Toronto: University of Toronto Press, 1985); F. Mihlar, *Regulatory Overkill: The Cost of Regulation in Canada* (Vancouver: Fraser Institute, 1996).
7. D.P. Emond, 'Environmental Law and Policy: A Retrospective Examination of the Canadian Experience', in Bernier and Lajoie, eds, *Consumer Protection, Environmental Law and Corporate Power*, 89.
8. A.R. Thompson, *Environmental Regulation in Canada: An Assessment of the Regulatory Process* (Vancouver: Westwater Research Centre, 1980).
9. Teubner, 'Substantive and Reflexive Elements'.
10. I. Ayres and J. Braithwaite, *Responsive Regulation: Transcending the Deregulation Debate* (New York: Oxford University Press, 1992).
11. There are over 35,000 chemicals currently manufactured or imported into Canada for commercial use and the number of these chemicals that pose a threat to human health or to the environment is unknown. See House of Commons Standing Committee on Environment and Sustainable Development, *It's About Our Health! Towards Pollution Prevention* (Ottawa: Canada Communications Group, 1995), 29.
12. Public concern for the environment has been an enduring and widespread feature of Canadian public opinion since the early 1970s. See Doug Miller, *Whither the Environment? Public and Expert Opinion Projected to 2010* (Toronto: Synergistics Consulting, 1996), 2.
13. See Mihlar, *Regulatory Overkill*.
14. 'Command and control' is the name attached to a traditional directives style of regulation that is employed in environmental regulation almost exclusively. 'Commands' in the form of government-initiated specifications, directions, and prohibitions are issued; 'control' is achieved by obtaining compliance with the commands. Non-compliance leads to penalties and sanctions for transgressors.
15. Among many other urgent environmental issues, there is a universally mounting level of anxiety about stratospheric ozone depletion, global warming, habitat degradation, acid rain, toxic chemical contamination, and hazardous waste management. The catalogue of potential environmental ills is too lengthy to set out in detail. For further reference, see World Commission on Environment and Development, *Our Common Future* (New York: Oxford University Press, 1987).
16. T. Schrecker, *Sustainable Development: Getting There From Here* (Ottawa: National Roundtable on the Environment and the Economy, 1993), 65.
17. See G. Teubner, L. Farmer, and D. Murphy, eds, *Environmental Law and Ecological Responsibility: The Concept and Practice of Ecological Self-Organization* (Rexdale, Ont.: John Wiley & Sons, 1994).
18. Ayres and Braithwaite, *Responsive Regulation*, 3.
19. R. Howse, 'Retrenchment, Reform or Revolution? The Shift to Incentives and the Future of the Regulatory State', *Alberta Law Review* 31 (1993): 457.
20. See L. Farmer and G. Teubner, 'Ecological Self-Organization', in Teubner, Farmer, and Murphy, eds, *Environmental Law and Ecological Responsibility*, 3-5.
21. G. Teubner, 'After Legal Instrumentalism? Strategic Models of Post-Regulatory Law', in Teubner, *Dilemmas of Law*, 305-7.
22. See, e.g., G.B. Doern, *The Environmental Imperative: Market Approaches to the Greening of Canada* (Toronto: C.D. Howe Institute, 1990); M.G. Wroebel, *Environmental Problems: Market-Based Solutions* (Ottawa: Library of Canada, 1990).

23. Teubner, 'Substantive and Reflexive Elements', 273–4.
24. Ibid., 257.
25. See P. Nonet and P. Selznick, *Law and Society in Transition: Towards Responsive Law* (New York: Harper, 1978); N. Luhmann, *The Differentiation of Society* (New York: Columbia University Press, 1978); J. Habermas, *Legitimation Crisis* (Boston: Beacon Press, 1975).
26. Teubner, 'Substantive and Reflexive Elements', 240.
27. Ibid.
28. Ibid., 241.
29. Ibid., 251-7.
30. Ibid., 240.
31. Ibid., 275.
32. The term 'procedural orientation of law' or 'proceduralization of law' is easily misunderstood. This is not to be confused with 'procedural law' and does not mean that reflexive legal strategies are without content. Teubner's theory of reflexive law does not eliminate the need for substantive legal norms; instead, it focuses on the limitations imposed on law by the 'differentiation' of society and limits itself to channelling decision-making processes and social communications. See Gunther Teubner, *Law as an Autopoietic System* (Oxford: Blackwell, 1993), 66.
33. Ibid., 96.
34. *McRae v. British Norwegian Whaling Co. Ltd.*, [1927-31] Newfoundland Law Reports 274 at 283-284.
35. E.W. Orts, 'Reflexive Environmental Law', *Northwestern University Law Review* 89 (1995): 1259.
36. Council Regulation (EEC) No. 1836/93 of 29 June 1993 allowing voluntary participation by companies in the industrial sector in a Community eco-management and audit scheme.
37. International Organization for Standardization, ISO 14000 Series, 'Environmental Management Series—Specifications', Chapter 1, Clause 4.
38. Council Regulation (EEC), 29 June 1993, Article 3(b) and 3(c).
39. Ibid., Article 3(d), 3(g), 3(h), Article 4.
40. The Standards Council of Canada is a federal Crown corporation with the mandate to promote efficient and effective voluntary standardization in Canada.
41. International Organization for Standardization, ISO 14000 Series, 'Environmental Management Series—Specifications'.
42. J. Mingers, *Self-Producing Systems: Implications and Applications of Autopoiesis* (New York: Plenum Press, 1996), 140-1.
43. Teubner, 'After Legal Instrumentalism?', 308.
44. See G. Teubner, ed., *Autopoietic Law: A New Approach to Law and Society* (New York: Walter de Gruyter, 1988).
45. G. Teubner, 'The "State" of Private Networks: The Emerging Legal Regime of Polycorporatism in Germany', *Brigham Young University Law Review* (1993): 566.
46. Teubner, 'Substantive and Reflexive Elements', 268.
47. Luhmann, *The Differentiation of Society*.
48. Autopoiesis is concerned with self-organization (or auto-determination) and derives from the work of Chilean biologists Humberto Manturana and Francisco Valera in the areas of neurophysiology and perception. The word itself is coined from the

Greek 'auto' (self) and 'poiesis' (creation; production) and is used to describe systems that maintain their defining organization despite perturbation and structural change while regenerating their components during the course of their operation. A biological example is the basic unit of a cell that is in a continuous symbiotic state of absorbing nutrients from its environment and expelling waste matter back into its environment, yet its system is unaltered by these exchanges. The hallmark of an autopoietic system is this relationship between its openness and closure. See H. Manturana and F. Valera, 'Autopoiesis and Cognition: The Realization of the Living', in R.S. Cohen and M. Wartofsky, eds, *Boston Studies in the Philosophy of Science* (Dordrecht: D. Reidel Publishing, 1980).
49. Teubner, 'Substantive and Reflexive Elements', 249. The theory maintains that all societal subsystems operate according to binary codes, for example, truth/falsity (science) or power/non-power (politics). The legal system deals only with the legality/illegality of particular acts or events. Particular laws (norms) are subject to change because the programs that contain the rules for applying the binary code can change according to a 'logic of normative development'.
50. Teubner, *Law as an Autopoietic System*, 74–5.
51. Ibid., 75.
52. Ibid., 77.
53. Ibid.
54. Teubner, 'After Legal Instrumentalism?', 309–10.
55. 'Directors' chill' is a term coined to describe the reaction of corporate directors to the relentless expansion of their areas of personal liability. It manifests itself in a number of ways, including refusal to serve as a director when asked, the resignation of directorships, and the practice of risk-averse management. For a more detailed explanation, see Toronto Stock Exchange, *Report of the Committee on Corporate Governance in Canada* (Toronto, 1994).
56. Teubner, 'Substantive and Reflexive Elements', 273–4.
57. Ibid., 242.
58. Ibid., 251.
59. Ibid., 275.
60. Ayres and Braithwaite, *Responsive Regulation*, 4.
61. Ibid., 106.
62. Ibid., 22.
63. Ibid., 93.
64. Ibid., 35–8.
65. Ibid., 36.
66. Ibid., 116.
67. D. Saxe, *Environmental Offences, Corporate Responsibility and Executive Liability* (Aurora, Ont.: Canada Law Book, 1990), 21.
68. T. Hodge, S. Holtz, C. Smith, and K. Hawke Baxter, eds, *Pathways to Sustainability: Assessing Our Progress* (Ottawa: National Roundtable on the Environment and the Economy, 1995), 44.
69. Ibid., 48.
70. Ibid., 45.
71. KPMG, *Canadian Environmental Management Survey* (Ottawa: KPMG, 1997), 9. Other sectors of the economy report as follows: natural resources and energy (10 per cent);

financial institutions (13 per cent); goods (13 per cent); municipalities (18 per cent); education (5 per cent); health (19 per cent); services (23 per cent).
72. Ibid.
73. L.A. Pal, *Public Policy Analysis* (Toronto: Methuen, 1987), 142–50. Pal sets out the complement of instruments of governing, which range from self-regulation (minimum) to public ownership (maximum).
74. C.C. Hood, *The Tools of Government* (London: Macmillan, 1983), 4.
75. Orts, 'Reflexive Environmental Law', 1326–7.
76. Hodge et al., *Pathways to Sustainability*.
77. I. Fredericks and D. McCallum, *International Standards for Environmental Management Systems: ISO 14000* (Toronto: MGMT Alliances, 1995), 1–3.
78. Ibid.
79. Ibid.
80. Orts, 'Reflective Environmental Law', 1123.
81. Environmental Assessment Panel, *Nuclear Fuel Waste Management and Disposal Concept* (Ottawa: Minister of Public Works and Government Services Canada, 1998).
82. Ibid., 34.
83. Ayres and Braithwaite, *Responsive Regulation*, 57–8.
84. Ibid., 54.
85. J.L. Cohen and A. Arato, *Civil Society and Political Theory* (Cambridge: MIT Press, 1992), 482.

Chapter 6

Democratic Environmental Governance and Environmental Justice

PETER SWAN

> Minimally, the new forms of regulation promise to improve the quality of our environment. At a maximum, they suggest a novel form of democracy that combines the virtues of localism and decentralization.[1]

Introduction

The use of law to further social goals through expanding legislative strategies has been problematized by thinkers drawing on disparate intellectual and political positions. The neo-liberal right has criticized the growth of the welfare state bureaucracy as an obstacle to economic freedom and innovation. This critique has also been taken up in a different form by theorists on the left who see bureaucracy as leading to the production of clientelism and passivity on the part of the recipients of state largesse.[2]

This contemporary critique of law as a means of regulation has also been articulated in a concrete and experiential manner by progressive social movement actors, who since the 1960s have placed more demands on law. Articulating these demands in the form of new rights claims, their expectations about what can be accomplished through legal intervention and reform have grown. The inability of law to respond effectively to these rising expectations has led to a questioning of the viability of law as a means of regulation.[3]

The polarized positions of the contemporary debate regarding law as a vehicle of social policy are replicated in more specific controversies around environmental law. Driven by an agenda of economic liberalization, the right has insisted that legal intervention to protect the environment interferes with the exercise of property rights, economic innovation, and growth. They counter with demands for deregulation or forms of regulation based on market mechanisms. Against this position, the environmental movement has relied heavily on strategies of legal intervention to advance its agenda of environmental protection. For the most part, however, environmentalists have been disappointed with the failure of traditional forms of command-and-control regulations to prohibit or control the main forms of economic activity responsible for the potentially catastrophic ecological risks. In addition, they stress the existence of a democratic deficit in the current modes of regulation—for the most part, participation in decision-making has been limited to governments and industry.

One of the major responses to the crisis of environmental law has been the active questioning of direct legal intervention. Accordingly, some academics and policy-makers have encouraged an attitude of legal restraint. Many believe that law must be reconceived to play a less direct role in achieving the objectives of contemporary governance, with the state co-operating with 'stakeholders' such as industry, local governments, and private associations and social movements. Such new modes of governance[4] suggest that law should stand further back and specify the procedures to allow interested participants to work out the content of rules and policies. One model that focuses on governance at a distance is found in 'informational' or 'performance-based' regulation. Information-based forms of regulation (IR), such as toxic release inventories, have been common in some North American jurisdictions since the 1970s. However, Charles Sabel and his colleagues in the United States have recently provided a new theoretical articulation for environmental governance.[5]

Assessing IR approaches through the lens of American pragmatist philosophy, Sabel and his collaborators emphasize their potential to contribute to innovative forms of environmental protection and to decentralize decision-making through the processes of social co-operation and social learning. Their analyses emphasize how IR approaches that incorporate the exchange of information among parties with interests in environmental decisions are able to provide for more effective environmental protection through better information that draws on a plurality of modes of knowledge. This claim with respect to the provision of more effective environmental protection is common to suggestions for environmental decision-making that emphasize partnerships between stakeholders as an alternative to top-down regulation from government. More significant for the present discussion is a second claim of Sabel and his colleagues. Unlike in models that call for self-regulation by industry, they also contend that the social co-operation entailed in deliberative information exchange has the potential to lead to novel forms of democracy through the enhancement of participation in decentralized decisions about environmental risks. Other recent theorists of governance are mainly concerned with making better and more efficient decisions, but Sabel and his collaborators also attempt to address the democratic deficit in existing forms of environmental regulation by articulating the basis for a democratic mode of environmental governance that has wider implications for democracy in general. This emphasis on the democratic promise of their performance-based environmental regulation will provide the main focus for this chapter.[6]

This chapter evaluates the assumptions informing Sabel's 'progressive' espousal of the potential of this co-operative form of environmental governance. Assessing the ideal of social reflexivity inherent in information-based approaches to environmental governance is first necessary, and then I offer an overview of some of the information-based programs that serve as models for Sabel and his colleagues. Following this, the central assumptions informing their views on democratic environmental governance are examined. Contemporary critical theory is then applied to problematize aspects of social co-operation between experts and the possessors of lay knowledge that remain unexamined by Sabel and his colleagues. The final section argues that the democratic potential for deliberation between laypersons and experts can be more effectively realized by relating these processes to the ideal of environmental justice, defined as equal access to decision-making in an emerging environmental public sphere. This concept of environmental justice also allows for a reconceptualization of the role of law in claims for environmental rights. Thus, through the process of claims-making and by using the moral/political language of rights, discourses of law can be redeployed at the centre of processes of environmental governance.

Reflexive Regulatory Alternatives

Writing in the 1970s, Roberto Unger described the phenomenon of the 'welfare-corporate state' in which the lines between public and private power were becoming increasingly blurred.[7] He described this situation of the late capitalist state as one in which bodies intermediate between the individual and the agencies of the state play an ever-increasing role in the life of society. In the 1990s this blurring of the public and private has taken on new forms at both domestic and international levels. This phenomenon has been described by theorists at the international level as constituting a form of 'governance without government'.[8] Such a characterization may erroneously convey the impression that the state is being pushed to the sidelines. However, it rightly emphasizes that many of the functions we normally associate with top-down regulation by government are increasingly being dealt with in a variety of practices and discourses linked to everyday life in the associations of the economy and civil society. This does not so much represent a departure from the state as it does a restructuring of its role in relation to the economy and society. Instead of being the major source of legal rules directly governing individual and collective behaviour, the contemporary state takes on a more general though no less significant role in the structuring of the rules, procedures, and practices of 'self-governing' economic and social entities.

In the last decade the priority attached to a growing role for the norms and practices of civil society has strengthened and there has been a move towards emphasizing the increasing responsibility of individuals for their own welfare. Nowhere is this shift towards the increasing responsibility of non-state agents more evident than in the area of environmental regulation.

Environmental regulation took on its modern configuration in North America during the late 1960s and early 1970s. Responding to a growing level of environmental awareness on the part of the public, policy-makers created the first legislation that had environmental protection as an explicit goal. Most such legislation took the form of direct and indirect commands from a central source that attempted to specify limits on liquid effluent or on quantities of emissions that would affect ambient air quality. Although this command-and-control approach has been successful in areas such as automobile emissions standards and has led to the development of new technology to control emissions, from the outset this form of environmental regulation has been subjected to severe criticism from all quarters. On one hand, industry has often resented such active government intervention, suggesting that environmental regulation is too intrusive and inflexible with regard to economic activity. On the other hand, environmentalists and citizens' groups have argued that existing regulation has not gone far enough. While recognizing that important gains have been made, they question both the adequacy of environmental protection that has been achieved and the lack of opportunity for democratic participation in decisions on environmental risks.

One response to the perception of the inadequacy of direct regulation by the state has been to suggest an alternative structure of regulation that attempts to mimic market behaviour. Rather than have the government intervene, the advocates of this approach have suggested that mechanisms be created to force industry to internalize the costs of abating environmental protection by attaching economic value to the environment.[9] According to advocates of such an approach, mechanisms such as tradeable pollution permits, property rights in the environment, and environmental taxation would provide both less intrusive and more effective forms of environmental protection. However, the attempts to implement market mechanisms have also proved to be far from successful. Industry has often successfully used its power to resist adapting adequate standards of environmental protection. In addition, such experiments, no less than command-and-control models, have depended on some centralized institution to define acceptable standards or to supply needed information.[10]

The continuing resistance by industry through threats of disinvestment and a lack of regulatory intervention have allowed the debate between those who would prefer the market to command-and-control methods to go on for more than 30 years without any resolution. Only in the last decade have novel experimental models begun to emerge that may have the potential to provide more effective environmental protection and also to make the power of industry and government more subject to democratic accountability. Some of these models focus on the organization of conditions and institutions of environmental governance that involve the co-operation of government, industry, and local communities and consumers rather than traditional approaches of top-down regulation. As such, these models are contributing to a reshaping of the regulatory terrain in which the state does not merely regulate from above but is involved in organizing the conditions for an active role in governance and self-governance by non-state institutions and agents.[11]

Two of the most significant alternatives are found in models of reflexive environmental regulation[12] and attempts at establishing information-based regulation or performance-based regulation[13] of environmental risks. The key to both approaches is the establishment of a more complete and 'accessible' information base on which to make decisions regarding environmental risks. They are designed primarily to encourage reflexive behaviour by actors involved in these decisions. These models suggest that by reflecting on the effects of their own economic and social activity on nature, as well as on their attempts to ameliorate environmental protection, actors should be able to revise their activities and goals in light of this new awareness and therefore promote more effective forms of environmental protection.

The theorists of reflexive law see it as a specific response to the use of law to intervene directly in social processes. Legal intervention, according to Gunther Teubner, fails to account for the increasing differentiation of society into semi-autonomous spheres or systems that are too dispersed and powerful to be readily subdued by law.[14] Once these limits on law are recognized, the pivotal question becomes one of how law can rationally structure social processes and procedures in the face of growing social complexity. Reflexive law answers this question by suggesting that rather than attempting to regulate social problems as a whole, law should make use of the self-reflective and self-governing capacities of other social institutions by providing the procedures to structure their decision-making processes.[15] Instead of trying to control the specific character or direction of social change, reflexive law aims at using legal procedures to establish a communicative strategy for the disclosure of information and for the provision of greater accountability.

As theoretical models, both reflexive law and informational regulation are posited as providing answers to the failure of command-and-control measures and market measures to use regulation to engender both social learning and problem-solving at the level of the regulated entities.[16] While these models are related, reflexive environmental law tends to be somewhat narrower in its primary focus on the internal operations within businesses and other 'intermediary social institutions'.[17] Its procedural approach is based on using legal procedure to structure the self-reflection of actors through communication. The goal of self-reflection through processes of disclosure and accountability, which is at the heart of reflexive environmental law, is mainly limited to the self-reflective processes within the very institutions that are the source of processes that cause most environmental degradation. One of the main purposes of informational regulation is to provide information from potential polluters to third parties. The legal procedures guiding informational regulatory systems, for example, may require that companies provide the public with information about regulatory compliance. They may also require that communities or workers be provided with information on potentially toxic substances used in industrial processes. At least in theory, systems aim to increase the reflexivity of a greater range of actors, who may or may not be encouraged to participate in wider processes of environmental governance.

Innovation and Environmental Governance

For the most part the main objective of both reflexive law and information-based approaches is to promote better environmental decision-making. In this context 'better' is defined in epistemological terms in the sense that communication within a firm or between firms, government, and local people can create a more comprehensive knowledge base. For our purposes we can categorize this more limited focus on the technique of regulation as a form of 'thin proceduralism' that downplays the moral and political dimensions of the regulatory process.[18]

However, Charles Sabel and his collaborators want to draw out the potentially more radical implications of informational regulation or performance-based regulation.[19] They do so by analyzing the traits shared by a number of existing variants of these forms of environmental governance[20] in the United States and by pointing out obstacles to the realization of their potential. More importantly, they clarify the assumptions underlying these regulatory forms that might promote both more effective and more democratic forms of environmental governance. Drawing on ideas about social learning and social co-operation derived from John Dewey's philosophy of pragmatism, Sabel and his colleagues reinterpret the epistemological dimensions of informational regulation. This approach emphasizes not only the production of knowledge through processes of social co-operation but also the significance of democratic participation in regulation or governance. Consequently, they attempt to reconceptualize information- or performance-based regulation in such a way as to emphasize what Julia Black has called 'thick proceduralism'.[21]

This theorization of the more radical potential of the informational regulatory approach illustrates an alternative both to the now dominant forms of regulation and to 'deregulation' and the existing forms of informational regulation. By emphasizing the problem-solving capacity and the democratic potential of informational regulation, Sabel and his associates focus on two issues that must be of central concern to those who want better forms of environmental protection that are also sensitive to social justice issues.

Sabel et al. begin their examination of modern environmental governance by evaluating the main contenders in the ongoing debate about the most appropriate forms of regulation.[22] They argue that neither command-and-control nor market models of environmental regulation are adequate in meeting the need for social knowledge or the conditions for democratic participation that they believe are necessary for contemporary environmental decision-making. According to their analysis, command-and-control mechanisms wrongly try to find lasting solutions by fixing, once and for all, permissible concentration levels of pollutants. In the face of rapidly changing epidemiological evidence, such expressions of regulatory omnipotence may well preclude the possibility of better technological solutions.[23] Nor do market mechanisms such as tradeable pollution rights and cost-benefit analysis fare any better in producing effective environmental protection. According to Sabel and his associates, devices that attempt to mimic the market fail to deliver on their promises of greater decentralization because they also require centrally organized information in order to have precise definitions and allocations of property rights and costs.[24]

Deregulation is a third contender and is also ruled out. The authors see that even with the growth of processes of economic liberalization at both national and global levels, 'the commitment of the "democratic public" to some form of stewardship for the environment is unquestioned.'[25] Any attempt to avoid a public commitment to environmental protection would therefore be potentially disastrous politically.

Sabel, Fung, and Karkkainen believe that the best existing alternative to the dominant models of environmental regulation are found in a number of experimental programs that have focused on a partial decentralization of decision-making to the local and regional levels. These performance-based models are grounded in the exchange of information between governments, regulated industry, and the broader community. As an example of law at a distance, they are premised on the idea of an increase in the autonomy of regulated entities in exchange for greater accountability.[26] In the place of centralized processes, the practices of information-based systems rely on the decentralization of decision-making to local communities, industries, and governments. Sabel and his associates assume that such processes are most capable of improving the quality of information through the promotion of democratic experimentalism.

They locate potential models for innovative or experimentalist environmental regulation in a number of existing IR-type programs in the US.[27] Some of their primary examples are the TRI (Toxic Release Inventory), TURA (Massachusetts Toxic Use Reduction Act), and the federal Environmental Protection Agency (EPA) Program for Regulatory Excellence (XL) and Common Sense Initiative (CSI).[28]

One of the best known of these programs is the Toxic Release Inventory, which emerged out of the Emergency Planning and Right to Know Act of 1986 (EPRA).[29] These programs resulted from public concern following a number of environmental catastrophes, such as Love Canal at Niagara Falls, New York, during the 1970s and 1980s. Emerging from both local and national demands for communities' right to know about the toxic chemicals released in their communities, networks of environmental activists began to lobby for the use of state coercion to compel public disclosure of information. Based on initial models of right-to-know legislation for workers, the TRI provides a national system for the benchmarking of self-reported releases of toxic substances.[30] It requires that all private or government facilities of a certain size must report to the EPA on the amounts of 650 specific substances that are transferred off site or deliberately or accidentally released on site. These reports are made available to the public and compare amounts released by substance, facility, industry, and location. While there are penalties for failure to report, there is neither independent verification of the accuracy of the report nor penalty for inaccurate reports. However, citizens may sue for failure to comply with the disclosure provisions and the data obtained may be used as evidence of failure to comply with other environmental legislation or used as a 'lever by which to apply public pressure for improvement'.[31]

Sabel and his colleagues recognize that the TRI represents only the beginning of adequate performance standards but maintain that public reporting of information can lead to the improvement of performance. They argue that publicity from such reports can force changes because it can lead both to the decline in share values of a bad performer and to competitive disadvantages with similar companies that see advantages to better performance. In addition, they argue that the TRI has been instrumental in inducing changes in industrial associations and the type of information that they circulate among firms. It also has been influential in the creation of the 'Responsible Care' program of the Chemical Manufacturers Association, which has led to published reports and improved standards and performance measures by its members.[32]

According to Sabel et al., Massachusetts' Toxic Use Reduction Act both broadens and extends the reporting requirements of the TRI.[33] This Act sets general reduction targets over a number of years and its reporting covers not only the release of toxic substances but also their use and generation. Reporting includes the generation and shipment of the by-products and a comparison with the performance from the previous year. Rather than specifying performance standards, the Massachusetts legislation relies on self-monitoring to encourage both firms and citizens to obtain information that can be used to reveal potential problems and their solutions. It also provides for the training of a peer inspectorate to review the usage reduction plans and of planners to provide a consulting service to help firms set and realize their objectives. To promote a capacity for self-reflection and self-revision in regulatory practices, the TURA also provides for a central governance structure to suggest the modification of state services and reporting requirements in light of performance reviews. The Act has had some success in reducing both the use of toxic chemicals and their by-products. Moreover, it has been well received by participating companies, who argue that they would continue to participate in the program even if they were not required to do so.

Project XL can be regarded as a further extension of the philosophy behind the TRI and the TURA. As a pilot project to promote and supervise other pilot projects, it authorizes the waiver of regular permitting requirements in return for better environmental performance. The bargain of greater autonomy for local actors in the determination of the precise ends and means of environmental regulation, in return for increased monitoring, is further extended in Project XL.[34] Unfortunately, in an effort to avoid the problems associated with command-and-control regulation, this model may concede too much to decentralization by failing to establish standards for granting waivers or for pooling experiential knowledge from the local actors. Without provisions for standards or the pooling of information, there can be no effective way to provide for self-reflective corrections in the system. Thus, Sabel et al. insist that in any effective IR system, the state must retain a crucial role by providing the standards necessary to guide decentralized regulatory systems. While these programs in no sense represent ideal forms of experimental environmental governance, they can be regarded as anticipations of such a regime.

Theorizing the Architecture of Deliberative Environmental Governance

While none of these approaches embodies all the qualities that Sabel, Fung, and Karkkainen regard as necessary for contemporary environmental governance, they are seen as having the potential to contribute to democratic reform in a revivified environmental public sphere. To clarify this potential, Sabel et al. maintain that a theoretically informed understanding of the principles behind an information-based regime is necessary. This is achieved through a conceptual reconstruction of existing programs that emphasizes elements with the capacity to promote co-operative decision-making by a plurality of social actors. This more radical model for a regime of environmental governance is based on two main objectives: improving the decision-making capacity of institutions and enhancing the opportunities for citizen participation in environmental decision-making. The main features of their 'architecture' for this regime are decentralization, reflexivity, social co-operation and social learning, accountability, and democratic deliberation.[35]

The new regulatory architecture rejects the idea of a 'central, panoramic knowledge' but maintains a role for centralized pooling of information.[36] Sabel and his colleagues agree that the decentralization of decisions is necessary for better and more democratic environmental decision-making, but at the same time they insist that neither of the main contending regulatory models can deliver on these objectives. Although the autonomy of local actors from centralized decision-makers is a significant element of the 'architecture' of their 'rolling-rule' or self-revising regime, the interaction of the 'local' and 'general' or centre is integral to a revitalized conception of reflexivity. While institutions friendly to local experimentation are needed, they also recognize that decision-making requires the institutionalization of links among these local units.[37] Knowledge based on concrete experience helps to account for local complexity, but such knowledge can only be effective if it is combined with general epidemiological evidence of the toxicity of substances in the environment.[38] Reflexivity thus is redefined as an active exchange and pooling of knowledge between a plurality of local actors and a centralized administrator to ensure that this knowledge is shared. Rather than there being fixed objectives that are difficult to change, environmental performance goals can be revised in the course of the ongoing sharing of knowledge among the relevant actors. Therefore, Sabel et al. contend, the state must continue to provide an oversight role to ensure that necessary knowledge is shared.

This sharing of knowledge should produce better environmental decisions because it allows for the possibility of a dynamic revision of the general framework of regulation in accordance with the revelation of unexpected risks at the local level. The sharing process therefore produces better information because it allows for its consolidation from a number of dispersed sources. If this were all there was to it, such models would not be much of an advance beyond the existing contenders in the regulatory debates. While problem-solving represents a valuable end in itself, the approach also appeals to the ideal of providing more accountable collective decision-making through more direct forms of democracy.

At the core of this model is an ideal of social learning. Sabel, Fung, and Karkkainen maintain that '[t]he philosophy of this architecture is pragmatist: while it rejects immutable principles, it keeps faith with the idea that we can always institutionalize better ways of learning from the inevitable surprises that experience offers us.'[39] Their objective is to foster mutual learning from local experimentation. Drawing on philosophical pragmatism, the authors suggest that sharing information is a form of social learning. An information exchange between periphery and core, wherein local entities and central regulators are both responsible for reflecting on and evaluating the results of local regulatory experiences, represents the recognition and revision of the contemporary division of labour. It is usually assumed that only centralized expert knowledge needs to be reflexive. Sabel et al. rightly recognize that the possessors of experiential local knowledge are also capable of revising that knowledge and their goals as a result of a co-operative exchange of information with scientific, technological, and economic experts.[40]

Following the pragmatist philosophy of John Dewey,[41] Sabel and his colleagues recognize that the 'circle of expertise' must be extended to an informed citizenry.[42] This process can only occur through co-operation among social actors to address common social problems. Therefore, to be effective, programs of experimentation must facilitate the conditions that make such social co-operation possible. They contend that co-operative activities in environmental governance can only be established if deeper and more extensive citizen participation is institutionalized. Environmental governance as problem-solving must also be a form of democratic governance.

Their institutionalization of this more radical ideal of participation emphasizes a specific type of democratic practice through deliberation. The examination and revision of choices regarding environmental risks is possible because citizens are able to consider their choices in light of the relevant experiences of others. Collective decision-making takes place, according to this 'pragmatist architecture', through direct deliberation with other social actors who are affected by the outcomes of those decisions. In this process of deliberation, citizens are expected to advance proposals and then to defend or justify them with public reasons. This form of justification of positions is regarded as essential to all collective democratic decision-making.[43]

Deliberative problem-solving 'is by its nature focused on addressing specific problems in local settings'.[44] Institutional fora for deliberation on environmental issues require that it be possible to debate the implications of general principles in light of the 'particulars of local experience' and to invite 'discussion of such experience in whatever terms suits participants'.[45] According to Joshua Cohen and Charles Sabel, deliberation by a diverse citizenry is capable of bringing out the strengths and weaknesses of different proposals.[46] Accordingly, all interested individuals and relevant units at local, state or provincial, regional, and national levels are to be encouraged to contribute to deliberation, thereby facilitating the comparison of information from relevant practices elsewhere. In addition, by promoting greater reflection on the definitions of problems and on proposed solutions, public discussion is intended as a means to facilitate the reflexive revision of existing proposals. Participants are expected to be accountable for their reasons and those regions or units that perform poorly are expected to revise their proposals in light of better results from other regions. Thus, deliberation about common problems by diverse participants can enhance not only social learning but also the problem-solving capacity of the entities involved in environmental governance. It also enhances the legitimacy of decisions by making all participants accountable through the justification and revision of their positions in public discourse.

Environmental Knowledge and Social Co-operation

The experimental approach to environmental regulation articulated by Sabel and his various associates holds much promise. Based on an ideal of mutual learning as a social process, it elaborates the connection between social co-operation and democracy. It also represents an attempt to redefine a new role for law in 'governing at a distance'. Within their pragmatic perspective, law does not govern directly by specifying the content of performance goals but rather provides a framework for a plurality of actors to exchange information and to realize goals that they establish in co-operative processes of public deliberation. Law in a sense becomes indistinguishable from the process of experimentation.[47]

Despite this potential, however, problems remain with the theoretical and practical articulations of these projects. The presuppositions of the pragmatic approach regarding the roles of the public or publics need further clarification. Democratic social co-operation as a goal is very important for the establishment and maintenance of an environmental public sphere, but it is necessary to recognize that the asymmetrical distribution of discursive power between technical experts and laypersons may make accessibility to the public sphere by citizens and social movements very difficult.

In the published version of their paper, Sabel, Fung, and Karkkainen emphasize the contribution of experimental regulation to environmental democracy.[48] While continuing to emphasize the role that information-based regulation plays in more effective problem-solving, they also focus on its significance as a contribution to an emerging environmental public sphere. They believe that mechanisms of reporting and information pooling that they find in their empirical studies of different information-based approaches have such potential. By providing for accountability through reporting, industrial facilities, local units, and scientific experts are forced to make their thinking and goals public. These mechanisms create opportunities for publicity and thus have the potential to increase the accessibility of the environmental public sphere for a much larger number of social actors. However, the demands of Sabel and his various associates go beyond compulsory accountability. The participatory requirements of their pragmatic approach are quite rigorous. Their ideal of experimental regulation also reflects a commitment to the political dimension of environmental regulation and demands a 'deep participation' by private as well as public actors.[49]

Reviving a theme of Deweyan pragmatism, Sabel and his colleagues insist that both accountability and democratic participation must be based on a transformation of traditionally antagonistic relationships between citizen and expert into 'partnerships for environmental protection'.[50] This involves the fusing of the expertise of the professional to the contextual intelligence of the citizens. In their opinion, such a fusion is possible and feasible because of evidence of a high degree of participation in the institutions of local environmental governance that they have examined. According to their analysis, not only has there been a consistent willingness on the part of citizens to participate but there also has been 'unexpectedly deep deliberative capacities among a surprisingly broad range of the citizenry'.[51] The latter observation suggests that, at least at the epistemological level, citizens have the competence to participate actively in decisions often defined as being technical in character. While I do not question the validity of this argument with respect to the participatory capacity of citizens to bring local knowledge to bear on decisions regarding environmental risk, I have some concerns about whether or not Sabel and his colleagues are being overly sanguine about the possibilities of mutual co-operation between scientific experts and citizens as the possessors of a locally based experiential knowledge.

Do these decentralized regulatory mechanisms of self-rule really have the potential to bring processes of technology and science under democratic control? Realism may dictate that we recognize that scientific expertise and democracy represent different and possibly conflicting forms of publicity.[52] Despite recognizing that individuals, groups, and communities cannot simply bracket and leave behind their interests and acculturation, Sabel and his collaborators do not sufficiently problematize the relationship between scientific experts and the possessors of local knowledge. While desirable as a social goal, an inclination to co-operate cannot be assumed, especially in regard to relationships between experts and non-experts.

One of the major problems with social co-operation that aims at creating communication across divergent interests and cognitive frameworks is maintaining the credibility and legitimacy of the processes in which it is embedded.[53] Such co-operation is difficult to achieve within the context of forms of public discussion wherein scientific experts and laypersons have their own values and assumptions that are embedded in particular cultural traditions and attitudes about what counts as valid knowledge. At the very least, it is necessary to recognize that this 'intercultural deliberation' between experts and laypersons raises issues about asymmetries of power.[54] As a result of 'institutionalized patterns of interpretation and evaluation'[55] that can be attributed largely to the demands of the capitalist economy, experts are able to neutralize and depoliticize processes of negotiation over technology and the amelioration of environmental risks by resorting to their own expert discourses to explain the character of problems and the character of solutions.[56] The hegemony of such expert discourses allows industry, governments, and experts to avoid social negotiation by resorting to technological fixes to redefine social uncertainties as technical ones. In effect, real access to the environmental public sphere is denied as moral or political issues around environmental risks are suppressed. Thus, lay publics face an extremely difficult task in getting their particular views on problems and their resolution onto the public agenda.

Within the context of their own conceptual framework theory, Sabel and his colleagues would undoubtedly recognize this as a denial of the more democratic regulation they seek. However, such institutionalized patterns of interaction between technical experts and lay publics raise more than issues of simply expanding environmental participation: they also raise profound issues of social justice.

Environmental Justice as Democratic Governance

One can best make sense of the problematic relationship between expertise and local knowledge by reinterpreting it within the context of an emerging concept of environmental justice. Within the last two decades, a third wave of environmentalism has been led by a grassroots movement of those most directly and severely affected by environmental problems. This movement has begun to conceptualize its claims about exposure to environmental risks in terms of demands for social justice.[57]

For the most part, the literature on this environmental justice movement has focused more narrowly on issues relating to discrimination based on race or class in decisions to locate polluting activities or hazardous waste facilities. Many of the demands of the movement focus on the need for a more equitable redistribution of the benefits of a safe and healthy environment and the burdens of exposure to ecological risks. However, despite this focus on racial and class discrimination in the distribution of environmental risks, environmental justice issues are neither confined to non-whites nor exclusively concerned with the distribution of environmental risks.[58]

A broader conception of environmental justice focuses more on the way that decisions about environmental risks are made than about the fairness of the distribution of risks.[59] As shown in the early work of Iris Marion Young, many members of communities exposed to environmental risks regard themselves as victims of injustice because they are not consulted and thus are excluded from participation in decisions that affect their future and that of their communities.[60] Their sense of injustice and demands for justice therefore seem to be motivated by their exclusion from the decision-making procedures that shape their lives. Environmental justice, according to this broader theory, should be measured not so much by a society's ability to promote a fair distribution of environmental burdens as by its ability to provide and support the institutional conditions necessary to realize a reflexive self-determination of acceptable levels of risks by individuals and groups in the relevant communities. It is within such a conception of environmental justice and injustice that the institutionalization of the subordination of local knowledge to the discursive power of scientific expertise is to be understood.

Interactions between laypersons as possessors of local environmental knowledge and scientific and technical experts can involve forms of identity politics and environmental injustice.[61] Science and scientific institutions and the traditional knowledge of occupational groups or of people living in a particular area are implicated in the perception of the identity of both technical experts and local people. As Brian Wynne has shown, the cultural traditions that are particular to scientific inquiry often are not open to questioning by people from non-expert cultures. He suggests that when attempts to articulate a local point of view on environmental conditions are ignored by scientific expertise, it is perceived as a form of disrespect that is experienced as a threat to the identity of these local groups.[62] Thus, as a result of an institutionalized pattern for the interpretation and evaluation of environmental risks, local groups are often denied a recognition of their status as full partners in social decision-making. Because technical experts often do not regard them as morally or epistemologically accountable, local groups are denied the 'participatory parity'[63] or political equality necessary to take part in the environmental decisions that affect their community. In order to be regarded as just, a state must not only guarantee the political autonomy of all its members but should also provide the conditions for their recognition as morally accountable persons capable of participating in the kind of public debate and communication envisaged by Sabel and his colleagues.

Institutionalized patterns of interpretation and evaluation that continue to privilege the epistemological advantages of expert knowledge come at the cost of democracy and thus are unjust. As long as alternative models of environmental governance, in their concern for 'practical problem-solving', speak in the technical language of 'baseline studies' and 'flexible innovation', as does Sabel's articulation of environmental experimentalism, they concede too much to the existing technocratic hegemony. They thus undermine the political, cultural, and moral character of deliberations between experts and laypersons, the recognition of which is necessary for democratic environmental governance.

In addition, any potential co-operation between expert and lay publics must take place in the resulting context of a distrust of science, which undermines its legitimacy and authority. The authority of science in environmental decision-making can only be constituted and renewed if the conditions of democratic accountability are established. To establish such conditions, we need to institutionalize the possibility for a discursive challenge to the authority of science and also the promotion of a political and legal culture that encourages participation.[64] The pragmatist architecture of the decentralized information-based regulation proposed by Sabel and his associates may contribute to the further development of these background cultural conditions. However, their informal character, even when institutionalized, does not go far enough to secure a role for lay knowledge in deliberative fora that acknowledge the continued priority of 'objective' scientific and technical knowledge. A more politicized response in the form of claims for environmental rights to challenge discursively the authority of science and the subsequent institutionalization of these claims in law may be the only way to stabilize the ideal of deliberative environmental regulation that underlies the regulatory innovations suggested by Sabel and his associates.

Within the environmental justice movement, many communities experience their exposure to environmental risks and the subsequent disrespect shown to their concerns as a systematic denial of their rights by more powerful social actors.[65] People in these communities have experienced their exposure to risk and the denial of their rights to participate in the environmental decisions affecting those communities as forms of exclusion from full membership in the community. Their political claims are based on demands for rights to a public process of open debate on the definitions of environmental risks and on the solutions to those risks. Thus, environmental rights are perceived as necessary to their recognition as equal citizens of the political community.

Such environmental rights emphasize the importance of political communication. They make sense primarily as rights of equal access to the public world.[66] Accordingly, we need to be able to participate on an equal basis in public affairs in order to articulate our opinions and have them listened to and to ensure that our actions may have some effect.[67] At the same time, however, environmental rights entail more than mere access to the public sphere. They also must refer to a broad concept of political equality that means that ordinary citizens and members of the environmental movement also can initiate action with others and initiate acts of deliberation with others about matters of common concern.[68] Without rights of access that can also address social inequalities that act as obstacles to the initiation of deliberation, people have little opportunity to define issues of common concern. Consequently, public deliberation may be reduced to the one-way discourse of expertise that allows the powerful to continue to marginalize other voices and to set the public agenda as they see fit. Without an expansive conception of environmental rights, there can be no deliberative equality and the public space for democratic environmental governance will be diminished.

What kind of special ecological rights are most appropriate for those who have suffered from environmental injustice? The exact character of such rights remains to be determined. However, a suggestive model is provided by the Mackenzie Valley Pipeline Inquiry conducted during the 1970s by Mr Justice Thomas Berger.[69] The purpose of the Berger Inquiry was to determine the environmental and social impacts of locating a natural gas pipeline in the Canadian North. Since the majority of the population in the proposed location of the pipeline were Aboriginal peoples, the Berger Inquiry had to provide special measures to ensure that their interests and those of the natural environment were represented. Among the significant and precedent-setting arrangements, which achieved a rights-like status within the Inquiry, were the provision of information on all aspects of the development to all parties; funding for the participation of needy groups; and the provision of special community hearings in remote locales that were sufficiently sensitive to the cultural traditions of the affected Native peoples to encourage a broader basis for participation. Future attempts to institutionalize 'participative parity' through environmental rights must demonstrate a similar sensitivity in order to promote meaningful communication and democratic deliberation between the possessors of lay knowledge and experts.

The pragmatist architecture of the reflexive regulation proposed by Sabel, Fung, and Karkkainen promises to provide fora for the participation of the people and communities most directly affected by environmental decisions. This process is important both for the constitution of an environmental public space in civil society that is separate from the state and for making that state more accountable. However, political communication within such informal bodies must necessarily remain unstable and subject to being diluted without a continual articulation of claims for participatory equality. Such claims can only be effective if they take the form of environmental rights as a language of political equality and also as demands for the institutionalization of these rights in law.

Experimental democratic processes, by their very nature, remain unstable. To stabilize the communicative power of publicity emerging from these efforts at co-operative governance, it is necessary to embody that power within institutions. Law, which is somewhat distant in the current forms of this experiment, may well have to be redeployed in order to provide the conditions that are necessary not only for experimentation but for the more just form of participation that is a central goal of such projects.

Conclusion

One of the main assumptions behind the alternative form of environmental governance proposed by Sabel, Fung, and Karkkainen is that the traditional antagonistic relationships between citizens and experts must be transformed into partnerships for environmental protection.[70] As I have suggested, this transformation may be much more difficult than Sabel and his colleagues would have us believe. Despite evidence of social co-operation in the experimental programs to which they try to give a theoretical and practical coherence, neither the cultural background of the relationships between expert and lay knowledge nor institutionalized patterns of the interpretation and evaluation of environmental risks are easily overcome.

We must also ask if such a transformation is even desirable if the main result of social co-operation is simply to be a better or 'more efficient' approach to environmental problem-solving rather than a furthering of democratic governance. If co-operative forms of environmental regulation promote mutual learning and encourage reflexivity on the part of both experts and laypersons, better environmental protection is likely to be a valuable result. However, as Brian Wynne has demonstrated in his cultural studies of the interaction between science and lay knowledge, there is a danger that social learning may become a one-way street in which only local people are willing to reflect on and change their assumptions about local environmental conditions.[71] Citizens are to become quasi-experts, according to Sabel, Fung, and Karkkainen. Without a similar transformation of the capacity of experts to re-examine and transform their own assumptions with respect to the 'objectivity' of technical knowledge, there is a real danger that a needed source of challenge to such knowledge will disappear. If this occurs, the political and moral dimension of democratic deliberation will continue to be hidden under the presumed neutrality of science and technology. The possibility of this depoliticization and neutralization of environmental governance may go hand in hand with the neo-liberal agenda within which some critics see performance-based regulation fitting.[72]

Whatever the validity of this specific argument about the compatibility of environmental experimentalism and neo-liberalism, the goal of achieving more complete forms of environmental knowledge, and at the same time promoting democratic deliberation through forms of social co-operation and learning, runs the risk of taming the political element in environmental governance. Only a more agonistic form of deliberation that emphasizes reflexivity in the confrontation of different opinions, and that is firmly anchored in the articulation of equal rights of access to the public sphere, can preserve the moral/political dimension of the negotiation about the kind of relationship with nature that we want.

Sabel and his associates rightly contend that we need institutions that can promote communication between different kinds of cultures and different kinds of knowledge. As I have argued, such institutions need to provide the conditions that allow those who have traditionally been marginalized to have a voice and to be listened to. To paraphrase the political philosopher, Hannah Pitkin, these claims to equal say and equal respect are then claims that must become 'negotiable by public standards'.[73] Such standards can only be reflexively constituted within the context of specific struggles for environmental justice.

Notes

1. C. Sabel, A. Fung, and B. Karkkainen, 'Beyond Backyard Environmentalism: How communities are quietly refashioning environmental regulation', *Boston Review* 24, 5 (1999). Available at: < http://bostonreview.mit.edu/BR24.5/sabel.html > .
2. See especially J. Habermas, *The Theory of Communicative Action*, vol. 2: *Lifeworld and System: A Critique of Functionalist Reason* (Boston: Beacon Press, 1987), 367–70.

3. See Chapter 3 in this volume, by A. Hunt.
4. In this paper, I will use the term 'governance' to describe the emerging regulatory terrain. Unlike 'regulation', which implies the existence of some central directing agent of regulation, 'governance' implies more general projects for ordering behaviour in which the state remains involved insofar as it specifies criteria for the rule-making activities of a plurality of actors but does not specify the substantive results of regulation. In addition, objects of regulation are not pre-existing but rather are constituted in the very process of rule-making and in the interchange among local, regional, national, and global sites of governance.
5. See Sabel et al., 'Beyond Backyard Environmentalism'; C. Sabel and M. Dorf, 'A Constitution for Democratic Experimentalism', *Columbia Law Review* 98 (1998): 267–473. Although these papers are obviously collaborative efforts, I will continue to refer to Sabel as the main author because both works represent the application of Sabel's ideas with respect to democratic deliberation and innovative forms of regulation that he has developed elsewhere.
6. For an analysis of the potential and limits of co-operative approaches in such performance-based regulation, see K. Harrison, 'Talking with the Donkey: Cooperative Approaches to Environmental Protection', *Journal of Industrial Ecology* 2 (1999): 51–72.
7. R.M. Unger, *Law in Modern Society* (New York: Basic Books, 1976).
8. J. Rosenau, 'Governance, Order, and Change in World Politics', in Rosenau and Otto Czempiel, eds, *Governance Without Government: Order and Change in World Politics* (Cambridge: Cambridge University Press, 1992), 4.
9. For an excellent overview of market-based models of environmental regulation, see Robert Howse, 'Retrenchment, Reform or Revolution? The Shift to Incentives and the Future of the Regulatory State', *Alberta Law Review* 31 (1993): 464.
10. See Sabel et al., 'Beyond Backyard Environmentalism', Section 1.
11. See T. Purvis, Chapter 2, this volume.
12. The most developed reflexive legal theory is found in the work of the German legal theorist, Gunther Teubner. See especially, G. Teubner, 'Substantive and Reflexive Elements in Modern Law', *Law and Society Review* (1983): 239–85. Another version of reflexive law may be found in the Jürgen Habermas's 'procedural paradigm' of law. See J. Habermas, *Between Facts and Norms* (Cambridge, Mass.: MIT Press, 1996). For a specific application of reflexive law to environmental law, see E. Orts, 'Reflexive Environmental Law', *Northwestern University Law Review* 89 (1995): 1227–1340; and D.J. Schneider, Chapter 5, this volume.
13. See Sabel et al., 'Beyond Backyard Environmentalism', Section 3. In the draft of the paper that forms the central focus of this chapter and was available on Sabel's personal Web site, the authors used the term 'information-based regulation'. In the published version they designate the same regulatory approaches, which are based on co-operative ongoing reflection and information exchange, as 'performance-based regulation'. I will use the two designations as being synonymous in this paper.
14. Teubner, 'Substantive and Reflexive Elements', 263.
15. Orts, 'Reflexive Environmental Law', 1254.
16. Ibid., 1264–8.
17. P.R. Kleinsdorfer and E.W. Orts, 'Informational Regulation of Environmental Risks', *Risk Analysis* 18 (1998): 156.

18. J. Black, 'Proceduralizing Regulation: Part 1', *Oxford Journal of Legal Studies* 20 (2000): 607.
19. See Sabel et al., 'Beyond Backyard Environmentalism', Section 3.
20. In both Sabel and Dorf, 'A Constitution for Democratic Experimentalism', and Sabel et al., 'Beyond Backyard Environmentalism', the more traditional term 'regulation' is preferred over 'governance'. However, they use the term within the context of a concept of an environmental regime that emphasizes the interaction of government with self-governing entities, such as corporations, local community organizations, and non-governmental organizations, in a complex exchange of information through mutual learning processes from the local to the centre. This usage is closer to the concept of governance articulated above than it is to traditional notions of regulation.
21. Black, 'Proceduralizing Regulation: Part 1', 597, 607.
22. Sabel et al., 'Beyond Backyard Environmentalism', Section 2.
23. Ibid.
24. Ibid.
25. Ibid.
26. Ibid., Introductory Section.
27. See Sabel and Dorf, 'A Constitution for Democratic Experimentalism', 373–86. In his collaboration with Fung and Karkkainen, Sabel also chooses to examine most of the same programs.
28. I have been selective by focusing on types of regulatory innovations that illustrate the possibilities of the deliberative framework that Sabel and his colleagues find so important.
29. 42 U.S.C. ss. 11001–11050 (1994).
30. Sabel and Dorf, 'A Constitution for Democratic Experimentalism', 375–6.
31. Ibid., 376.
32. Ibid., 376–8.
33. Ibid., 379–81.
34. Sabel et al., 'Beyond Backyard Environmentalism', Section 3.
35. This list of elements is based on a consolidation of elements in recent articles by Sabel and various associates. See especially ibid.; Sabel and Dorf, 'A Constitution for Democratic Experimentalism'; J. Cohen and C. Sabel, 'Directly Deliberative Polyarchy', available at: < http://www.law.columbia.edu/sabel/papers/DDP.html > .
36. Sabel et al., 'Beyond Backyard Environmentalism', Section 2.
37. See also Cohen and Sabel, 'Directly Deliberative Polyarchy'.
38. Sabel and Dorf, 'A Constitution for Democratic Experimentalism', 374.
39. Sabel et al., 'Beyond Backyard Environmentalism', Section 2.
40. For an argument suggesting that 'traditional' communities may have a greater capacity for such critical self-reflection than do scientific experts, see B. Wynne, 'May the Sheep Safely Graze? A Reflexive View of the Expert-Lay Divide', in Scott Lash et al., eds, *Risk, Environment & Modernity: Towards a New Ecology* (London: Sage Publications, 1996).
41. See J. Dewey, *The Public and its Problems,* in John Dewey, *The Later Works*, vol. 2 (Carbondale: Southern Illinois University Press, 1988).
42. Sabel et al., 'Beyond Backyard Environmentalism', Section 3.
43. This description of democratic deliberation is only implicit in the description of the

'pragmatist architecture' of experimentalist environmental regulation. However, it is more explicitly and systematically developed in Sabel's recent work with Joshua Cohen. See Cohen and Sabel, 'Directly Deliberative Polyarchy'. The following description of deliberative problem-solving is derived from this work.
44. Ibid., Section 4.
45. Ibid.
46. Ibid.
47. Sabel and Dorf, 'A Constitution for Democratic Experimentalism', 387.
48. Sabel et al., 'Beyond Backyard Environmentalism'.
49. Ibid., Section 2.
50. See ibid., Introductory Section.
51. C. Sabel, A. Fung, and B. Karkkainen, 'Sabel, Fung and Karkkainen Respond to their Critics', *Boston Review* 24, 5 (1999). Available at: < http://bostonreview.mit.edu/BR24.5/sabel2.html >.
52. Publicity may be defined as the political exercise of public reason to establish a public space for the exercise of citizenship. See James Bohman, 'Citizenship and Norms of Publicity: Wide Public Reason in Cosmopolitan Societies', *Political Theory* 27 (1999): 187.
53. J. Bohman, 'Democracy as Inquiry, Inquiry as Democratic: Social Science and the Cognitive Division of Labor', *American Journal of Political Science* 43 (1999): 599.
54. Bohman, 'Citizenship and Norms of Publicity', 187.
55. N. Fraser, 'Social Justice', *The Tanner Lectures on Human Values, 1998* (Salt Lake City: University of Utah Press, 1999).
56. The institutionalization of expert or technical discourses finds one of its clearest expressions in the concept of science-based regulation. Government and industry favour basing policy decisions on the authority of science, which is largely derived from the ideal of value-neutrality and objectivity. The appeal to the authority of science and technology thus is used by governments 'to legitimate policy making processes that have no public mandate'. Decisions based on science are thought to be less open to challenge because of their 'objectivity'. Hence, science is used to depoliticize decisions and ensure that the real decision-makers in government and industry are not accountable. Sabel and his associates downplay the extent to which expert discourses remain entrenched in policy decision making. For an excellent analysis of science-based regulation, see Lucy Sharratt, 'Deconstructing a "Science-Based" Regulation: Towards Rendering the Risks of Genetic Engineering Visible' (MA thesis, Carleton University, 2000).
57. L. Cole, 'Empowerment as the Key to Environmental Protection: The Need for Environmental Poverty Law', *Environmental Law Quarterly* 19 (1992): 634-6.
58. Without a consideration of who makes environmental decisions or of the legitimacy of the procedures for making these decisions, theories of environmental justice must remain incomplete at both the conceptual and the practical levels.
59. S. Čapek, 'The "Environmental Justice" Frame: A Conceptual Discussion and Application', *Social Problems* 40 (1993): 7-9.
60. I.M. Young, 'Justice and Hazardous Waste', in M. Braidie et al., eds, *The Applied Turn in Contemporary Philosophy* (Bowling Green, Ohio: Bowling Green University Press 1983), 175.
61. See B. Wynne, 'Misunderstood Misunderstandings: Social Identities and Public

Uptake of Science', in A. Irwin and Wynne, eds, *Misunderstanding Science?* (Cambridge: Cambridge University Press, 1996).
62. Ibid.
63. Fraser, 'Social Justice', 25. 'Moral accountability' refers to whether people are recognized as having the cognitive capacity for viable participation. In this sense, 'cognitive' refers both to the moral capacity to offer good reasons to justify their positions and to the quality of their knowledge.
64. M. Warren, 'Deliberative Democracy and Authority', *American Political Science Review* 90 (1996): 56.
65. Čapek, 'The "Environmental Justice" Frame', 7.
66. J. Bohman, 'The Moral Costs of Political Pluralism: The Dilemmas of Difference and Equality in Arendt's "Reflections on Little Rock" ', in Jerome Kohn and Larry May, eds, *Hannah Arendt: Twenty Years Later* (Cambridge, Mass.: MIT Press, 1996). While Bohman is referring primarily to a more general moral/political right that underlies specific articulations of legal rights, the general demand for environmental rights can be explained in a similar way as political claims for inclusion and recognition that are necessary for participatory equality.
67. Ibid., 63.
68. Ibid., 63-4.
69. T.R. Berger, *Northern Frontier, Northern Homeland: The Report of the Mackenzie Valley Pipeline Inquiry* (Ottawa: Queen's Printer, 1977).
70. Sabel et al., 'Beyond Backyard Environmentalism', Section 1.
71. See Wynne, 'May the Sheep Safely Graze?'; Wynne, 'Misunderstood Misunderstandings'.
72. For an example of such a critique, see T.J. Lowi, 'Frontyard Propaganda: A Response to "Beyond Backyard Environmentalism" by Charles Sabel, Archon Fung, and Bradley Karkkainen', *Boston Review* (1999). Available at: < http://bostonreview.mit.edu/BR24.5/lowi.html >.
73. H.F. Pitkin, 'Justice: On Relating the Private and Public', *Political Theory* 9 (1981): 347.

PART II

Labour, Capital, and the State

Chapter 7

Corporate Governance

R. LYNN CAMPBELL

The nature and structure of a corporation present unique challenges for regulation. This uniqueness stems from the metaphysical qualities of the corporate persona, its structure, and the fact that the corporation can do no act by itself. In addition, the court's application of regulatory legislation complements an extensive body of common law and equity that has been employed to control corporate directors. This chapter explores the challenges of regulating the corporation and its directors.

The Corporation and Governance

What is a corporation? In law, a corporation is a separate legal person.[1] Of course, a corporation cannot do everything that an individual can, but in areas of contract, tort, and crime, the law treats a corporation as if it were an individual. What this means is that a corporation, in its own right, can be party to a contract, can sue to have a contract enforced, and can be sued to have a contract enforced against it. A corporation owes, and is owed, common-law duties in tort and can likewise enforce them or have them enforced against it. The corporation is responsible, not the shareholders, for its debts. In tort, the corporation is vicariously responsible for the acts of employees and agents.[2] In crime, a corporation can be convicted as long as the offending act has been done by one of its agents with sufficient authority.[3] Regulatory legislation, for the most part, treats corporations as persons in terms of regulation and enforcement. A corporation has an existence of its own and offers limited liability to those who own its shares. This means that the shareholder is not financially responsible for more than the value of the share.[4]

How does one know when one is dealing with a corporation? The corporation, after all, is really a legal myth or an abstraction in terms of physical presence.[5] From an outsider's point of view, a business may be carried on in the same manner as before incorporation. After incorporation, then, the corporation is responsible for contracts entered into on its behalf, not the shareholders, directors, or corporate agents. This is important because, in law, another person may be responsible or be party to a contract. Usually one knows that a corporation is involved by its name. An incorporated name will end with such words as 'Limited', 'Ltd.', 'Incorporated', or 'Inc.'. It takes merely an administrative act of the state to create a corporation. Persons who want to carry on business for profit by way of a corporation must apply to either Industry Canada, Corporations Branch, or the provincial counterpart for a Certificate of Incorporation.[6] In most cases, relatively simple standardized forms, called Articles of Incorporation, must be completed after a name has been selected and cleared. After the name search report, application forms, and the prescribed fee have been submitted, the applicants (or incorporators) are entitled to a Certificate of Incorporation.[7] The date on this certificate is the effective date of incorporation.

There are many types of corporations—for-profit, not-for-profit, private, public, and even statutory.[8] All corporations contribute to the economy of a state and have considerable impact on individuals. A corporation may be an employer, consumer, producer of goods and services, and investment instrument for shareholders. A corporation may also be a polluter, a bad debtor, and even a not-so-good entity that closes down or makes many or all of its employees redundant through downsizing. This chapter will focus on the private for-profit corporation.

'Governance', by dictionary definition, means the manner of governing, controlling, directing, or regulating.[9] This includes the method of management and system of regulation. An integral part of the 'method' or 'system' includes the authority or power accorded to an office or its permission to govern. Also included in this definition are words not normally considered part of governance, such as behaviour, demeanour, and conduct. As with other dictionary definitions, meanings evolve and develop through variables of time and expectation. Expectation of a corporation itself has not only changed over time, but now the expectations of others can legitimately be considered by the directors. Today, employees, consumers, creditors, and even community standards must be routinely taken into account by directors.

Corporate governance can be divided into two modes—internal and external. Internal governance refers to control exerted upon the powers of management by the structure and organization of the corporation. External governance refers to legal accountability of management through the regulation of the exercise of their corporate powers. Constraints and limits on the exercise of power are primarily imposed by legislation or the extension of existing common-law equitable duties by the courts. The latter mode of governance has been far more effective in terms of controlling management behaviour.

Internal Governance: The Corporate Structure

Today, legislation entitles a corporation to exist with a minimum of one shareholder. A corporation has two 'organs' or bodies that have power to act—the general meeting of shareholders and the board of directors. The shareholders own the shares of the corporation and elect the directors. The board of directors has all the power to manage the business and affairs of the corporation.[10] In terms of governance, the corporate structure, as set up by statute, provides the legal mechanisms through which authority and control are exercised. The shareholders in general can only exercise their powers in meetings duly called to transact the business of the corporation.[11] To convene a formally constituted meeting there has to be notice given to each shareholder, the directors, and the auditor.[12] The notice must state the time, date, and place of the meeting, which can be virtually anywhere in Canada that the directors determine. Also, a quorum, that is, the majority of the voting shareholders unless otherwise provided in the bylaws, must be present at the beginning of the meeting.

The business of the corporation normally transacted by shareholders includes the election of directors, approval of the financial statements, appointment of the auditors, the ratification of any bylaws or special resolutions, and any other business put on the agenda by the board of directors.[13] Decisions in the shareholders' general meeting are made by a majority of votes cast unless the bylaws require a special majority.[14] The shareholders' general meeting is a forum in which the shareholders are entitled to question the directors' actions and policies. In reality, even though individual shareholders may attempt to make the general meeting an accountability session, the majority of the shareholders will ultimately make any decision with respect to corporate business of the meeting.

Like the shareholders, the directors can only exercise their powers in a meeting that has been duly called. There must be notice and a quorum, otherwise any business transacted is void and has no effect.[15] The notice need not specify the purpose of or business to be transacted at the meeting unless specifically mandated by the bylaws. A quorum for the board meetings is the majority of the directors unless otherwise stated in the bylaws.[16] The directors are authorized not only to manage the business and affairs of the corporation, but also to make, amend, or repeal any of the bylaws with the approval of the shareholders in general meeting.[17] Director decisions are determined by majority vote.

Older statutes required there to be seven directors. However, current legislation only requires one director unless the corporation is offering its shares to the public, in which case there must be three.[18] Directors hold office for a fixed term determined by the bylaws and can be removed from office at any time by a majority vote of the shareholders in general meeting. There are few restrictions on who may become a director. A director must be 18 years of age, of sound mind, and not bankrupt.[19] Thus, there are no special qualifications to become a director.

In theory, residual corporate power rests with the shareholders in general meeting because, as in a democratic state, they elect the officials who govern for a period of time. Also, shareholders have an opportunity to question the directors at election time. Policies and decisions of the board of directors can be challenged, and if the shareholders are unhappy they can refuse to elect or re-elect individual directors. As in democracies, a majority decision determines any matter in the shareholders' meeting. However, in terms of nature and purpose, democratic and property rights can conflict. The exercise of the voting right is not restricted merely to the election of directors. Not only are corporate policies determined by majority vote of shareholders, but also the actions of individual directors can be whitewashed or censured by the shareholders, for directors sometimes enrich themselves or others to the detriment or exclusion of other shareholders.[20] This is particularly true if one or more directors hold the majority of voting shares in a general meeting.

The courts have traditionally deferred to the doctrine of majority rule and rarely have assisted aggrieved minority shareholders. A partnership principle of non-interference in internal disputes has been applied to corporations because the courts have been of the opinion that differences among shareholders could be resolved by a majority vote.[21] Also, except in extreme cases the courts have been reluctant to intervene to review imprudent management.[22] Carrying on business involves certain risks that, so courts generally have determined, are best left to the directors to assess. In addition, the courts traditionally have been reluctant to second-guess commercial matters with the benefit of hindsight. Therefore, directors have not had to perform corporate duties with a greater degree of skill and care than might reasonably be expected from a person of their knowledge and experience.[23] A very low level of skill and care left little room for shareholder comfort. Being liable for errors of judgment was not generally considered to be part of a director's duties to the corporation.

Thus the corporate structure has not provided an effective check on management powers. Direction of corporate policies and operations has clearly rested in the hands of controlling shareholders, whose actions could seldom be checked by other shareholders. Indeed, the law through the doctrines of agency and separate legal personality assisted those with effective control in exploiting their unfettered power. Controlling shareholders owed duties only to the corporate persona that could routinely ratify abuses. Aggrieved minority shareholders were a frustrated lot. Corporate governance was more procedural than substantive. The corporation could be used not only to conduct business, but also to shield controllers, who created extra wealth at the expense of minority shareholders, from accountability. Corporate governance, until the mid-twentieth century, was clearly in its early stages of development.

The Seeds of Change

Today, the corporate form has become the norm for carrying on business. Sole proprietorships and partnerships, while important forms of conducting business, have given way to the corporation rather quickly. Limited liability had attractive features for investors. Raising capital was made easy and economical by issuing shares of a corporation. Corporate shares could be transferred more easily than partnership interests, particularly upon the death of a partner. However, shares could be sold to an unwary public because until the early 1930s the equity market was largely unregulated.[24]

With the explosive growth in the public's desire to become investors, corporations grew in size. Likewise, corporate shareholdings became more widespread and rarely did someone holding a few shares show concern or seek direct involvement with the operation of the corporation. Shareholders were primarily interested in return on and growth of their shares. As a consequence, shareholder meetings were rarely attended by all shareholders, but their proxies became valued instruments of control to whoever held them. In most cases, the holders of these proxies were directors who sent out management circulars and notices calling the meetings with proxy forms attached.

As noted in the seminal 1932 work of Berle and Means,[25] share ownership of corporations became separated from management. Managers, who were appointed and authorized by the board of directors, now directed the operations of the corporation. Through proxy solicitation and management circulars they also controlled the shareholders' general meetings. With this control came dominance of the board of directors because managers controlled the proxies for director elections. Further, managers controlled valuable information that shareholders either did not receive or did not care to receive, provided that the corporation continued to produce profits. Maximizing shareholder value, then, became a goal of management. However, management techniques and methods of doing so were mostly unregulated and left to the conscience of ill-informed directors. Their position was fraught with difficulty and conflict because those the directors were obliged to monitor were normally the same ones who put them on the board.

Despite management control of day-to-day operations and operational information, the board of directors was still legally vested with the authority to manage the affairs of the corporation.[26] Thus, any legal check on the operation of the corporation had to start with the directors because they had delegated corporate authority to act. The board had always had the power to delegate its authority even if it had not done so. The difficulty, then, was that the actual corporate control was vested in a managerial group that owed no legal duties to the shareholders nor to the corporation, other than normal employee duties. However, if corporations were to be accountable to anyone, the board of directors was the trigger point because it was the legal focus of corporate power. Therefore, legislation to regulate some aspect of corporate activity invariably provided additional penalties for the directors, who had authority to prevent a breach.[27]

External events also left their impact on the corporation. The Great Depression left its scars on many skeptical shareholders during the late 1920s and the 1930s. International state conflicts during the 1940s transformed open market economies into war-oriented economies with central government control and dominance. New business opportunities related to supplying the needs of war resulted in greater corporate size. Despite these events, however, the corporate form continued to flourish and grow in transnational and international operations. Indeed, large corporations situated in industrial countries became instruments of foreign investment for the benefit of many Third World countries.

Securities legislation followed the Great Depression and the collapse of the equity markets.[28] This legislation controlled the exchange of shares of corporations listed on a public exchange. The purpose of the legislation was to prevent financial gain due to abuse and unfairness in the trading of shares. Thus a shareholder was precluded from using information that was not known to others to his or her advantage. In particular, corporate insiders were not allowed to use confidential corporate information to their advantage by selling or purchasing corporate shares before all shareholders were informed. Insiders, who were always defined to include directors, had to register share transfers, and important information affecting share value had to be released to all shareholders in a timely and uniform manner. Commissions were set up to enforce trading rules and to administer exchange policy.[29] Since 1966, securities legislation could be viewed as a mechanism of external control of corporate governance.

As well, during the 1960s and 1970s changes in legislation in Canada at both the federal and provincial levels impacted on corporate operation. New legislation set the ground rules at the place of work. Employee standards, safety, and human rights legislation imposed new duties on both corporations and directors.[30] Labour legislation allowed employee organizations to bargain collectively and strike if necessary.[31] Modern ground rules for distribution and marketing of corporate products had been established. Combines, packaging, labelling, and consumer protection legislation had set new and higher standards.[32] Governments had retained control of corporate operations by requiring that certain business activities must be licensed.[33] Thus, oil, natural gas, forestry, transportation, and communications sectors of the economy were governed by legislative as well as administrative regulation.[34] Fiscal reporting and remitting of source deductions required diligent supervision by directors.[35] These types of statutes not only provided a greater respect for the public interest, but also forced directors to take into account new considerations when acting in the boardroom. Through legislation, therefore, the state forced accountability for action and also provided the tools for others to enforce and expand new duties of directors. This legislation could also be viewed as an external mechanism of corporate governance.

There were also major changes in the legislation under which corporations came into existence. Since reform during the 1970s, procedural and substantive law provisions were enacted. Limits on the authority of the managing director or committee of directors became enumerated.[36] Thus, specified acts such as a management proxy circular or bylaw proposals had to be approved by the full board before being sent to the shareholders. A process for disclosing a director's interest in contracts to which the corporation was party was established to avoid subsequent rescission.[37] Insider trading rules for corporations distributing shares to the public were set up.[38] Civil liability could result if a director used confidential information to an advantage when transacting in corporate shares. Processes for derivative actions, appraisal, and oppression remedies were updated.[39] Further, a security holder or the Director of the Corporations Directorate of Industry Canada could apply to a provincial supreme court to launch an investigation in certain circumstances of fraud or oppression.[40] The court had a wide arsenal of orders as an effective remedy to the complaint.

Duties of directors became codified in the governing legislation. The common-law and equitable duties that required a director to act honestly and in good faith in the best interests of the corporation were explicitly set into legislation.[41] However, the legislation in the 1970s changed the internal mode of corporate governance and accountability. For example, the requirement of skill, care, and diligence was changed from the subjective to the objective standard. Directors now had to exercise a degree of skill, care, and diligence that a reasonably prudent person would exercise in comparable circumstances.[42] Thus, the standard of conduct became more refined but fell short of a professional level of competence. Further, these duties were imposed not only on directors but also officers of the corporation. Neither director nor officer could contract out of these statutory duties. Nor could the corporation indemnify directors or officers who had not acted honestly and in good faith in the best interests of the corporation.[43]

Directors who voted for or consented to certain resolutions contrary to the governing legislation could be personally responsible for any loss to the corporation, such as payment of a dividend, an indemnity, or a shareholder loan that contravened the legislation.[44] Likewise, when a corporation was at or near insolvency, a purchase, redemption, or acquisition of shares could result in personal liability for directors who voted in favour of such a resolution.[45] Directors could also be personally liable to employees for six months' wages in the event that the corporation failed to pay.[46] These statutory liabilities were joint and several, which meant a plaintiff could make the total claim against one director. The onus was then on this director to seek contribution from other directors, provided they were agreeable and solvent.

Corporate governance was clearly changing through legislative direction. Controlling shareholders no longer had open-ended discretion when exercising corporate powers. Corporate policies and operations were influenced by the state through diverse legislative proclamations. Duties of directors became more articulated and were monitored closely by minority shareholders. They now had a statutory process to enforce these duties. Also, such action often received considerable press coverage that reflected negatively on the corporation. Thus corporate governance had become more substantive than procedural in nature. The corporation as a vehicle for creating wealth became more regulated and was answerable to the state, shareholders, creditors, employees, consumers, and the general public for its actions. More importantly, corporate directors and officers were just as accountable to these persons.

External Regulation

Change in corporate governance by way of regulatory legislation set down new parameters within which corporate operations had to be carried on. New or varied rules of conduct had to be instituted and corporate operational practices had to be brought into line with legislative pronouncements. If necessary, corporate policy also had to be changed to accommodate legislative and regulatory direction. From the corporation's perspective, change in corporate governance, when imposed externally, invariably meant additional expense. These added costs reduced profits, which in turn affected the competitiveness of a corporation in its industry. Indeed, because competition would now be measured in global terms, additional costs due to legislative mandates could also impact on a state's ability to run its economy.

The courts were another important external force in determining the scope of legislative-driven regulation. When legislation was challenged by corporations or the state, the courts had to determine the extent to which it applied. Legislation and regulation might not cover a particular case, or the courts themselves might be persuaded that the legislation should not apply to a given set of circumstances. Legislation might be found to be ultra vires or be declared contrary to the Charter of Rights and Freedoms.[47] Whatever the legal argument, the courts had the final say in applying the law as set down in legislation. Thus, any legislation that dealt with corporate governance could be challenged.

Here we will consider how the courts have dealt with governance issues in three broad areas: (1) enforcement provisions of legislation that have dealt directly with the operations of the corporation; (2) legislation that has offered minority shareholders the tools to enforce duties of directors and officers; and (3) the court's application of common-law and equitable doctrines to corporate directors and officers to ensure their honesty and integrity.

Legislative Intervention

Before we look at how regulatory legislation has been enforced against corporations and their directors, it may be helpful to examine legislation that applies to society in general—the Criminal Code. As with natural persons, corporations must conduct their transactions in a manner consonant with society's values as reflected by the Criminal Code. Therefore, as persons, corporations must transact business in accordance with the Criminal Code and can be punished for any contravention of its provisions. However, how can a corporation commit fraud or conspiracy because it cannot act by itself, let alone form a guilty mind?

The courts have applied the law of agency to impute the actions of senior corporate agents to be those of the corporation itself. In agency, an agent could bind its principal if the agent acted within its scope of authority. The actions and intentions of directors and officers were, the courts have reasoned, the corporation's because they were its 'alter ego', and therefore original or personal liability would follow.[48] The courts have applied the identification doctrine to determine whether a corporation would be criminally responsible. This doctrine stated that the actions and intentions of senior agents, such as directors and officers, were considered to be those of the corporation because they were its 'directing mind and will'.[49]

The identification doctrine has applied a hierarchical test to determine whose acts would render a corporation criminally liable. Directors and officers are obviously not a problem because they have sufficient authority and exercise corporate power. However, other positions, such as managers and supervisors, are not always seen as clothed with sufficient authority.[50] Also, as corporations have grown in size, their structures have become more complex and knowledge is not as easily imputed to the top. The courts then developed the delegation doctrine to expand the scope of people whose actions could make the corporation criminally responsible. This doctrine has applied a functional test to determine liability.[51] If an agent had total control over the assigned duties and responsibilities in a particular area, then that person was considered to be the 'directing mind and will' of the corporation. This formulation resulted in a much wider net of persons whose actions and intentions would render the corporation criminally responsible. However, it seemed to blur the common-law doctrine that a principal could not be vicariously liable for the criminal acts of its agents.[52]

Offences contained in regulatory legislation also provided an exception to this common-law doctrine. The object of such legislation was to prevent acts detrimental to the public interest but otherwise devoid of moral turpitude.

However, if prevention was to be effective, it had to be administered through the criminal courts. Most acts in contravention of regulatory legislation were done by employees who did not have sufficient authority to bind the corporation to original or personal liability. Thus, if the legislation was to be significantly effective, vicarious liability had to be imposed for regulatory offences. This liability arose directly by way of legislation. Regulatory legislation developed a common formula and provided that it was sufficient proof of the offence to establish that it had been committed by an employee or agent of the corporation, whether or not that person could be identified.[53]

Three types of offences were created by regulatory legislation: *mens rea*, strict liability, and absolute. Even though the courts had been reluctant to impose vicarious liability for *mens rea* offences in other areas, they have been willing to infer intent from the doing of the prohibited act in regulatory legislation. In the 1978 Supreme Court of Canada decision, *R. v. Sault Ste Marie*,[54] it was held that the doing of the prohibited act was sufficient for conviction of a strict liability offence unless the accused could prove due diligence. Therefore, after the Crown had brought forth factual evidence to show that an offence had been committed, the onus then moved to the accused to bring forth evidence to show that it had exercised due diligence. No doubt the Crown had an advantage, but only the accused had possession of facts required to prove the exercise of due diligence. Also, the Court in *Sault Ste Marie* suggested that absolute liability offences might well be abolished.

Directors and officers could also be targeted for prosecution when the corporation had been accused. In crimes, the actual wrongdoer could always be charged. Thus a charge of fraud or conspiracy could be laid against a director or officer if that person had done the offending act. In addition, any person who aided or abetted another person in committing an offence could be a party to the offence.[55] Further, every person who counselled another person to be party to an offence was a party to that offence. An offence was defined to include ones under regulatory legislation. Thus, there was broad liability for crime that could result not only in fines, but also in incarceration.[56]

While sanctions may have been less onerous under regulatory legislation, there have been an enormous number of incidental statutory liabilities of directors. Most statutes contain provisions of liability when a director directs, authorizes, assents to, acquiesces in, or participates in an activity that gives rise to an offence.[57] Other statutes provide for liability if a director has permitted, acquiesced to, or knowingly concurred in an offence.[58] Still other statutes impose liability on directors whenever a corporation has been convicted of an offence, unless the court concludes that they had not 'authorized, permitted or acquiesced in' the offence.[59] In regulatory offences, the exercise of due diligence is invariably a defence.

In many sectors, regulatory legislation contains administrative boards that also regulate corporate conduct. Corporations in many industries have been required to obtain a licence from either an administrative board or a government ministry to conduct business. For example, alcohol production, distribution, and consumption in public places have required licensing.[60] The forestry, petroleum, communications, transportation, and manufacturing sectors of the economy have also been heavily regulated by direct regulation and licensing.[61] Licences have required the licensee to comply with certain conditions for approval. Once issued, if any provision has been breached, the licence normally contains a provision for penalty of the licensee. In the extreme, a licence can be revoked, which normally results in the licensee no longer carrying on business. Administrative boards have often been empowered to issue orders or levy fines if regulatory legislation has been contravened. These Acts, being considered 'administrative' in nature, have not always been subject to court review.

Shareholder Action

Shareholders have become important monitors for making the directors, officers, and management more accountable in the exercise of their corporate duties. Shareholders, particularly minority shareholders, have a vested interest in ensuring that these people comply with their duties because any breach could be detrimental to maximizing profits and share values. Further, shareholders have traditionally been more familiar with corporate operations because they have a right to attend shareholder meetings and have rights to receive information outsiders cannot demand.[62] In the past, the board of directors controlled general meetings and retained a firm grip on corporate operations and policies. Indeed, for the most part, shareholder general meetings had become mere formalities that had to be endured once a year.

Today, however, shareholder activism in certain economic sectors is alive and paving new ground for management accountability. Individual shareholders in certain instances can insist on an item being placed on the agenda so that it can be fully discussed at a general meeting.[63] Also, recognition of the minority position has blossomed because the nature of that shareholder has changed. Many minority shareholders can be in the form of investment, pension, or registered funds that are large in size and have considerable voice in the power structure of the corporation. These funds also have the resources and expertise to challenge decisions made by corporations. For example, 'poison pills', or management's defence to a takeover bid, now have to be ratified by the shareholders in general meeting before becoming effective.[64] Funds have even questioned the level of remuneration of a chief executive officer.[65] Further, shareholders today have effective remedial tools for more accountability.[66]

In common law, the corporate derivative action was the primary remedial action to address wrongs of directors.[67] However, the common law refused to take the position that the controlling shareholders, when exercising corporate powers, owed any duties to the minority.[68] *Foss v. Harbottle* (1843) also stood in the way of the minority taking action. This case held that when the directors had done a wrong, then the proper person to remedy the wrong was the corporation itself because the directors owed duties only to the corporation. The courts considered such wrongs as internal disputes and matters to be worked out by the shareholders themselves. Only in certain exceptions could a shareholder bring action. These exceptions were narrowly defined but include ultra vires actions, a fraud on the minority, or some act that could not be ratified by the majority.[69] These exceptions were also riddled with procedural obstacles that made the wrong very difficult to pursue. As a result, the common-law corporate derivative action was not an effective remedy to make directors accountable for their wrongs.

To better protect the legitimate interests of minority shareholders, current legislation has attempted to remove the procedural difficulties of *Foss v. Harbottle*. The new legislation of the 1970s has allowed a complainant, on compliance with specified conditions, to bring an action on behalf of the corporation to enforce a duty owed to it.[70] In order to weed out unnecessary harassment of management, leave of the court has to be obtained to commence an action. Also, the complainant has to show that every reasonable effort has been taken to cause the corporation to bring action. This has required a refusal of the directors to take action or proof that such a request to do so would be futile.[71] Further, the court has to be satisfied that the complainant is acting in good faith and that the commencement of the action would be in the best interests of the corporation.[72] Thus, it is necessary for an applicant to ascertain the views of the majority of shareholders and to establish a prima facie case by affidavit material before leave will be granted.

Although well intentioned, legislation that codified the common-law corporate derivative action has not afforded the minority shareholder an effective remedy. Procedural hurdles have again been significant for the person taking action. The bona fides of an applicant has been a questionable requirement. Also, the 'interests of the corporation' has been viewed with skepticism if the object of the legislation is to enhance managerial accountability. What is needed are new criteria upon which a corporate derivative action is to be evaluated.[73]

Modern Canadian legislation has revolutionized the oppression remedy. Today, this may be invoked by a complainant whenever the affairs of the corporation or the powers of the directors have been exercised in a manner that has been 'oppressive or unfairly prejudicial to, or that unfairly disregards the interests of' one or more of the shareholders, creditors, directors, or officers.[74] The complainant, a person who has taken control of the action, has included more than a minority shareholder. A current or former director or officer has been included, as well as any other person deemed 'proper' by the court to make an application.[75] Thus, any person who has had a contractual relationship with a corporation can be a potential applicant when legitimate expectations have not been met because of corporate conduct.[76]

To determine whether conduct has been 'unfair or prejudicial' so as to constitute oppression, the courts have regarded not only the actions of the directors, but also the expectations of the applicants.[77] Expectations have been determined as questions of fact and have had to be reasonable. However, they may have changed and even have evolved over time. Circumstances giving rise to an allegation of oppression can be far-reaching. Amalgamation, a takeover bid, a tender for bids, and buy-sell provisions are examples of when legitimate expectations might be unfairly or prejudicially affected. Manoeuvres resulting from shareholder family disputes might also give rise to oppression. However, focusing on the impact that management conduct has had on others' expectations has been a fundamental shift in the resolution of corporate disputes.[78]

Expectations have been also vital to crafting a remedy in the event of a finding of oppression. Remedial orders can be broad and all-encompassing in nature. The court might make whatever interim or final orders that it believes will fit a particular situation.[79] The court has been empowered to intervene by making orders that could assist the complainant in self-help within the existing power structure of the corporation, that could realign the existing power structure, or that could change the ground rules of the corporation.[80] The court also has been given the power to deal directly with any conduct or transaction that has caused the oppression. However, in drafting specific orders, the court has taken into account the expectations of the applicants and has even returned the corporation, directors, and applicants to their positions prior to any oppression.[81] This has included setting aside contracts freely entered into or forcing the sale of shares of one of the respondents in the action.[82]

The oppression remedy may well radically reshape corporation law. Management must now consider the interests of any potential applicant when exercising powers. This is in addition to the traditional duty of considering only the best interests of the corporation. Also, directors need not have conscious knowledge that what they were doing was unfair to the applicant because it is the effect of the conduct that is relevant to oppression.[83] Thus, corporate conduct that would otherwise be done in good faith can still be judged to be oppressive if legitimate shareholder expectations have been abridged. It is these expectations that are protected by the oppression remedy. Further, the oppression remedy has the potential of striking down the corporate persona and ignoring the limited liability doctrine. The court has the power to impose personal liability on directors or officers to compensate an aggrieved person. Even the time-honoured business judgment rule that shields directors from court interference may be attacked.[84] Independent legal advice, outside financial advisers, and independent or special board committees are essential supports for the business judgment of the directors in circumstances wherein duties might conflict.[85] Clearly, the spectacle flowing from the oppression remedy has barely begun.

The oppression remedy may well be used as an effective tool for making management more accountable for corporate action. Also, it is an efficient action because the court has not let procedural or technical claims become obstacles either to a finding of oppression or to imposing remedies to deal with the source of the unfairness or prejudice. Further, the person taking the action is not the state, but someone connected to the corporation. However, the courts have guarded against debasing the nature of the oppression remedy by refusing to employ it to address trivial complaints. Needless to say, this remedy is becoming the action of choice to cope with unjust management conduct.

Judicial Control

The courts have applied principles of common law and equity to check the exercise of powers of management and the directors. Each director and senior management official owes fiduciary duties of loyalty and good faith to the corporation.[86] This includes always acting in the best interests of the corporation, never fettering one's discretion, exercising powers for the purpose for which they have been conferred, and never allowing one to be in a position of potential conflict between duty and self-interest.[87] The fiduciary duty also requires a certain level of skill, care, and diligence. To a lesser extent, the courts have also applied common-law doctrines to control management action. For example, piercing the corporate veil or applying the tort of inducing breach of contract have been used to remedy questionable actions of the directors.

Even though corporate powers can only be exercised in properly constituted meetings, fiduciary duties are owed by directors and officers individually. The courts have consistently stated that the best interests of the corporation are to be determined by the board, not by the court.[88] Thus, business judgment of the board is rarely reviewed in terms of its soundness or insight. Risk-taking and corporate policy have been entrusted to the board of directors, and their decisions determine the profitability of the corporation. However, the courts have reviewed the manner and process by which these decisions have been made. The powers of directors and officers are held in trust for the corporation, thus business decisions can be tainted by taking improper considerations into account. Therefore, only the interests of present and future shareholders can be considered, even though exclusive focus on these interests might result in inappropriate or poor policy in other areas of the business.[89] Interests of the shareholders have to include all of the shareholders and not only what a controlling shareholder has demanded of nominee directors.

The courts have also applied the collateral purpose doctrine to control management. Directors have to exercise their corporate powers for the purpose for which they have been conferred.[90] Purpose has been given a reasonable interpretation and any ambiguity in definition has been resolved by examining the corporate constitution. Thus, issuing shares for the purpose of maintaining control of the corporation has been found to be in breach of the directors' fiduciary duty.[91] However, a sale of corporate assets to purchase sufficient shares to maintain control has been found not to be an exercise of powers for an improper purpose because the directors took all reasonable steps to maximize value for all shareholders.[92] In the future, the process employed in making such decisions will have to be carefully examined by the courts and scrutinized in conjunction with other remedies, such as oppression, that have allowed the expectations of complainants to be taken into account. This will particularly be the case when the directors have refused an attractive takeover bid.

The courts have shown great ingenuity in controlling director and officer conduct in the application of the residual fiduciary duty. Equity has long precluded any person in a position of trust to benefit personally from the trust. This rule has been rigidly applied for the protection of the person who was the beneficiary of the trust, or for whom the trust powers have been exercised.[93] Thus, the concept of equity has meant that no trustee could ever be in a position of potential conflict between personal interest and duty.[94] Duty in this sense has meant the duty to consider exclusively the best interests of the trust when exercising trust powers. Any benefit personally received by the trustee had to be returned because the trustee was in a 'conflict of interest' position. The only defence to a breach of this duty was consent of all beneficiaries after full and informed disclosure.

Directors and officers, like trustees, have held the powers of the corporation in trust and, when exercising them, have had to take into account only the best interests of the corporation. The equitable rule against potential conflicts of interests therefore has been applied in full rigour to directors because they are in a fiduciary position.[95] Senior management and employees placed in positions of responsibility have also owed fiduciary duties.[96] No fiduciary could make any secret profit without full disclosure and consent of the majority of shareholders in general meeting. Also, a fiduciary could not take up any lucrative employment contract at a competitor or divert any corporate opportunity so that a personal benefit occurs.[97] If a fiduciary received information of an opportunity that could be taken by the corporation, the courts have stated that the information has been received in the capacity of the fiduciary position and thus belongs to the corporation. It was irrelevant whether the corporation would or could have taken the opportunity.[98] What the court was looking for was use of a fiduciary position as such to benefit oneself.[99] Even if the benefit had been diverted to another person, there was a breach of duty. Further, fiduciary duties extend beyond the termination of the office or after a director leaves a corporation.

The Supreme Court of Canada has carefully examined the appropriate remedy for breach of fiduciary duty. Traditionally, damages or rescission has been awarded to an aggrieved litigant. However, damages had to be calculated and often did not reflect the true loss to the plaintiff. Rescission may not have reflected the true loss to a plaintiff either. In *Lac Minerals v. International Corona*,[100] Justice La Forest, writing for the majority, granted a constructive trust by way of the equitable proprietary claim. By this remedy, the asset that had been acquired through a breach of fiduciary duty and breach of confidence was restored to the party who would have purchased it but for the breach. Allowing a defendant to retain a specific asset obtained through conscious wrongdoing would be, in the Court's opinion, offensive. Thus the moral quality of the defendant's act was a consideration in determining whether to grant the proprietary remedy.

In *Lac Minerals v. International Corona*, the plaintiff had discovered a gold vein in northern Ontario. The plaintiff, being a junior in the mining industry, required a larger mining corporation, a senior, with sufficient capital to mine the gold. As was the custom in the mining industry, the plaintiff disclosed information to the defendant with the view of concluding an agreement to undertake this venture together. However, the defendant ultimately acquired the land containing the vein and mined the gold itself. To acquire the property the defendant used public information as well as information disclosed by the plaintiff. Justice La Forest concluded that even though the commercial exploration agreement had not been concluded, the plaintiff was vulnerable and a fiduciary relationship had been created. He stated:

> the essence of the imposition of fiduciary obligations is its utility in the promotion and preservation of desired social behaviour and institutions. Likewise with the protection of confidence. In the modern world the exchange of confidential information is both necessary and expected. Evidence of an accepted business morality in the mining industry was given by the defendant, and the Court of Appeal found that the practice was not only reasonable, but that it would foster the exploration and development of our natural resources. The institution of bargaining in good faith is one that is worthy of legal protection in those circumstances where that protection accords with the expectations of the parties.[101]

Thus, fiduciary duties have had a major impact on director, officer, and management conduct. These duties now have required the utmost good faith and honesty and consideration of others' interests in decision-making beyond those merely of shareholders. Shareholders, directors, creditors, and any other person who can assert a fiduciary relationship with the corporation are owed fiduciary duties. The categories of fiduciary relationships have not been closed and courts have continued to broaden them. The standard of conduct is to be determined by general principles because there have been no guidelines established to sanction specific actions. The remedies flowing from a fiduciary breach, including the equitable proprietary claim by way of constructive trust, are better suited to aggrieved plaintiffs. Judgments may be more easily executed in the event that the corporate defendant has become insolvent. Also, the benefit of any accretion to the property subject to the constructive trust would belong to the plaintiff. Further, the court would not have to become involved in the tedious task of calculating damages.

In addition to loyalty and good faith, the fiduciary has been required to exercise a degree of skill, care, and diligence. The common law did not require a director to exercise the same degree of skill and care as a trustee.[102] However, legislation has changed the subjective standard applied in the common law to one of a reasonably prudent person in comparable circumstances.[103] Diligence has required attendance at regular board meetings. Blind reliance on information from corporate officials could result in breach of the duty of diligence. Active steps to dissent from a board resolution now have to be taken by a director.[104]

Traditionally, the courts have not interfered with business judgments of the directors by reviewing their competence or skill in making decisions about the business. Corporate policy and decisions relating to risk-taking were clearly vested in the board and not to be reviewed, with the benefit of hindsight, by the courts. Today, the courts have adopted a similar approach, but they have more carefully scrutinized the process adopted in arriving at a decision. In takeover, offering, and insolvency issues, diligence has now required that the board establish independent committees to recommend a course of action, to consider outside financial and legal advice, and to take into account and to document advice received.[105] If the directors have believed what they were doing was reasonable and have had grounds upon which to base this belief, then it would be difficult to attack the decision by breach of fiduciary duty of skill, care, and diligence.[106] However, the courts could review business decisions in other areas, such as oppression.

The required standard and level of skill, care, and diligence have been extensively tested by the courts in cases dealing with the failure of corporations to remit source deductions. Section 227.1(1) of the Income Tax Act has made a director jointly and severally liable for unremitted source deductions unless the director has exercised diligence to prevent the default.[107] The statutory wording for the 'skill, care, and diligence' has been virtually copied from other Canadian corporate legislation and has been enacted as a defence to a claim by Revenue Canada under this section of the Income Tax Act. A high level of diligence and prudence has been required for successful defence.[108] Reassurance that payment has been made from a director responsible for finances has been insufficient diligence. A director has had to take positive action to prevent a default from occurring. Prudence has required that all deductions must be held in trust and processes put in place to signal an early failure to remit. The defence has had little success when a bank has refused to honour remittance cheques due to financial difficulties of a corporation. Potential personal liability for these corporate obligations has changed directors' attitudes towards debts to the Crown, and as a consequence most corporations have had procedures set up to ensure that remittances have been paid in a timely manner.

The courts have used common-law actions such as the tort of inducing breach of contract or piercing the corporate veil to review questionable corporate behaviour. Interfering with contractual relations have resulted in damages for the tort of inducing breach of contract.[109] A director might be personally liable in damages for stripping a corporation of assets so that a plaintiff could not successfully enforce a claim. If a director was acting under the compulsion of a duty to the corporation, then there might be no interference.[110] However, when no compulsion to act on behalf of the corporation has existed, there would be no justification and thus a director would be personally liable. Acting with the intention of depriving a plaintiff of the benefits of a contract with a corporation could be evidence of bad faith and also result in personal liability of a director.[111] In this case, the corporation would be insulated from liability because the director has acted outside the scope of authority.

In certain circumstances, the courts have pierced the corporate veil to attach personal liability to a director or controlling shareholder.[112] Normally, the corporation has shielded directors and shareholders from personal liability for corporate debts because it was the contracting party. Directors and officers have acted as mere agents for their principal—the corporation. In the same manner, the corporation has been responsible for torts and even crimes committed by its agents. Fundamental to corporate identity has been the principle stated in *Salomon v. Salomon*—the corporation has had to be regarded as a separate legal person. This has been the only consistent principle, but there have been circumstances in which the courts have disregarded the corporate veil.

The corporate veil has been pierced to prevent a fraud or to curtail improper conduct of corporate agents.[113] The veil has also been discarded when a corporation has been deliberately used to evade contractual obligations.[114] Likewise, a corporation has been considered an agent of another corporation that has had effective and constant control.[115] In other areas, the courts have pierced the corporate veil to recognize a group enterprise or the economic realities of two corporations. Tax law has often ignored the separated legal existence of related corporations when one has become an instrument of a controlling parent.[116] Conduct and control have been considered important considerations to determine instrumentality. Inadequate capitalization has also allowed creditors to pierce the corporate veil to attach liability to shareholders.[117]

There has been no guiding principle to predict when the corporate veil will be pierced. Economists have argued that the corporate veil could be pierced to promote efficiency.[118] Management would then be precluded from externalizing costs that rightly belonged to the corporation. Another argument has been to avoid misleading creditors. Misrepresentation has been regarded as a mechanism of increasing creditors' information costs that have been viewed as wasteful. No fixed policy or legal principle has been developed to determine when the veil will be disregarded, other than that the courts will do so in the interests of justice. This lack of clear policy, however, could be regarded as a check on the behaviour of management, officers, and directors.

Judicial flexibility has been important in making directors accountable for their actions. The prompting for directors to become more accountable may well have been a product of actions by public institutions and society itself during the 1960s and 1970s. Society began to view persons in positions of power with greater skepticism. It was a time of challenging and changing existing values. Society was demanding that persons in positions of trust and power be more accountable for their actions. This transformation did not go unnoticed by the courts. While deciding a case dealing with the filching of a corporate opportunity by the directors, Supreme Court Justice Laskin, as early as 1972, recognized a need for change. He noted in a powerful, perceptive statement:

> What these decisions indicate is an updating of the equitable principle whose roots lie in the general standards that I have already mentioned, namely, loyalty, good faith and avoidance of a conflict of duty and self-interest. Strict application against directors and senior management officials is simply recognition of the degree of control which their positions give them in corporate operations, a control which rises above day accountability to owning shareholders and which comes under some scrutiny only at annual general or at special meetings. It is a necessary supplement, in the public interest, of statutory regulation and accountability which themselves are, at one and the same time, an acknowledgment of the importance of the corporation in the life of the community and of the need to compel obedience by it and by its promoters, directors and managers to norms of exemplary behaviour.[119]

Recent Developments

In the early 1990s Canada experienced a recession. Investors and creditors of corporations that either failed or had to be restructured were often dissatisfied with the performance of management and the board of directors. In response, the Toronto Stock Exchange formed a committee to review the state of corporate governance specifically at the board of directors' level.[120] The committee launched its work from the perspective that well-functioning corporations were vital to wealth creation and social progress. The committee also wanted to design proposals to meet the growing expectations concerning how boards are constituted and the relationships between boards, management, and shareholders. After consulting with interested parties, the Toronto Stock Exchange published a report, *Where Were the Directors?*,[121] that spelled out guidelines for improved corporate governance in Canada.

The report identified principal responsibilities of a board. They included the stewardship of the corporation, the strategic planning process, the identification and monitoring of the major risks of the business, the appointment, development, and succession of senior management, the implementation of an effective communications policy, and the adoption of internal systems to ensure the fulfillment of these responsibilities. The principal recommendations focused on increasing the effectiveness of the board by addressing the constitution of the board and by prescribing board committees to carry out governance-related functions. The report expressed concern with the extensive system of director liability for corporate conduct and invited provincial and federal governments to review legislation imposing personal liability on directors.

The report emphasized the importance of outside directors and recommended that the majority of directors be unrelated ones who would be independent of management and free to act in the best interests of the corporation. The report also recommended that the majority of board committee members be composed of outside directors, but that the audit and nominating committees be composed only of outside directors. Orientation and education programs should be developed for new board members. The size of the board should be reviewed to determine an appropriate number for effective decision-making. Appropriate structures and procedures to ensure a board can act independently of management ought to be in place. To this end, individual directors ought to be able to hire outside advisers at the expense of the corporation. The board as a whole should have the responsibility of defining its relationship with the chief executive officer. Responsibility for developing the corporation's approach to governance issues should be assumed either by the board of directors or assigned to a board committee.

Where Were the Directors? has stimulated a great deal of high-level discussion and debate. The study of corporate governance has prompted interesting cross-disciplinary work in legal journals.[122] The negative impact of personal liability of directors has often appeared in this commentary, with some arguing that the capacity of corporations to generate wealth necessary to fund the goals of the welfare state might be undermined if this liability continued. Unchecked personal liability, it has been claimed, could curtail the corporation's need to be more creative in response to globalization pressures. However, the federal Working Group on Directors' Liability did not find sufficient evidence to warrant a critical review of federal legislation.[123] The risk directors faced could be managed by the introduction of appropriate prevention and accounting systems, use of comprehensive indemnification contracts, and the purchase of better liability insurance. The Working Group also concluded that directors had not been resigning in significant numbers to avoid liability, but were taking a more active role in managing. Further, the group concluded that the costs imposed by directors' liabilities have not resulted in any noticeable harm to Canada's international competitiveness.

The guidelines adopted by the Toronto Stock Exchange did not provide a penalty for non-compliance. However, each corporation listed on the TSE would be required to describe in its annual report its system of corporate governance with reference to the guidelines. This disclosure would also explain any difference between the corporation's system and the guidelines. Thus, all listed corporations would have to examine the adequacy of their own systems of governance and develop systems reflecting their own circumstances.

Conclusion

Long gone are the times when directors considered only shareholders' interests in making corporate decisions. Nor can shareholders now ratify director action that may be prejudicial or unfair. Legislatures, together with the courts, have transformed the manner and care with which corporate decisions must be made. Today, the impact and effect of decisions on others must be considered and an appropriate balance of competing interests must be struck. Indeed, interests of the Crown, creditors, consumers, shareholders, and other parties must be taken into account and the advice of external experts such as lawyers, financial consultants, and industry specialists must be carefully considered. What this indicates is that corporate decision-making has become more process- and substance-driven rather than procedural. The shareholders in general meeting can no longer provide a protective shield for directors.

Superimposed on traditional internal corporate governance, wherein the directors managed and the shareholders ratified, is a complex process of new considerations at the behest of legislatures, courts, and shareholders. The challenge has become a delicate act of balancing many interests that often compete with those of shareholders and others. This task has become even more difficult in a competitive economic environment that is measured globally, as opposed to simply nationally.

Competition, however, has not been the only driving force behind the transformation of corporate governance. Society has also left its mark. The bottom-line test, or economic value, of the 'best interests of the shareholders' has been transformed to include society's value of accountability of persons in positions of power. These persons include not just directors but also officers of corporations. Accountability has been imposed and expanded by a plethora of legislation that provides personal liability of directors, and sometimes officers, when governing legislation has been breached. The courts have also imposed high standards of conduct on directors and officers by applying an exacting fiduciary duty to their actions. These duties will also continue to expand to new circumstances when injustice occurs or when the legitimate expectations of others have been cut short by the actions of senior corporate persons.

What will happen to corporate governance in the future? Crystal gazing never has been an exact science. However, corporate governance will be affected by values that society demands at any given point in time. The original requirement to act in the best interest of the shareholder has had to accommodate the protection of the minority. The protection of the minority has, in turn, yielded to a greater desire to make corporate officials, particularly directors, accountable for their actions. Society's values will continue to change. As in the past, the change will not come from the corporation itself, but from legislatures and the courts. The reason for this is because the cost of implementing societal values is incompatible with the corporate duty to maximize profits for the shareholders. Effective systems of corporate governance also carry a cost, but they should be viewed as complementing, not conflicting with, bottom-line profits.

What will these values be in the future? No doubt they will be a product of society's demands. It would be futile to attempt to articulate them. Some may guess the areas of environment, human rights, and human security will be of growing concern. Others may point out that global competition might force corporations to focus more on profits to maximize share value. However, states are becoming more co-operative in trade and in the collective condemnation of states that do not act according to the norms of international behaviour. Societies are demanding that their governments take action whenever any state falls below minimum standards. Corporations and the factors that influence their governance must follow the course set by national, or perhaps even international, society.

Notes

1. *Salomon v. Salomon & Co.*, [1987] A.C. 22, 1985–9 All E.R. Rep 33; *Macaura v. Northern Assurance Co.*, [1925] A.C. 619; *Lee v. Lee's Air Farming Ltd.*, [1961] A.C. 12. Today, corporations have the powers of a natural person: Canada Business Corporations Act, R.S.C. 1985, c. C-44, s. 15 (hereinafter cited as CBCA).
2. *Einhorn v. Westmount Investments Ltd.* (1969), 6 D.L.R. (3d) 71. Also see *McFadden v. 481782 Ontario Ltd.* (1984), 47 O.R. (2d) 134.
3. *Rex v. Fane Robinson Ltd.*, [1941] 2 W.W.R. 235, 76 C.C.C. 196, [1941] 3 D.L.R. 409; *Regina v. St. Lawrence Corp. Ltd.*, [1969] 2 O.R. 305, 7 C.R.N.S. 265, [1969] 3 C.C.C. 263. See also *Regina v. Waterloo Mercury Sales Ltd.* (1974], 4 W.W.R. 516; *Regina v. P.G. Marketplace & McIntosh* (1980), 51 C.C.C. (2d) 185.
4. Unless the corporate veil is pierced by the courts. For example, see *B.G. Preeco I*

(Pacific Coast) Ltd. v. Bon Street Holdings Ltd. et al. (1989), 60 D.L.R. (4th) 30; De Salaberry Realties Ltd. v. Minister of National Revenue (1974), 46 D.L.R. (3d) 100; Clarkson Co. Ltd. v. Zhelka et al. (1976), 64 D.L.R. (2s) 457; Re Canadian Commercial Bank; Canada Deposit Insurance Corp. v. Canadian Commercial Bank (1987), 67 C.B.R. 136.
5. Lennard's Carrying Company Ltd. v. Asiatic Petroleum Co. Ltd., [1915] A.C. 705.
6. See CBCA, ss. 6–8.
7. CBCA, s. 9.
8. Private corporations have few (usually less than 25) shareholders. Public corporations have many shareholders and usually are listed on a stock exchange. Non-profit corporations are incorporated under Part II of the Canada Corporations Act, R.S.C. 1985, c. C-18. Statutory corporations normally have a particular public interest, i.e., Canada Post Corporations Act, R.S.C. 1985, c. C-10, or regulate a particular industry, i.e., Atomic Energy Control Act, R.S.C. 1985, c. A-16. Cities, boards of education, colleges, universities, and health delivery centres are provincial statutory corporations.
9. *The Oxford English Dictionary*, 2nd edn (Oxford: Clarendon Press, 1989), 710.
10. CBCA, s. 102. The directors are not subject to direction by the shareholders: *Automatic Self-Cleansing Filter Syndicate Co. Ltd. v. Cunninghame*, [1906] 2 Ch. 34. See also *Quin & Axtens Ltd. v. Salmon*, [1909] A.C. 422.
11. CBCA, s. 139.
12. CBCA, ss. 135, 137. See *Barron v. Potter; Potter v. Berry*, [1941] 1 Ch. 895.
13. CBCA, ss. 106, 155, 162 (2).
14. CBCA, s. 139.
15. CBCA, s. 114(5).
16. CBCA, s. 114(2).
17. CBCA, s. 103.
18. CBCA, s. 102.
19. CBCA, s. 105.
20. *North West Transportation Co. v. Beatty* (1887), 12 App. 589; *Burland v. Earle*, [1902] A.C. 83; *MacDougall v. Gardiner* (1875), 10 Ch. App. 606. See also B. Welling, *Corporate Law in Canada: The Governing Principles*, 2nd edn (Toronto: Butterworths, 1991).
21. *Edwards v. Halliwell*, [1950] 2 All E.R. 1064. See also *In re Jury Gold Mines Ltd.*, [1928] 4 D.L.R. 735.
22. *Re Brazilian Rubber Plantations & Estates Ltd.*, [1911] 1 Ch. 425.
23. *Re City Equitable Fire Ins. Co.*, [1925] Ch. 407; *Re Owen Sound Lumber Co.* (1917), 38 O.L.R. 414; *Re Denham & Co.* (1883), 25 Ch. D. 752. Diligence today requires an increased level of attendance, attention, and care. See *Re Standard Trustco* (1993), 6 B.L.R. (2d) 241.
24. The U.S. Securities Act first appeared in 1933. For a historical overview, see L. Loss, *Securities Regulation*, 2nd edn, vol.1 (Boston: Little, Brown and Company, 1961). The first comprehensive Canadian securities legislation was the 1945 Ontario Securities Act. See Securities Act, R.S.O. 1990, c. S.5.
25. Adolfe A. Berle and Gardiner C. Means, *The Modern Corporation and Private Property* (New York: Macmillan, 1932; rev. edn, 1967).
26. CBCA, s. 102.
27. This is in addition to normal offence provisions dealing with corporate liability. For

example, see Canadian Environmental Protection Act, S.C. 1999, c. 33, s. 280.
28. Securities Act, R.S.O. 1990, c. 55.
29. Securities Act, ss. 2, 3.
30. Canada Labour Code, R.S.C. 1985, L-2; Canadian Human Rights Act, R.S.C. 1985, H-6.
31. Ibid.
32. Competition Act, R.S.C. 1985, C-34; Consumer Packaging and Labelling Act, R.S.C. 1985, C-38.
33. Canadian Radio-television and Telecommunications Commission Act, R.S.C. 1985, C-22; Canada Transportation Act, S.C. 1996, c. 10.
34. Canada Petroleum Resources Act, R.S.C. 1985, C-36; Canada Oil and Gas Operations Act, R.S.C. 1985, O-7; Canadian Wheat Board Act, R.S.C. 1985, C-24; Forestry Act, R.S.C. 1985, F-30.
35. Canada Income Tax Act, S.C. 1980-81-82-83, c.140, s.124(1).
36. CBCA, s. 115(3).
37. CBCA, s. 120.
38. CBCA, ss. 126-31.
39. CBCA, ss. 239, 190, 241.
40. CBCA, ss. 229-37.
41. CBCA, s. 122(1)(a).
42. CBCA, s. 122(1)(b). The common law provided only a subjective test of skill. See *Re City Equitable Fire Ins. Co.*, [1925] Ch. 407. Also, the common law intervened only in extreme cases. See *Re Brazilian Rubber Plantation & Estates Ltd.*, [1911] 1 Ch. 425.
43. CBCA, s. 124.
44. CBCA, s. 118.
45. CBCA, s. 118(2).
46. CBCA, s. 118(2).
47. The customary legal argument to challenge legislation was that it was outside the jurisdiction (ultra vires) of the body that enacted it. See *Citizens Insurance Co. v. Parsons* (1881), 7 App. Cas. 96. See also Bora Laskin, *Laskin's Canadian Constitutional Law*, 5th edn (Toronto: Carswell, 1986), 511. Today, the Charter of Rights and Freedoms (ss. 1-34 of the Constitution Act, 1982) is another legal argument to attack legislation.
48. See R.L. Campbell, *Legal Framework of Business Enterprises* (Toronto: York University Captus Press, 1995), 272.
49. *Lennard's Carrying Company Ltd. v. Asiatic Petroleum Co. Ltd.*, [1915] A.C. 705. See also *R. v. Fane Robinson Ltd.* (1941), 76 C.C.C. 196; *R. v. St. Lawrence Corp. Ltd.*, [1969] 3 C.C.C. 263.
50. *Tesco Supermarkets Ltd. v. Nattrass*, [1972] A.C. 153.
51. *R. v. St. Lawrence Corp. Ltd.*
52. *Regina v. P.G. Marketplace & MacIntosh* (1980), 51 C.C.C. (2d) 185.
53. For example, Canadian Environmental Protection Act, S.C. 1999, c. 33, s. 282.
54. *R v. Sault Ste Marie* (1978), 40 C.C.C. (2d) 353.
55. Criminal Code of Canada, ss. 21-2.
56. *Canadian Dredge & Dock Co. Ltd. v. The Queen* (1985), 19 D.L.R. (4th) 314, affirming (1981), 56 C.C.C. (2d) 193.
57. Consumer Packaging and Labelling Act, R.S.C. 1985, C-38, s. 20; Bankruptcy and

Insolvency Act, R.S.C. 1985, B-3, s. 204; Transportation of Dangerous Goods Act, S.C. 1992, c. 34, s. 39.
58. Ontario New Home Warranty Plan Act, R.S.O. 1990, c. O.31, s. 22; Real Estate and Business Brokers Act, R.S.O. 1990, c. R.4, s. 50.
59. Business Practices Act, R.S.O. 1990, c. B.18, s. 17; Employment Standards Act, R.S.O. 1990, c. E.14, s. 79
60. Liquor Control Act, R.S.O. 1990, c. L.26; Liquor License Act, R.S.O. 1990, c. L.19.
61. Forestry Act, R.S.O. 1990, c. F.26; Broadcasting Act, S.C. 1994, c.11; Canadian Transportation Act, S.C. 1996, c.10; Petroleum Marketing Act, R.S.A. 1980, c. p.-5; Mining Act, R.S.O. 1990, M.14.
62. CBCA, ss. 134(2), 135(6), 150.
63. *Michaud v. National Bank of Canada*, [1997] R.J.Q. 547. Michaud asserted his shareholder's right to compel the Royal Bank and the National Bank of Canada to circulate to shareholders a series of proposals designed to increase managerial accountability and reduce levels of executive pay. For comment, see B. Cheffins, 'Michaud v. National Bank of Canada and Canadian Corporate Governance: A "Victory" for Shareholder Rights?', *Canadian Business Law Journal* 30 (1998): 20-72.
64. *347883 Alberta Ltd. v. Producers Pipelines Inc.* (1991), 80 D.L.R. (4th) 359; *United Grain Growers Ltd. v. 3339351 Canada Ltd.*, [1997] M.J. No. 111. See also S. Wishart, 'Are Poison Pills Illegal?', *Canadian Business Law Journal* 30 (1998): 105-39.
65. TD Asset Management, as minority shareholder of Repap Enterprises Inc., successfully challenged a lucrative contract employment of the Chairman of the Board of Directors. The investment fund used the oppression remedy. See *Globe and Mail*, 15-18 June 1999.
66. CBCA, ss. 239-41.
67. S. Beck, 'The Shareholders' Derivative Action', *Canadian Bar Review* 52 (1974): 159-208.
68. *Percival v. Wright*, [1902] 2 ch. 421; *North-West Transportation Co. Ltd. v. Beatty* (1887), 12 App. Cas. 589. *Percival v. Wright* was distinguished in *Allen v. Hyatt* (1914), 17 D.L.R. 7 and considered to be wrongly decided in *Coleman v. Myres*, [1977] 2 N.Z.L.R. 297. See also *Brant Investments Ltd. v. KeepRite Inc. et al.* (1991), 3 O.R. (3d) 289.
69. *MacDougall v. Gardiner* (1875), 10 ch. App. 606; *Burland v. Earle*, [1902] A.C. 93; *Menier v. Hooper's Telegraph Works* (1874), 9 Ch. App. 350; *Cook and Deeks*, [1916] 1 A.C. 554; *Pavlides v. Jensen*, [1956] Ch. 565. Also see L.C.B. Gower, *Principles of Modern Company Law*, 6th edn (London: Sweet & Maxwell, 1997), 647. Since the development of the oppression remedy, fiduciary breaches do give rise to action. See *82099 Ontario Inc. v. Harold E. Ballard Ltd.* (1991), 3 B.L.R. (2d) 113.
70. CBCA, ss. 238, 239.
71. *Re Doan Development Corporation* (1984), 54 B.C.L.R. 235.
72. *Re Marc Jay Investments Inc.* (1974), 50 D.L.R. (3d) 45; *Abraham v. Prosoccer Ltd.* (1981), 119 D.L.R. (3d) 167.
73. See M.A. Moloney, 'Whither the Statutory Derivative Action?', *Canadian Bar Review* 64 (1986): 310-41.
74. CBCA, s. 241. For background, see *First Edmonton Place Ltd. v. 315888 Alberta Ltd.* (1988), 40 B.L.R. 28. In common law, a shareholder had to show conduct that was

'burdensome, harsh and wrongful' and also a lack of probity or fair dealing. See *Scottish Co-operative Wholesale Society Ltd. v. Meyer*, [1959] A.C. 324, [1958] 3 All E.R. 66.
75. See J.A. Van Duzer, 'Who May Claim Relief from Oppression: The Complainant in Canadian Corporate Law', *Ottawa Law Review* 25, 3 (1993): 463-84.
76. *Deluce Holdings Inc. v. Air Canada*, [1993] 12 O.R. (3d) 131; *Naneff v. Con-Crete Holdings Ltd.*, [1993] O.J. No. 1756; *Westfair Foods Ltd. v. Watt* (1991), 5 B.L.R. (2d) 160.
77. *820099 Ontario Inc. v. Harold E. Ballard Ltd.* (1991), 3 B.L.R. (2d) 113; *Naneff v. Con-Crete Holdings Ltd.*
78. *Deluce Holdings Inc. v. Air Canada*.
79. *820099 Ontario Inc. v. Harold E. Ballard Ltd.*
80. B. Welling, *Corporate Law in Canada: The Governing Principles*, 2nd edn (Toronto: Butterworths, 1991).
81. *820099 Ontario Inc. v. Harold E. Ballard Ltd.*; *GATX Corp. v. Hawker Siddley Canada Inc.*, [1996] O.J. No. 1462.
82. *820099 Ontario Inc. v. Harold E. Ballard Ltd.*
83. *347883 Alberta Ltd. v. Producers Pipelines Inc.*, [1991] 4 W.W.R. 577, (1991), 80 D.L.R. (4th) 359.
84. *Rogers Communications v. MacLean Hunter Ltd.* (1994), 2, C.C.L.S. 233; *Olympia & York Enterprises Ltd. v. Hiram Walker Resources Ltd.* (1986), 59 O.R. (2d) 254; *C.W. Shareholdings Inc. v. WIC Western International Communications Ltd.* (1998), 39 O.R. (3rd) 755.
85. *Palmer v. Carling O'Keefe Breweries of Canada Ltd.* (1987), 37. B.L.R. 316.
86. *Canadian Aero Service v. O'Malley* (1973), 40 D.L.R, (3d) 371; L.S. Sealy, 'Fiduciary Relationships', *Cambridge Law Journal* (1962): 69-81; P.D. Finn, *Fiduciary Obligations* (Sydney: Law Book Company Ltd., 1977).
87. Gower, *Principles of Modern Company Law*, 647.
88. *Re Smith & Fawcett Ltd.*, [1942] Ch. 304; *Re W. & M. Roith Ltd.*, [1967] 1 All E.R. 427.
89. Gower, *Principles of Modern Company Law*.
90. *Punt v. Symons & Co. Ltd.*, [1903] 2 Ch. 506; *Hogg v. Cramphorn*, [1967] Ch. 254.
91. *Bamford v. Bamford*, [1970] Ch. 212; *Teck Corp. Ltd. v. Millar* (1973), 33 D.L.R. (3d) 288; *Exco Corporation v. Nova Scotia Savings and Loan Company* (1987), 35 B.L.R. 149.
92. *Re Olympia and York Enterprises Ltd. and Hiram Walker Resources Ltd.* (1986), 59 O.R. (2d) 254.
93. *Keech v. Sandford* (1726), Sel. Cases. ch. 61.
94. *Boardman v. Phipps*, [1967] 2 A.C. 46; *Regal Hastings Ltd. v. Gulliver*, [1942] 1 All E.R. 378.
95. *Canadian Aero Service v. O'Malley*.
96. *Industrial Developments Consultants Ltd. v. Cooley*, [1972] 2 All E.R. 162.
97. *Cook v. Deeks*, [1916] 1 A.C. 554; *Canadian Aero Service v. O'Malley*; *Industrial Developments Consultants Ltd. v. Cooley*; *Peso Silver Mines Ltd. v. Cropper*, [1966] S.C.R. 673.
98. *Ex parte James* (1803), 8 Ves. 337; *Boardman v. Phipps*; *Regal Hastings Ltd. v. Gulliver*.
99. *Regal Hastings Ltd. v. Gulliver*.

100. *Lac Minerals v. International Corona* (1989), 61 D.L.R. (4th) 14.
101. Ibid., 47.
102. *Re City Equitable Fire Insurance Co.*
103. CBCA, s. 122(1)(b). See also *Re Standard Trustco* (1993), 6 B.L.R. (2d) 241.
104. CBCA, s. 123.
105. *Brant Investments Ltd. et al. v. Keeprite Inc. et al.*; *Re Olympia & York and Hiram Walker Resources Ltd.*
106. *Teck Corporation v. Millar* (1972), 33 D.L.R. (3d) 288; *Howard Smith v. Ampol Petroleum Ltd. et al.*, [1974] A.C. 821.
107. Canadian Income Tax Act. S. 227. 1 (3).
108. R.L. Campbell, 'The Fiduciary Duties of Corporate Directors: Exploring New Avenues, *Canadian Tax Journal* 36 (1988): 918–47; Campbell, 'Directors' Diligence Under the Income Tax Act', *Canadian Business Law Journal* 16 (1990): 480–501.
109. *McFadden v. 481782 Ontario Ltd.*; *Morgan v. Saskatchewan* (1985), 31 B.L.R. 173.
110. *Said v. Butt*, [1920] 3 K.B. 497.
111. *Kepic v. Tecumseh Road Builders* (1985), 29 B.L.R. 85.
112. See note 4 above.
113. *Lockharts Ltd. v. Excalibur Holdings Ltd. et al.* (1988), 47 R.P.R. 8; *Pacific Rim Installations Ltd. v. Tilt-Up Construction Ltd.* (1978), 5 B.C.L.R. 231; *B.G. Reeco I (Pacific Coast) Ltd. v. Bon Street Holdings Ltd. et al.*
114. *Jones v. Lipman*, [1962] 1 All E.R. 442; *Saskatchewan Economic Development Corp. v. Patterson-Boyd Manufacturing Corp. et al.*, [1981] 2 W.W.R. 40; *Clarkson Co. Ltd. v. Zhelka et al.*
115. *Smith, Stone and Knight Ltd. v. Birmingham*, [1939] 4 All E.R. 116; *D.H.N. Food Distributors Ltd. v. Tower Hamlets*, [1976] 1 W.L.R. 852; *Wallersteiner v. Moir*, [1974] 1 W.L.R. 991.
116. *DeSalaberry Realties Ltd. v. Minister of National Revenue.*
117. *Arnold v. Philips* 117 F. 2d 497; *Re: Canadian Commercial Bank* (1987), 67 C.B.R. 136. See also *Henry Browne & Sons Ltd. v. Smith*, [1964] 2 Lloyd's Rep. 476.
118. R.A. Posner, *Economic Analysis of Law*, 3rd edn (Boston: Little, Brown and Company, 1986); P. Halpern, M. Trebilcock, and S. Turnbull, 'An Economic Analysis of Limited Liability in Corporation Law', *University of Toronto Law Journal* 30 (1980): 117–50. But see *Walkovsky v. Carlton* 223 N.E. 2d 6.
119. *Canadian Aero Service v. O'Malley* at 384.
120. The chair of the committee of 13 members was Peter Dey, QC. The draft report was released in 1994 and submitted to the Toronto Stock Exchange.
121. *Where Were the Directors? Guidelines for Improved Corporate Governance in Canada* (Toronto: Toronto Stock Exchange, 1994).
122. For symposium papers on corporate governance, see *Canadian Business Law Journal* 26 (1996). See also K.E. Montgomery, 'Market Shift—The Role of Institutional Investors in Canada', *Canadian Business Law Journal* 26 (1996): 189–201; B.R. Cheffins, 'Corporate Governance in the United Kingdom: Lessons for Canada', *Canadian Business Law Journal* 28 (1997): 69–106.
123. Industry Canada, Working Group on Directors Liability, 1996 (unpublished report).

Chapter 8

Governing Employment

MICHAEL MAC NEIL

Introduction

The 1996 Canadian census showed that 12.4 million of 14.3 million working Canadians were employees.[1] More than 85 per cent of Canadian workers, and the dependants who rely on them for economic security, count on an economic relationship in which the worker sells his or her labour in a market transaction. Employment, not state-sponsored redistribution programs, is the primary source of economic security in our society. Not only is employment important to workers for the economic security it provides, but for many the workplace constitutes an extremely important community within which social attachments are formed and opportunities for self-fulfillment realized. It is also a place of considerable danger—of physical injury, illness, and psychological trauma. That employment is far more than a pure economic relationship was captured early in the twentieth century by the declaration that 'labour is not a commodity', contained in the part of the Treaty of Versailles establishing what is now the International Labour Organization.[2]

The employment relationship is not only of intense economic and personal importance to the employee. It is also one of the foundations of a capitalist economy. Capitalism depends on combining material resources and labour, under the control of capitalist investors, to produce products and services, ultimately, for a consumer society. Because the vast majority of citizens have insufficient capital resources or property to be able to maintain an autonomous productive regime, they are forced by economic exigency to sell their labour services on a labour market, becoming employees of enterprises that have been able to pool sufficient capital and obtain access to sufficient natural resources to engage in productive undertakings. Not surprisingly, employment is a matter of considerable public policy importance for the state, and the successful organization of work and the management of employees are necessarily crucial in determining the well-being of capitalist enterprise.

The governance of the employment relationship is the subject of this chapter. In particular, given that employers and employees do not necessarily share a common interest, the dynamics by which conflicting interests are managed and controlled are of fundamental importance.[3] This is especially worth examining at the level of controlling employee behaviour and the process by which productive work is accomplished. Employment has been notable for the extent to which it has been marked by hierarchy and subordination, inequality of power, and the exertion of control by employers over employees. A wide variety of forces influence the process of governance in the workplace, ranging from ideological beliefs to the balance of economic and political power, the use of technology, and organizational techniques. That the relationship has been marked by inequality and employer control is also in part attributable to the forms of legal regulation applied to employment.

Contract law and property rights serve as the regulatory basis for a majority of employment relationships in Canada. These common-law regulatory regimes have built on historical master-servant relationships, and notwithstanding the contractualization of employment, they have managed to preserve extensive control by employers. A movement towards statutorily based regulation of many aspects of the employment relationship has served as much to legitimize the hierarchy of control as to acknowledge the claims of workers to justice in the employment relationship. Finally, extensive efforts of workers to introduce governance regimes to ensure a greater collective voice through trade union representation and collective bargaining have been significantly hampered by labour laws that are more concerned with the control of industrial conflict than with the democratization of the workplace or the empowerment of workers.

Recent discourse about employment and labour markets in Canada has reflected concerns about increased efficiency and productivity, about flexibility in responding to challenges from a global economy, and about timely adaptation to new technologies.[4] Employers are continually seeking more effective ways of maximizing the productive efforts of workers. Earlier discourses of scientific management are giving way to talk of employee empowerment, worker involvement in decision-making, job enrichment, and equitable distribution of the gains likely to be realized from transforming the organization of work.[5] This chapter posits the thesis that those legal regimes within which employees and employers govern themselves more willingly acknowledge workers' claims to human dignity and justice, without fundamentally altering the hierarchical nature of the employment relationship, and thus without fundamentally altering the ability of employers to control the work process. Hence, the discourses of employee empowerment and involvement have not led to a fundamental restructuring of the legal regime that regulates employment.

The chapter demonstrates, as well, that any attempt to assess governance mechanisms must look beyond the state alone. While the many statutory regulatory initiatives appear to suggest that the state is at the centre of regulating the work relationship, in fact, the operation of the market, supported by contract, property, and customary rules, plays a major role in structuring the balance of power within the employment relationship. This analysis supports the contention that the employment relationship is itself a site of regulatory practice, albeit one in which the aspiration to 'thick proceduralism' and its commitment to pluralistic and deliberative processes of norm formation is unlikely to be met.[6]

The Contract of Employment

It is important to think about the common law of the contract of employment as a system of regulation of the employment relationship.[7] In the debates about juridification of labour relations[8] and the need for removal of inappropriate, ineffective, and inefficient regulatory rules governing the employment relationship, the assumption is not that a regulatory vacuum would be created by the deregulation, but rather that the ordering of the relationship effected by appropriate rules of contract and property law, which operate in the absence of statutory intervention, provides a better, market-based, private scheme of regulation.[9] However, it is also important to recognize that the contract of employment does not normally result from the full negotiation of terms by individuals with equal bargaining power. Many of its terms are in fact filled in as default rules, either as spelled out by the employer in contracts of adhesion or, equally importantly, as a result of terms implied by the courts. These terms are sometimes said to be based on the intent of the parties, but more often they are a reflection of judicial perceptions about what is needed to ensure that the employment relationship is a workable one in a modern, complex, industrialized society.

The contract of employment establishes the legal framework within which privatized forms of self-regulation can take place. A key element of the contract of employment is that it allows considerable managerial discretion to establish the conditions under which employment is carried out. Employers are able to use technology, bureaucratic structures, and extensive control mechanisms to co-ordinate the productive effort. Employers have supervisory control over employees, and have the legal authority to specify a wide variety of workplace rules controlling employee behaviour, including attendance, grooming and attire, and safety measures. In many cases, such as grooming and conflict-of-interest policies, this control extends to the off-duty conduct of employees. The employer's rule-making function is recognized as a legitimate exercise of managerial prerogatives, ultimately enforceable through the imposition of discipline, especially through the power of dismissal.[10] The exercise of managerial authority may be subject to review on grounds of reasonableness, for example, where employees claim to have been wrongfully dismissed for breach of the rules or where a union has negotiated a collective agreement that submits the employer to a just-cause standard for discipline. Nevertheless, the privatized, contractual ordering of the employment relationship continues to provide the foundation that other forms of ordering, such as collective bargaining or legislated standards, take as their point of departure.

The modern contract of employment crystallized as nineteenth-century courts accepted market arrangements as the most efficient and just means of organizing productive efforts. It marked a transition from paternalistic regulation rooted in concepts of status, to individualist emphasis in the structuring of relationships. The paternalist emphasis is evident in pre-industrial England, as demonstrated by the analysis of the leading eighteenth-century systemizer of English law, Sir William Blackstone. He described the master-servant relation (the precursor of modern employment) as one of the three great relations of private law, the others being husband-wife and parent-child.[11] The relationship is situated within the domestic sphere, with corresponding concepts of patriarchal authority, deference of inferiors to superiors, and obligations on the master for the welfare of the servant. Implicit in Blackstone's language is a moral economy of fair play within the relationship, enforced by law.[12] However, in practice, law often sanctioned rather harsh treatment of servants by masters, with masters having the privilege to subject servants to arbitrary and legally enforced physical punishment, fines, and dismissal with the forfeiture of payment for very minor infractions of the master's expectations. Workers who quit before the end of a term would not be entitled to payment for work done. Given that it was frequently the practice not to pay workers any or a major portion of their wages until the end of the term of contract, this served as a major control device on the actions of many classes of servants. These powers of the master derive from a vision of hierarchical relationships in which the superior master is seen as having authority over the servant to correct and discipline, deriving not so much from the consent of the servant as from the subservient status of the servant.

The process of industrialization brought about tremendous changes in the actual nature of the employment relationship, transforming it from one of personal, domestic service to one that had many more commercial aspects. The growth of mechanization, factories, and urbanization and the decline of agriculture all contributed to a reconceptualization of employment. Pentland, in his analysis of the development of labour and capital in Canada, notes that from 1830 on 'employers were clearly conscious of the fact that the old world of scarce labour and paternalistic relations was disappearing', being replaced by a strong appeal to free and individualistic labour markets.[13] The contractualization of the relationship emphasized its impersonal nature, with the primary nexus between worker and employer being commercial and cash-based. The employer was no longer regarded as having any moral responsibility for the welfare of the worker. Labour came to be regarded as a fungible commodity, with the worker free to leave, but also with the employer free to replace the worker with another on the best terms that could be obtained.

The courts in the nineteenth century played a major role in transforming the legal understanding of the relationship to one more in accord with the individualistic ideology and its emphasis on freedom of contract that flourished during the period. This placed upon individuals a responsibility for making their own obligations, and concepts of justice emphasized the freedom of individuals to choose to do as they wished, without focusing on the resources that might be available to allow an individual scope of choice. The contract that applied to the employment relationship in many instances was not negotiated, but was one to which the employees adhered to terms unilaterally imposed by the employer. To the extent that the contract was silent on most issues, it was assumed that the employer retained extensive discretion to order the worker to perform work in the manner the employer desired. Especially important in maintaining the employer's power was the right to dismiss an employee without notice for just cause, and with only a minimal amount of notice where there was no cause.

Although it depended on agreement for its creation, the employment contract in fact incorporated customary norms and rules. A movement to contractualization of employment was fraught with dangers for emerging capitalism. Because the vision of contract includes the ideas of limited commitment and the freedom of individuals to define for themselves the terms of the relationship through bargaining, there was a potential threat to the capitalists' unfettered control over labour resources. This was avoided, employing Fox's terminology, 'by marrying contractualism to the traditional master-servant notions'.[14] Courts implied into the contract of employment duties of obedience and loyalty, granting to the employer full authority to direct and control the employee, and thereby enabled employers to extract the maximum productive labour from the labour power that was purchased. Most importantly, the master/employer retained the power to dismiss an employee for breach of the employment contract, which meant, generally, a failure to follow the lawful orders of the employer. This gave the master enormous power of control over the actions of the worker. Dismissal operated outside the institutional framework of the law— a self-help remedy that has been characterized as the capital punishment of the industrial disciplinary system.[15]

The regulation of individual employment relationships continues to be greatly influenced by its contractual origins. In Canada, there is considerable litigation over the contract of employment, in the area of law known as wrongful dismissal. Non-unionized workers are generally presumed to be hired on individual contracts of indefinite duration, unless the contract otherwise specifies. An employer is entitled to fire the employee if there is just cause. Technically, just cause exists if the employee has breached the employment contract in a substantial way. If there is no just cause, the employer may lay off an employee by giving reasonable notice or pay in lieu of notice. A worker who is fired without reasonable notice where there is no cause for the firing can sue the employer for wrongful dismissal. The courts are then called upon to determine the issues of cause and what constitutes reasonable notice. Courts do not order reinstatement of an employee who has been fired without cause. Damages are the normal remedy, and they are usually measured by the amount of wages that would have been earned during a period of reasonable notice. In other words, under the contractual analysis, the wrong that is done is not the firing but the failure to give notice. The contract of employment has been structured in such a way that the employer retains the ability to get rid of an employee at any time and for any reason,[16] so long as the requisite notice is given.

To some extent there is a disjuncture between the judicial recognition of the unique nature of the employment relationship, including its relational nature and its non-economic attributes, and the underlying legal structure, which continues to support a hierarchical, employer-controlled governance approach to employment. The judicial recognition of employment as a special relationship has been enunciated in a number of Supreme Court of Canada decisions. The Court recognizes that the contract of employment has many characteristics that set it apart from the ordinary commercial contract.[17] The Court has emphasized the inequality of bargaining power in the relationship and characterized employees as a vulnerable group in society.[18] It has also noted that work is one of the fundamental aspects in a person's life, providing individuals with a means of financial support and an avenue to contribute to society. Employment, according to the Court, is essential to an employee's sense of identity, self-worth, and emotional well-being.[19] As a result of recognizing these features of the employment relationship, the Court has made a number of decisions tempering the apparent harshness of a contractual regime that allows employers to dismiss at will. The courts have imposed obligations to provide a substantial amount of notice, especially in the case of managerial or professional employees, and have imposed obligations of good faith with respect to the manner of dismissal.

However, the intervention has not fundamentally challenged the right of the employer to dismiss, even in the absence of just cause. As one Supreme Court decision put it:

> In the context of an indeterminate employment contract, one party can resiliate the contract unilaterally. The resiliation is considered a dismissal if it originates with the employer and a resignation if it originates with the

employee. If an employer dismisses an employee without cause, the employer must give the employee reasonable notice that the contract is about to be terminated or compensation in lieu thereof.[20]

The Supreme Court has refused to impose a regime that would prohibit dismissal in the absence of just cause. Indeed, it has refused even to impose a general obligation of good faith in dismissal, stating that doing so would:

> deprive employers of the ability to determine the composition of their workforce. In the context of the accepted theories on the employment relationship, such a law would, in my opinion, be overly intrusive and inconsistent with established principles of employment law, and more appropriately, should be left to legislative enactment rather than judicial pronouncement.[21]

The refusal to provide reinstatement as a remedy for wrongful dismissal is based on a belief in the necessity to uphold managerial prerogatives to utilize resources as management sees fit in the pursuit of the enterprise. Traditional justifications claimed that courts could not properly supervise reinstatement, and that it would be unfair to allow workers to demand specific performance of the contract where employers could not do so.[22] The court has not yielded from this position, despite collective bargaining regimes in which reinstatement of unjustly dismissed employees is accepted as normal and statutory provisions in a number of areas that empower adjudicators to reinstate improperly dismissed workers. The judicial reluctance to change is all the more debatable in the light of studies showing that reinstatement, although far from a perfect remedy, in many situations offers a viable and more complete form of protection for employees.[23]

A more extensive analysis of the law of wrongful dismissal is beyond the scope of this chapter. The factors determining what the courts perceive to be the reasonable notice that must be given by an employer, judicial treatment of express contractual terms seeking to minimize notice periods, the development of a law of constructive dismissal that tempers the ability of employers to 'force' an employer to quit and to unilaterally make major modifications in job duties or benefits, and the development of the jurisprudence on punitive and exemplary damages all serve to highlight the tension between acknowledging the special nature of the employment relationship and the continuation of a legal framework that reinforces a governance regime giving considerable control and unilateral discretion to the employer. This chapter now turns to an examination of two other key components of the external laws shaping the governance regime in employment: collective bargaining and employment standards legislation. The question is the extent to which these legal frameworks have substantially modified the governance regimes within employment relationships.

Collective Bargaining

The traditional justifications for collective bargaining law rest on the assumption that collective bargaining can ameliorate the inequality of bargaining power that exists between employers and employees and that is in part responsible for the employer being able to formulate the terms and conditions of employment. Collective bargaining is also espoused as a means by which employees can have a voice in the formulation of workplace rules and policies, a form of democratization of the workplace notably absent in regimes based on individual contracts of employment. Finally, collective bargaining is promoted as a fundamental right of workers, arising from a commitment to freedom of association and the right of workers to join together in the pursuit of collective interests. While collective bargaining can be assessed from a wider variety of perspectives, such as effectiveness, efficiency, and its contribution to attaining justice, however defined, the focus here is on the extent to which collective bargaining fundamentally alters the employer's ability to control the work process and make decisions affecting the continuation of the employment relationship.

Governance issues arise in a wide variety of employment contexts, but this analysis focuses on two in particular. The regulatory provisions of collective bargaining law are of two main types: the first sets out constitutive rules whereby access to collective bargaining is gained and protection is given to those seeking to obtain such access. The second sets out rules that broker power between employers and workers and their unions. These rules include those regulating when a strike may occur and the nature of the duty to bargain.[24]

The extent to which collective bargaining as a regulatory regime empowers workers to participate in fundamental decisions about the direction of the enterprise is central to understanding how this regime has reshaped the workplace relationship. These decisions are obviously important for workers, because they ultimately determine what products or services will be produced, where they will be produced, and what amount of labour will be required. If one has little or no voice in the big questions about the undertaking at which one is employed, it can be fairly argued that governance remains essentially in the control of capital investors and/or management. A second area of inquiry concerns the employer's ability to define and control day-to-day work practices, through the organization of the productive process, the establishment of workplace rules, the use of monitoring and other techniques to maximize productivity, and the use of disciplinary processes to enhance and enforce workplace rules. To the extent employers are required to obtain employee agreement on practices and rules, and to the extent employer decisions may be reviewed by a neutral adjudicator against standards of reasonableness or justice, it is arguable that employees have gained an important role in the governance of the workplace. The analysis below concludes, however, that while collective bargaining regimes subject employers to some significant limitations on their ability to govern the workplace unilaterally, regulation through collective bargaining falls considerably short of granting employees full democratic voice in the governance of the workplace.

In the Canadian collective bargaining regulatory regime, once a union has been certified to represent the employees of an employer, as defined by the certificate of bargaining authority granted by a labour relations board, an employer is under a duty to bargain in good faith with the union and to make a reasonable effort to agree to terms and conditions of employment. On the face of it, this appears to ensure that a union has a major governance function in the workplace. However, a closer examination of the duty to bargain in good faith belies to some considerable extent the initial first impression. The duty to bargain is not a duty to agree.[25] The analysis of governance authority must take into account both the nature of the agreement reached between the employer and the union and what happens in the absence of agreement. The general rule in Canada is that while bargaining is taking place, there is a statutory freeze on the power of an employer to modify the terms and conditions of employment. This, too, would appear to confirm that the power of the employer to make unilateral decisions affecting workplace arrangements is thereby significantly compromised. This is not really so, however. First of all, at some point in time, if the parties have bargained to an impasse, the employer is entitled to make changes to terms and conditions.[26] The union is left in the position of either submitting or engaging in a strike in an attempt to pressure the employer to come to an agreement. The union does not have the option of having work continue on the terms and conditions that it proposes, or even on the basis of the status quo, while the parties seek a way out of the impasse.

Further, even during the period that a statutory freeze is in place, it is not really a freeze on the power of the employer to make decisions. Rather, it has been interpreted merely as a requirement that business may be carried out as usual, leaving the employer to make many decisions without having to obtain the consent of the union, including engaging in layoffs and imposing discipline on employees.[27] The employer is still able to exercise the traditional managerial prerogatives, based on property rights and customary managerial privileges. As a result, the employer can continue to govern the workplace without extensively sharing that responsibility with workers or the union.

Where the employer and the union do enter into a collective agreement, a stronger case can be made that a fundamental shift in governance structure has taken place. Collective agreements cover a wide range of terms and conditions of employment, including such issues as wages and benefits, hours of work, occupational health and safety, layoff procedures, and the introduction of seniority as a major criterion both for the determination of level of some benefits and for decisions about who should be laid off or promoted. Equally important, collective agreements typically introduce a regulated approach to discipline and dismissal, imposing a standard of just cause against which to measure the legality of an employer decision. The collective agreement also introduces a sophisticated grievance arbitration system by which to settle disputes about the interpretation and application of the collective agreement. However, several major caveats are in order.

Most collective agreements contain clauses recognizing a wide scope for continuing employer control in the workplace. These management rights clauses in effect allocate to the employer the power to do that which it is not expressly prohibited from doing by the terms of the collective agreement.[28] Thus, the fundamental decisions about capital investment, organization of work, and so on are typically left for management to make. While arbitrators and courts have introduced some restraints on the exercise of managerial authority, in some limited circumstances requiring that decisions be made fairly or reasonably, these limitations are at best a narrow restriction on the exercise of managerial authority. Employers typically can make major decisions such as to close a plant without obtaining the consent of workers, and indeed without even having to bargain with the union about the closure.[29]

Even if the employer's ability to make unilateral decisions about major entrepreneurial initiatives is not substantially curtailed when workplace governance mechanisms are subjected to the regulatory influence of collective bargaining, it could be argued that at least in the area of workplace discipline, matters have substantially changed. Unlike the regime engendered by individual employment contracts, the employer is subject to a just-cause standard in the absence of which the employer is not entitled to dismiss. Unilateral control by giving notice of dismissal is no longer possible for the employer. Furthermore, the grievance arbitration system introduces what is supposed to be a quicker and cheaper way for employees to challenge an employer decision, and arbitrators have the remedial authority to reinstate employees where they find the employer has failed to prove just cause for dismissal. Furthermore, the just-cause standard has been interpreted as giving arbitrators the right to review unilaterally established workplace rules, the breach of which may lead to discipline, to ensure that they are reasonable.[30] On the face of it, these provisions make it possible to argue that a key justification for the promotion of collective bargaining is that it introduces the rule of law into the workplace, with 'legislated' standards and a 'judicial' system for the enforcement of those standards. The employer's power to govern through the disciplinary system appears to be significantly hampered. Yet, an employer's disciplinary control is still very significant.

The arbitral conception of just cause is still firmly rooted in the customary employee obligations acknowledged by nineteenth-century courts to be essential elements of the employment relationship. The employee is still bound to obey orders and to comply with employer-established rules.[31] The very continuation of the language of obedience, rooted in a hierarchical master-servant relationship, serves to accentuate the nature of governance in the workplace. There is an ongoing obligation of loyalty owed by the employee to the employer,[32] as well as a duty to avoid conflicts of interest. The employer is still in a position to establish dress codes and grooming requirements[33] and, generally, to organize the means of production in such a way that the employee wishing to remain employed is without choice in submitting to these dictates. While much modern human resources literature may stress the importance of employee participation, the encouragement of teamwork, and the acknowledgement of employee empowerment, in fact the regulatory scheme within which the governance structure in the workplace is constituted by no means requires or guarantees that employees will have much voice or participation in governance at the day-to-day level of employee operations. Collective agreements may provide for joint employer-employee participation in a variety of committees, but more often than not these committees are more consultative in nature, and they seldom acquire broad decision-making power.

In the final analysis, collective bargaining can be seen as considerably increasing the opportunity for employees to be more active participants in the governance of the workplace. However, the legal framework is such that these opportunities must constantly be bargained for by the union, and in the absence of collective agreement language, employers and their managers are left with considerable unilateral control over the governance of many of the basic elements of the day-to-day work of the enterprise.

Statutory Standards

Did the introduction of statutory standards, through a variety of legislative initiatives such as minimum employment standards, occupational health and safety laws, and human rights codes, fundamentally transform the governance regime in the workplace? In some regards, certainly, this is true, but in other ways the employer and management still hold the reins of governance.

Even in the heyday of nineteenth-century laissez-faire approaches to the market and labour, legislatures were starting to intervene to regulate the employment relationship. Some of the earliest interventions aimed to ameliorate the appalling working conditions many workers were forced to endure and sought to limit some of the worst abuses of children in the workplace. In the twentieth century, this regulation of workplace conditions extended beyond health and safety issues to the regulation of a wide variety of terms and conditions of employment, including minimum-wage standards, hours of work, holidays, entitlement to vacation pay, and notice of layoff or dismissal. Workers' compensation schemes were introduced in recognition of the social responsibility to provide for workers injured on the job. More recently, a sensitivity to claims of equality has led to extensive statutory regulation of discriminatory practices. Most of these statutory interventions involve not only the establishment of standards, but the introduction of novel enforcement mechanisms that in theory are designed to increase access of workers to means of enforcing their rights. A closer examination of some of these statutory schemes, however, demonstrates that they still have not fundamentally altered the ability of the employer to dominate the governance processes in the workplace.[34] Several short illustrations of these claims follow.

In the modern regulation of occupational health and safety, two major reforms are pointed to as evidence that workers have been considerably empowered in the workplace. Many jurisdictions have moved to the promotion of an internal responsibility system, through which employees become entitled to a great deal of information about workplace risks and are entitled, indeed required, to be represented on joint committees charged with responsibility for the pursuit of workplace safety. Furthermore, workers are empowered to take responsibility for their own safety by refusing to work in unsafe conditions, free from employer retaliation.[35] These reforms appear to increase considerably employee voice in governance decisions, limiting arbitrary or unilateral employer decision-making power about how to organize work. Yet, the extent to which these reforms make inroads on the traditional domination of employers in governance of the workplace must be questioned.

Participation in the internal responsibility system undoubtedly gives employees an added dimension of voice. They now have forums in which complaints can be made; they receive greater training in recognizing risks and prevention of accidents. However, they have not received a great deal more power to influence the structuring of work processes. For instance, if one examines the powers of joint health and safety committees in Ontario, half of whose members must be non-managerial employees, one sees that these committees have a variety of functions, which include identifying hazards, obtaining information, making recommendations, and being represented when testing takes place.[36] While the acquisition of knowledge does provide one form of power, it is an essentially limited one, and the crucial decisions about what is to be done to ameliorate hazards in the workplace are left to management to make; the joint committee's power to recommend is not a power to implement changes.[37]

The one point at which it appears that workers are given significant additional power is the right to refuse unsafe work. This is explicitly protected by statutory provisions that prohibit the employer from retaliating against a worker who has exercised her or his right.[38] This right is strictly regulated, however, and it challenges the underlying hierarchical, authoritarian relationship between employer and employee only to a limited extent. Hence, we find a member of the Ontario Labour Relations Board stating:

I am not persuaded that the Act creates a statutory right to substitute methods of safety protection in the face of an employer's express and often repeated directions to the contrary. Such actions are regarded by the Board as insubordinate conduct rather than the exercise of an independent discretion, and do not draw the Act's protections.[39]

An employee's ability to act independently in the face of employer orders is only possible where the employee has reason to believe that workplace equipment or conditions are likely to endanger himself or herself or another worker.[40] The 'likely to endanger' standard is a high one, such that many workplace problems cannot be addressed by the worker pursuant to this right. The real problem in exercising the right is that a whole host of forces make it practically difficult to withstand pressure from the employer to continue to work. Evidence clearly indicates that the right to refuse unsafe work is more likely to be exercised effectively in workplaces where the workers are represented by a union. As a result, it can be fairly argued that these provisions do not fundamentally alter the framework by which work processes are organized through the employer's ability to control and co-ordinate.

A similar analysis can be made of employment standards legislation, which regulates a wide variety of employment conditions, ranging from hours of work, minimum wages, and holidays to the termination of employment. It is on this last feature that I wish to focus, to determine if the statutory standards are likely to alter significantly the disciplinary power of management and to limit the unilateral power of management to make fundamental decisions leading to the layoff or dismissal of large numbers of workers.

The regulation of termination of employment in employment standards legislation takes several forms. Such statutes typically provide that minimum periods of notice must be given to workers, usually based on the length of service or, where a mass layoff is implemented, on the number of employees being laid off. Employers are normally able to provide pay in lieu of notice, so that the notice provisions work in a manner similar to the requirement of reasonable notice in the individual contract of employment. These provisions place a limit on the power of the employer to define contractually the notice period at less than the minimum standard, and they provide a means by which the notice provisions (or the requirement to make payments in lieu of notice) can be enforced without a court action. However, the employer is still entitled to dismiss an employee without notice where there is cause. More importantly, the employee is not protected from an unjust dismissal, only from a dismissal without notice. Similarly, the provision for longer periods of notice where group layoffs are occurring does not prevent an employer from implementing a decision to close a plant or relocate work, although it does force the employer to internalize some costs for doing so. Furthermore, the employer may be required to provide assistance to workers in helping them to adjust, such as aiding them in looking for new work.

A few Canadian jurisdictions go further and actually extend the concept of just-cause dismissal as developed in the collective bargaining regime to non-unionized workers.[41] Under the Canada Labour Code, which applies to about 10 per cent of Canadian workers who are within the regulatory sphere of the federal Parliament, a worker claiming to have been dismissed without just cause can have the claim adjudicated by a neutral adjudicator appointed by the government.[42] That adjudicator, like an arbitrator pursuant to a collective agreement, but unlike a court in a wrongful dismissal case, can order the reinstatement of an unjustly dismissed employee.[43] Nonetheless, it would be difficult to claim that this has resulted in a fundamental alteration of the governance regime in the workplace. The decisions of adjudicators substantially parallel the decisions of collective agreement arbitrators,[44] with the result described above that the fundamental right of the employer to make orders and co-ordinate the work process is not fundamentally affected. Furthermore, the Labour Code specifically exempts from review those dismissals arising as a result of lack of work or the discontinuance of an employer function. Unilateral employer discretion in organizing the labour process is thereby maintained.

Conclusion

There is little doubt that extensive changes have been made to the regulatory regime governing the employment relationship since the development of the modern labour market took place in the nineteenth century. All three regulatory regimes examined—wrongful dismissal law, collective bargaining, and legislated employment standards—demonstrate some considerable recognition of the claims of workers to be treated with dignity and justice in the workplace. Many employers are required to provide a minimum period of notice when laying off employees and to involve employees in an internal responsibility system to better promote health and safety in the workplace. Some employers may be prohibited from dismissing an employee except for just cause or where there is no work to be done. Non-unionized employers whose relations with employees are governed by individual contracts of employment may be required to give longer, reasonable periods of notice, especially to senior and long-serving employees, and may be constrained as to how they end the employment relationship by an obligation to act in good faith. Unionized employers may be required to develop rules that are reasonable and they are required to extend a measure of due process in evaluating employee conduct. Nevertheless, at their core, these regulatory regimes continue a venerable tradition of legitimizing the predominant role employers play in controlling work processes and making fundamental decisions about the direction of the work enterprise. The promotion of workplace justice and employee dignity is not premised on a fundamental right of employees to exercise control in their working lives. It is not based on a commitment to democratic governance.

Notes

1. Data derived from Statistics Canada, *Labour Force 15 years and over by Class of Worker, 1996 Census*. Available at: <http://www.statcan.ca/english/Pgdb/People/Labour/labor43a.htm>. Accessed 6 Sept. 1999.
2. For a full development of this theme, see D. Beatty, 'Labour Is Not a Commodity', in B. Reiter and J. Swan, eds, *Studies in Contract Law* (Toronto: Butterworths, 1980), 313–55.
3. R. Edwards, *Contested Terrain: The Transformation of the Workplace in the Twentieth Century* (New York: Basic Books, 1979).
4. See, e.g., P. Kumar, *Unions and Workplace Change in Canada* (Kingston: IRC Press, Queen's University, 1995). Consider also recent revisions to Ontario employment standards legislation, increasing the maximum hours per week an employee can work from 48 to 60. Employment Standards Act 2000, S.O. 2000, c. 41, s. 17(2). The Ontario government justified this increase on the basis that it would 'accommodate the needs of employers to run their businesses in a way that meets market pressures, new manufacturing processes or the intensive requirements of high technology industries.' Ontario Ministry of Labour, *Time for Change: Ontario's Employment Standards Legislation* (Toronto: Ontario Ministry of Labour, 2000). Available at: <http://www.gov.on.ca/LAB/es/00escpe.htm>. Accessed 5 June 2001.
5. Ibid., 11.
6. See Peter Swan, Chapter 1, this volume.
7. M. Freedland, 'The Role of the Contract of Employment in Modern Labour Law', in L. Betten, ed., *The Employment Contract in Transforming Labour Relations* (The Hague: Kluuwer Law International, 1995), 17–27.
8. S. Simitis, 'Juridification of Labor Relations', in Gunther Teubner, ed., *Juridification of Social Spheres* (Berlin: Walter de Gruyter, 1987), 113.
9. One of the most forceful proponents of this position is R. Epstein. See Epstein, 'In Defense of the Contract at Will', *University of Chicago Law Review* 51 (1984): 947–82; Epstein, *Forbidden Grounds: The Case Against Employment Discrimination Laws* (Cambridge, Mass.: Harvard University Press, 1992).
10. The employer has a right to dismiss an employee who has been guilty of 'serious misconduct, habitual neglect of duty, incompetence, or conduct incompatible with his duties, or prejudicial to the employer's business, or [who] has been guilty of wilful disobedience to the employer's orders in a matter of substance.' *R. v. Arthurs, Ex parte Port Arthur Shipbuilding Co.* (1967), 62 D.L.R. (2d) 342 at p. 348 (Ont. C.A.). For a recent example of a court upholding the summary dismissal of an employee for misconduct, disobedience, and neglect of duties, see *Guenther v. Saskatoon Livestock Sales Ltd.*, [1999] S.J. No. 22 (Q.B.).
11. *Ehrlich's Blackstone* (San Carlos, Calif.: Nourse Publishing, 1959), categorizes the master-servant relationship into four categories: menial or domestic; apprentices; labourers; and superior servants. These categories and the examples given by Blackstone demonstrate his vision as one centred on an agricultural economy domi-

nated by landowners who employed both agricultural and domestic workers. It is doubtful that this was an accurate vision of the range of employment relationships that existed in Blackstone's own time, when the mercantile growth and industrialization of Great Britain were already underway. See Kahn-Freund, 'Blackstone's Neglected Child: the Contract of Employment', *Law Quarterly Review* 93 (1977): 508–28. The assimilation of master-servant relationships to familial ones is evident from Blackstone's description of the duties of the master, including the rule that menial servants were engaged for a term of a year, on the ground that it is appropriate for the master to provide for the servant both when there is and when there is not work, and the right of the master to punish the servant.

12. P. Karsten, '"Bottomed on Justice": A Reappraisal of Critical Legal Studies Scholarship Concerning Breaches of Labor Contracts by Quitting or Firing in Britain and the U.S., 1630–1880', *American Journal of Legal History* 34 (1990): 213–61.
13. H. Clare Pentland, *Labour and Capital in Canada, 1650–1860* (Toronto: James Lorimer, 1981), 188.
14. A. Fox, *Beyond Contract: Work, Power and Trust Relations* (London: Faber, 1974), 187.
15. G. England, 'Recent Developments in Wrongful Dismissal Laws and Some Pointers for Reform', *Alberta Law Review* 16 (1978): 470–520.
16. There are now statutory limits placed on the reasons an employer may have for dismissing workers. Human rights statutes prohibit discrimination on a number of grounds, and other statutes prohibit employer retaliation against employees for exercising their rights under the statutes.
17. *Wallace v. United Grain Growers Ltd.*, [1997] 3 S.C.R. 701 at para. 91.
18. *Slaight Communications Inc. v. Davidson*, [1989] 1 S.C.R. 1038.
19. *Reference Re Public Service Employee Relations Act (Alta.)*, [1987] 1 S.C.R. 313 at 368.
20. *Farber v. Royal Trust Co.*, [1997] 1 S.C.R. 846 at para. 23.
21. *Wallace v. United Grain Growers Ltd.* at para. 76.
22. If an employer were able to obtain specific performance against an employee who quit, this would amount to coercing an employee to work for a particular employee against her or his will, which is tantamount to slavery and thus prohibited. It is worth noting, however, that courts are willing to grant injunctions to prohibit workers from collectively refusing to work by engaging in strikes, with no qualms about this amounting to forced labour.
23. G. Adams, *Grievance Arbitration of Discharge Cases* (Kingston: Queen's University Industrial Relations Centre, 1979).
24. K. Van Wezel Stone, 'Labor and Corporate Structure: Changing Conceptions and Emerging Possibilities', *University of Chicago Law Review* 55 (1988): 82–4; G.W. Adams, 'Towards a New Vitality: Reflections on 20 Years of Collective Bargaining Regulation', *Ottawa Law Review* 23 (1991): 139–75.
25. *CUPE v. Nova Scotia (Labour Relations Board)*, [1983] 2 S.C.R. 311.
26. *CAIMAW v. Paccar of Canada Ltd.*, [1989] 2 S.C.R. 983.
27. *Bank of Nova Scotia v. Retail Clerks International Union*, [1982] 2 Can. L.R.B.R. 21; 82 C.L.L.C. 16,158 (Canada Labour Relations Board).
28. See *Re United Steelworkers of America and Russelsteel Ltd.* (1966), 17 L.A.C. 253 (Arthurs), a leading case establishing the dominant arbitral position that in the absence of a clause in the collective agreement prohibiting it, the employer has the right to contract out work. For some theoretical discussions of the scope of manage-

rial prerogative, see P. Weiler, 'The Role of the Labour Arbitrator: Alternative Versions', *University of Toronto Law Journal* 19 (1969): 16-45; B. Langille, 'Equal Partnership in Canadian Labour Law', *Osgoode Hall Law Journal* 21 (1983): 496-536; D. Beatty, 'The Role of the Arbitrator: A Liberal Version', *University of Toronto Law Journal* 34 (1984): 136-69; A Goldsmith, 'The Management-Control Bargaining Relationship: Three Models', *Ogsoode Hall Law Journal* 24 (1986): 775-832.

29. The collective bargaining regime has been supplanted to some extent by legislative requirements that an employer give notice of an intention to close a plant and to bargain in some circumstances, not about the decision to close, but about assisting the dislocated wokers to adjust to the plant closing.
30. *Metropolitan Toronto (Municipality) v. C.U.P.E.* (1990), 74 O.R. (2d) 239 (C.A.).
31. See one recent case in which an arbitrator describes an employee's refusal to follow an order as an 'act of gross and premeditated insubordination that baldly contested the right of the employer to manage its operation'. *Re Caligo and Canadian Auto Workers, Local 1285* (1998), 73 L.A.C. (4th) 365 (Roberts).
32. One recent case described the duty as 'one of loyalty or trustworthiness which obliges the employee to always act in the best interests of the employer'. *Re Canada Safeway Ltd. and U.F.C.W., Local 2000 (Schmidt)* (1997), 61 L.A.C. (4th) 1 (Larson).
33. Employer rule prohibiting employees from having beards held to be reasonable: *Re Canada Safeway Ltd. and U.F.C.W., Local 1518* (1998), 74 L.A.C. (4th) 306 (Kelleher).
34. This analysis is consistent with the work of Eric Tucker, *Administering Danger in the Workplace* (Toronto: University of Toronto Press, 1990). Tucker examined the history, law, and politics of health and safety regulation in late nineteenth- and early twentieth-century Ontario, concluding that formal state regulatory intervention did little to improve safety conditions, in part because of the lessening of worker control of the labour process throughout the period studied.
35. See, for example, Ontario Health and Safety Act, R.S.O, 1990, c. O.1, s. 50.
36. Occupational Health and Safety Act, R.S.O. 1990, c. O.1, s. 9(18).
37. For a case study demonstrating the tragic failure of internal responsibility systems, see Harry Glasbeek and Eric Tucker, *Death by Consensus: The Westray Story* (Toronto: Centre for Research on Work and Society, York University, 1992).
38. See, e.g., Occupational Health and Safety Act, R.S.O. 1990, c. o.1, s. 50.
39. *Re Gonder and Pan Oston Ltd.* (1996) O.L.R.B.D. No. 820.
40. See, e.g., Occupational Health and Safety Act, R.S.O. 1990, c. O.1, s. 43.
41. Canada, Quebec, and Nova Scotia each have such protection, although the length of time one must have worked to be entitled ranges from one year in the federal jurisdiction to three years in Quebec and 10 years in Nova Scotia.
42. Canada Labour Code, R.S.C. 1985, c. L-2, s. 240.
43. Canada Labour Code, R.S.C. 1985, c. L-2, s. 242(4). However, reinstatement is less often ordered by adjudicators because of the practical difficulty of supervising such orders without a trade union present in the workplace. See The Labour Law Casebook Group, ed., *Labour and Employment Law: Cases, Materials and Commentary*, 6th edn (Kingston: Industrial Relations Centre, Queen's University, 1998), 798.
44. G. Simmons, 'Unjust Dismissal Under Federal Jurisdiction', *Labour Arbitration Yearbook* 2 (1991): 41-54.

Chapter 9

Law, Regulation, and Becoming 'Uncivil': Contestation and Reconstruction within the Federal Administrative State

ROSEMARY WARSKETT

Introduction

Theories of the modern liberal democratic state and its internal administration have been in crisis for at least the last 15 years. Traditional perspectives that emphasize the rational-legal characteristics of the modern bureaucratic state are met with skepticism and also evidence that points to the contradictory and chaotic manner in which state administration is undertaken.[1] The focus of this chapter is the governance of state personnel within the liberal democratic state, in other words, it examines how the state regulators themselves are regulated. The argument is made that the traditional Weberian approach to state bureaucracy, administration, and regulation cannot explain the complex set of relations and conflict that takes place within the state. While contradictions and conflicts were always present within the early liberal democratic state, since the post-war period these have multiplied, resulting in the inapplicability of traditional theory.

The chapter begins with a brief examination of the theory of state bureaucracy and state personnel and recent attempts to deal with the crisis of ideas. It then focuses on the case of Canadian federal state administrators and how 'civil servants' were legally constructed in the first half of the twentieth century. This is followed by an examination of the class and gender conflicts and contestation that cut through the 'unity' of the bureaucracy in the latter part of the twentieth century and how these sets of relations have influenced the reconstruction and regulation of Canadian federal 'civil servants'.

Contradictions and Conflict within the Liberal Democratic State

Traditional approaches to the administrative state assume a unity of law and regulation in governing the personnel within the state. Max Weber's view of bureaucracy projects a vision of a set of politically neutral experts, scientifically applying their skills to construct a rational, efficient set of state regulation.[2] Characteristics of the administration include appointment by merit, impersonal, standard operating procedures, and a hierarchical, specialized division of labour. Employees of state bureaucracies are said to operate under politically defined financial controls, and they are involved in degrees of confidentiality and secrecy that have the effect of limiting public scrutiny of internal decision-making.[3] Neo-Marxist criticism of this perspective focuses primarily on the way in which the state in capitalist societies reproduces the dominant class hegemony.[4] This approach, however, also projects a unity of purpose within the state, with those in subordinate positions loyally carrying out the commands from above.[5] In this sense the state personnel are treated as a unified part of the rational functioning state. The unequal structure of representation within the state was addressed by Rianne Mahon, who concluded that conflicting functions of departments and agencies are hierarchically ordered and so act to preserve the state's 'contradictory unity'.[6] More recently, Bob Jessop adopted a strategic-relational approach that conceives state structures as 'given from the past' and providing possibilities for action and strategy by societal and state actors.[7] Jessop does not deal with conflicts among personnel within the state, but his approach does leave open the possibility of examining conflicts and contradictions occurring within the administrative state. In this chapter I adopt a strategic-relational approach to examine conflicts within the Canadian federal state.

The notion that the state is internally unified might fit to some extent the evidence of early state administration, but since World War II there have been radical changes. All liberal democratic states experienced rapid expansion in the period of the 1950s and 1960s with the growth of social welfare provision.[8] This resulted in substantially increased state responsibilities together with a rapid growth of fiscal and personnel resources. In all liberal democracies the state's legal, regulatory, and other kinds of interventions multiplied. All of these factors led to 'internal pluralization and fragmentation of departmental perspectives within the administration, an escalation of the respective rivalries, and on the whole, an increasing unpredictability of the resulting long-term and "synergetic" effects of individual policies which are nearly impossible to coordinate'.[9]

While it is true that many states in the last decade of the twentieth century cut back services and privatized some of their functions, nevertheless, the size of bureaucracy remains large as compared to the early modern state. Furthermore, the reduction of the number of civil servants without substantial reduction of the amount of work they perform has led to further conflict and dissatisfaction among state employees. It is clear from a substantial number of studies that civil servants are not a uniform, neutral set of personnel. The evidence calls out for an approach that can explain the divisions within state bureaucracies and the militancy of lower-level personnel that developed within a wide range of states in the late twentieth century.[10] Such an explanation needs to capture the distinctiveness of those working within the administrative state and yet be capable of explaining the recent conflict and contestation by those in subordinate positions, including women and other identities.

The approach that treats civil servants as a social category that was constructed discursively and historically through legal regulation is a useful perspective for understanding the 'civil servant' discourse that developed in the first part of the twentieth century and its dissolution in the second half. Nicos Poulantzas developed the idea of the state administrative bureaucracy as a social category.[11] Drawing on the work of both Marx and Weber, Poulantzas pinpointed how state administration is a 'social category deriving from bureaucraticism as a normative ideological model of organization'.[12] In this view personnel within liberal democratic state bureaucracies are constructed and governed, in varying degrees, according to the characteristics identified by Max Weber. These characteristics are an important part of the attempt to mould the bureaucracy into a unity apart from other social groups, and into a social category set apart not only in the world of work but also from the market economy. Certainly the discourse of civil service as an efficient, functioning set of personnel, appointed by merit rather than through patronage, expertly and loyally carrying out commands from above, was a salient part of the liberal democratic state prior to the post-war period. But this aspect of the state began to come apart with the rapid growth of the welfare state. Cracks began to appear in the unity of the social category and relations based on class and gender, and in the case of some states on race and ethnicity, began to come to the fore. This resulted in a significant change in the conditions for those working at all levels of state hierarchies. Demands by those in the lower echelons of the service for the right to be unionized and to bargain collectively appeared early on in the post-war period. Later, demands were made by women and various minorities for a range of equal rights. The disunity of the civil service continued and deepened in the latter part of the twentieth century, particularly in those states that adopted a neo-conservative discourse that resulted in the privatization of many state functions and cutbacks and downsizing in the social services.

Civil servants are now found in a hierarchy of classes within the bureaucracy: management, the new middle class, and the working class. The state bureaucracy is also divided in terms of gender, race, language, and other differences. In this chapter the focus is primarily on class and gender cleavages, but at certain moments other divisions have been salient within the Canadian administrative state.[13] This means that how the bureaucracy and civil servants are constructed has an impact both on the formation of class and gender and on other 'differences'. In turn, how the social category of bureaucrats is constructed is affected by the internal divisions of class, gender, race, and other differences.

The Early Canadian Administrative State and the Political Construction of Civil Servants

In keeping with the characteristics of a rational-legal administration, the transformation of the Canadian patronage state to a modern liberal democratic state bureaucracy had taken place in large part by 1919.[14] Appointment was by merit rather than patronage; the bureaucracy was hierarchically structured, with control emanating from the top down; the division of labour became more specialized, with a 'modern' classification system and job descriptions based on the principles of the new scientific management.[15] Furthermore, the neutral, non-arbitrary application of legal rules was emphasized, specialist expertise became more highly valued, and systems were put in place to increase financial control and accountability.[16] All of these elements were part of an effort to 'modernize' the federal civil service, to make it efficient and effective, and to remove it from political patronage. These changes took place under the direction of the Conservative government of Robert Borden, which was influenced both by the demands of the upper levels of the bureaucracy for a professionalized service, free of patronage, that could command the respect of its citizens,[17] and by 'capitalist supporters who demanded a strong and effective state apparatus to act as their agent, especially in external trade and the search for foreign markets'.[18]

As part of the changes a new Canadian Civil Service Act was passed by the Dominion Parliament in 1918, extending the powers of the Civil Service Commission. The Commission was given power to make appointments on the basis of merit; section 32 of the legislation explicitly stated that employees must not engage in partisan political activities.[19] The Commission used its new powers to make a sweeping reorganization of the service and contracted Arthur Young and Company of Chicago to devise and implement a new classification plan. After Parliament accepted the plan, the Civil Service Commission pronounced that the unscientific methods of the past had been replaced by the principles of 'scientific classification'.[20] The overall effect of the plan was to standardize tasks and to establish a rigid hierarchy of duties with assigned levels of position and pay. The service was organized with the intention of applying 'Taylorist' principles of efficient management by separating the conception of the tasks from their execution.

As part of the reorganization, the Civil Service Commission introduced the concept of indeterminate appointments, thereby distinguishing between permanent career civil servants and those employed on a contractual or casual basis. This was an important step in creating a professionalized service and constructing a social category of career civil servants. Through this process it became clear that women, in general, were not to be part of the category. In 1921 severe restrictions were placed on the appointment of married women to indeterminate positions. They were only to be appointed when qualified men were not available.[21] This meant that women already appointed to the service had to resign. They were then rehired as temporary employees and paid the lowest rate for their classification.[22] This violation of the merit principle appeared to escape the commissioners but was quite in keeping with the dominant view of the time. The proper place for married, middle-class women was thought to be in the home, even though prior to marriage the role of secretary and clerical worker had become acceptable as an occupation for young, respectable females by the 1920s.[23] The reorganization of the Dominion Civil Service in the early twenties, based on the principles of scientific management, resulted in the creation of routine clerical jobs that were for the most part filled by women.[24]

The rationalization and reorganization of the civil service in this period set in place a hierarchical system of work organization that continued until the introduction of collective bargaining in 1967. Career civil servants during the interwar period were mainly drawn from the upper- and middle-class ranks of white, Anglo-Saxon men. General clerical positions, at the lowest levels of the service, were the entry points for young men who potentially could advance rapidly up through the ranks. Women appointed as stenographers either left the service to be married or stayed on as higher-level secretaries and clerks.[25] Later on, in the post-war period, this gender differentiation in employment policy fed the growing dissatisfaction of 'civil service' women.[26]

Blue-collar men, those in the trades and general labour categories, were also marginal to the social category of civil servants, although they were members of the staff associations. Many of these jobs were term or casual appointments and patronage continued to play an important part in the awarding of public works contracts. Those who did have permanent positions wanted to earn at least the same wages and benefits as their counterparts in the private sector, plus they had the security of an indeterminate appointment. In this sense, although they were not able to move up through the bureaucracy, they too experienced the respect and benefits of being civil servants and gave their loyalty to the state.

For the most part, however, the social category of civil servants was constructed on the basis of young, male, clerical workers entering at the bottom of the service with the expectation that they could rise through the ranks to become part of the upper-level bureaucracy. With very few exceptions women remained as administrative support workers. The highest position they could aspire to was secretary to a deputy or assistant deputy minister. They achieved this by 'rising' in rank with their male supervisor much the same way that a politician's wife experiences the material benefits of her husband's promotion. This system of 'rug ranking' came under attack from women in the early 1970s.[27] The unity of the social category depended on the perception of privilege and respect accorded to its members. In the interwar period, therefore, women, blue-collar males, francophones, and other minorities were in effect excluded from the permanent civil service. This did not mean that cleavages of class, gender, and ethnicity did not exist within the service, but they were minimized as a result of the exclusions from the category. In this sense the Canadian civil service appeared at least on the surface to conform to the hierarchical unity depicted in traditional bureaucratic theory.

Civil servants were set apart from workers in the private sector and constructed as a special category in part because of the doctrine of sovereignty. Employees of the state were perceived as different from other employees because of the general perception of a neutral state standing above or apart from society. The idea that servants of the state must be politically neutral, acting only in the interests of the state, meant, in practice, that the state was perceived to have 'peculiar prerogatives over its employees'.[28] The unity of the social category meant that commands were issued from above and obedience rose from below. For those young men who perceived themselves as potentially being part of the command at some future date, this worked well. For others—tradesmen, postal workers, and veterans returning from the war—this posed a problem. Since women secretaries and clerks were required to leave the service on marriage until the mid-1950s, they were invisible to the dominant actors in the civil service and hence discounted, unable to be active agents of change.

During the war, the number and proportion of female appointments to the civil service increased sharply. Restrictions on the appointment of married women were relaxed and as a consequence more than 50 per cent of all civil service appointees were women. Restrictions, however, were reintroduced in 1947 with the return of the war veterans.[29] Once again the merit principle was sidestepped and veterans were given preference for appointment. This measure was one small part of the Liberal government's policy on jobs laid out in the White Paper on Employment and Income of 1945 and was part of the transformation of the economy from war to peace. Even though the White Paper laid out a policy of creating jobs by encouraging investment in the private sector, the post-war experience resulted in a rapid expansion of government programs and thus in substantial growth at all levels of the state. Federal government employees increased in number from 146,257 in 1946 to 258,281 in 1966, an increase of 77 per cent.[30] The increases at the provincial and municipal levels were even greater, as provinces and cities began to assert independence from the federal level and as regional social programs related to health, education, and welfare expanded.

The growth of the Canadian federal state was both horizontal and vertical, and resulted from added programs and departments and from a burgeoning of bureaucratic structures. In 1955 the restrictions on appointment of married women to the federal civil service were revoked in response to the need to fill newly created lower- and middle-level positions, especially clerical and typist positions. Women were appointed to the lower-level administrative support positions, while white-collar men continued to rise rapidly through the hierarchy.[31] Rapid growth of both departments and positions, however, increased the distance between the highest levels of control and front-line workers[32] and resulted in a divergence between the rationality of the internal bureaucracy and the requirements of the social groups served by various levels of the state.[33] These contradictions began to be felt by front-line civil servants attempting to provide services while acting within the legal-rational system. Demands by the staff associations for more input into the determination of their terms and conditions of employment were becoming more serious by the mid-1950s. Cracks were starting to appear in the unity of the social category of civil servants.

Contestation by Workers within the Administrative State

In the 1950s federal civil servants, active in both the Civil Service Federation of Canada (CSFC) and the Civil Service Association of Canada (CSAC), were at a crossroads. Should they attempt to press their claims for greater salaries and benefits through consultation with the higher levels of the service or should they reconstruct their identity, demand legal collective bargaining rights, and learn to become part of the Canadian trade union movement?[34] The Canadian Postal Employees Association (CPEA), which was part of the CSFC at that time, strongly supported the latter approach, as did many other blue-collar groups within the CSAC. Most civil servants, and certainly the leaders of the two staff associations, however, wanted to keep one foot in the traditions and principles of the civil service and the other in a limited form of collective bargaining that could resort, at most, to binding arbitration if a dispute arose between the employer and the employees.[35]

In 1966 the Public Service Alliance of Canada (PSAC) was formed by federal civil servants from the two associations. They carried with them the traditions of the earlier staff associations, which had decided at this point to merge in large part because of the collective bargaining structure proposed by the bill to create the Public Service Staff Relations Act (PSSRA) in 1967. The PSAC was at that time the largest union within the service and it remains so today. Its leaders in 1966 were committed to remaining part of the social category of civil servants and maintaining the characteristics that defined that category. The majority of these employees were white, Anglo-Saxon men, many of them returning veterans of World War II.[36] By 1966, however, the material basis of this category was already undergoing change. Women could now be appointed to permanent positions and, with the rapid expansion of the Canadian state and its administration, they rushed to fill many of the newly created positions in the 'administrative support' category. Blue-collar males, at that time comprising approximately 30 per cent of the PSAC membership, were under-represented in its leadership.[37] Furthermore, among this group there was mounting dissatisfaction with the disparities in conditions of work and compensation between civil servants and their private-sector counterparts. Prevailing rate comparisons, which had allowed blue-collar PSAC employees the same rates of pay as those in the private sector, had been abandoned by the employer.[38]

The Public Service Staff Relations Act placed a number of important restrictions on collective bargaining, but at the same time it provided legal union recognition for civil servants. The Act, therefore, structured and restricted the development of labour relations while providing some openings for union resistance. Binding arbitration as an option, as opposed to the strike/conciliation route, provided for a different kind of collective action for the PSAC than that practised by the postal workers. It allowed the newly formed union to continue its harmonious relationship with Treasury Board. Binding arbitration depends, in large part, on the expertise of union staff in presenting pay and benefit data before arbiters and convincing them of the justice of their case. It does not necessarily require the collective participation or mobilization of the membership.

Leaders of the newly formed union strongly supported and policed the merit principle; they wanted to maintain political neutrality; they were proud of their distinctive employee status and did not want to be treated like other workers in the private sector. They supported the idea of separate legislation that recognized their distinctiveness and the important role they played in maintaining a neutral state administrative bureaucracy far from 'the madding crowd'.[39]

The merit principle and its practice were essential (and still are in many respects) in reproducing the category of civil servants and cementing its unity. Appointment by merit was the guiding principle of the the Public Service Employment Act, 1967. In keeping with Weber's characteristics of a rational-legal bureaucracy, appointment and promotion on the basis of merit is an attempt to ensure that the civil service is staffed by personnel who act neutrally and apolitically, accepting their role as servants of the general will. Merit as a principle embodies the idea that positions should be allocated to those with the highest qualifications and demonstrated skill for performing the required duties. It is meant to assign individuals a place in the job hierarchy without any consideration of a person's gender, race, ethnicity, or class position in the society. In other words, the 'unjust hierarchy of caste . . . [is] replaced by a "natural" hierarchy of intellect and skill'.[40]

Criticism about the application of the merit principle and how practices and conceptions of meritorious behaviour are biased against those who are in a subordinate position within the dominant culture have been wide-ranging. Within the Canadian context, Rosalie Abella[41] undertook such a critique of employment practices and was instrumental in developing the principle of employment equity found in the present federal Employment Equity Act. Abella criticized systems of merit that discriminate against certain groups and proposed that bias and discrimination be eliminated by producing a set of practices that are truly based on the merit principle. In this view, employment equity perfects the merit principle. Criticism from outside the liberal democratic paradigm, however, portrays the merit principle as a myth that cannot be reconstructed in an objective and unbiased manner. The merit principle is an important part of the reproduction of the status quo and its practices operate to keep others subordinated. Merit, in practice, is normally constructed to value those of a certain class, gender, ethnicity, and ableness and thus reflects dominant cultural predispositions.[42] The principle of equal pay for work of equal value (now generally referred to as pay equity) approaches the devaluing of women from a different direction and requires that women's work be valued using a gender-neutral evaluation. It was this latter principle that women in the PSAC strongly adopted in the 1980s, and while they did not reject the merit principle outright they did demand that it be reconstructed and redefined.

Despite the large numbers of women in clerical positions, the male-dominated white-collar groups dominated the PSAC in the beginning. The female-dominated clerks' strike in 1980 was the defining moment of the present union and its present estrangement from the discourse of 'civil service'. Initially, the strike was not perceived as a woman's struggle within the PSAC. The analysis of this strike as an important moment in the feminization of the union came later, once the discourse of women's equality was well developed in the PSAC in the late 1980s.[43] Nevertheless, the strike played an important role in allowing women clerks new possibilities for organizing within the union. Its importance lay in the fact that it cracked apart the social category of 'civil servant' as it was constituted within the union. This resulted in the rejection of many practices of consultation and co-operation that characterized the relationship with the employer up to that point. Clerical workers, without any tradition or experience of striking, 'voted with their feet', rejecting the advice, the tradition, and the regulations that the PSAC had inherited from the old staff associations. Thus began the process of constructing a union based on collective solidarity and action.[44] The clerks' strike, in this sense, was the beginning of the end for the discourse of civil service within the PSAC.

After the clerks' strike, in the 1980s, the union turned to using the conciliation/strike route as the normal way of dealing with the employer's unwillingness to negotiate. By the mid-1980s the leadership took the decision to place all bargaining units, with Treasury Board as the employer, on that route and work towards a master collective agreement. Gradually, the PSAC changed its orientation, so that its executive leaders' main task became how to mobilize the membership around certain issues, rather than how to keep conflict at a minimum.

By 1991, the Alliance was ready to undertake the massive organizing task of a general strike against Treasury Board. Ten years of strategizing and learning through mobilizing the membership brought results, and the 1991 strike proved a resounding success in terms of organizing even though few material gains were made. The reconstruction of collective bargaining within the PSAC resulted in a remoulding of negotiations with Treasury Board and changes in the practices emanating from the Public Service Staff Relations Act. The regime and regulation of collective bargaining within the federal state radically changed from a system where arbitration of disputes was the norm to one where strikes and picket lines are not unusual.

As we saw earlier, women were excluded and then later marginalized within the traditional civil service. The status of women within the federal administrative state began to be seriously challenged through collective bargaining and with the clerks' strike. The contestation over pay equity, however, symbolizes the changing status of women in the service more than any other event.

During the 1970s the federal Canadian state became open to the adoption of a policy of equal pay for work of equal value (commonly referred to as pay equity), a policy with the potential to increase significantly the value and cost of women's wage labour. The human rights orientation of the Liberal government in the 1970s, together with Canada's ratification of International Labour Organization (ILO) convention 100 in 1972 provided a basis on which a change in law could be contemplated and then justified later on. This convention referred to work of equal value, going beyond the principle of equal pay for equal work. Once the principle of pay equity was given substance by its incorporation into the Canadian Human Rights Act of 1978, it became the focus of strategizing by women within the PSAC.

Initially, certain members of the PSAC in female-dominated jobs were able to benefit from the equal value law by adopting tactics that extended their bargaining power. This was achieved through using the wage discrimination sections of the Canadian Human Rights Act in concert with collective bargaining legislation, in this case the Public Service Staff Relations Act.[45] Given the initial success of this strategy, by 1984 the PSAC had filed complaints with the Canadian Human Rights Commission on behalf of approximately 65,000 members.[46] The filing of group complaints changed the effect of the legislation from an individual complaints process to one that challenged the systemic nature of women's lower pay. In this sense the PSAC and other unions participated in reconstructing the practice of complaint-based legislation.

Treasury Board, operating from a powerful position in the administrative hierarchy of the federal system, and also as employer of the largest group of women workers in Canada, moved decisively to limit the success of the PSAC's group complaint strategy and thereby to regain control over pay policy. Treasury Board, as the employer, was concerned not only about its wage bill but also about those of employers within the private sector. As representatives of Canadian corporate interests within the Canadian state, Treasury Board and the Department of Finance play a significant role in making labour relations policy. By dint of its power within the departmental hierarchy, Treasury Board was able, for a certain period of time, to emasculate the power of the Canadian Human Rights Commission (CHRC). It did this initially by establishing the Joint Union/Management Initiative (JUMI). Through this strategy Treasury Board moved the conflict to a new terrain and set off a further set of strategizing and learning for the PSAC. This stage of the process ended in 1990 with the PSAC withdrawing from the study and the unilateral imposition of Treasury Board's own version of the equal pay study's results.[47]

When it became clear, however, that wage discrimination still existed within the departmental system, the Canadian Human Rights Commission moved to investigate the PSAC's complaints by establishing a tribunal. It was at this point that Treasury Board, using its power within the administration, challenged the Commission's legal authority to act. While Treasury Board is powerful within the federal government bureaucracy, this power does not necessarily extend to the judiciary. At the level of the Federal Court the challenge was quashed and the CHRC established a tribunal to investigate the PSAC complaint.

After many years of hearing evidence, the Canadian Human Rights Tribunal finally decided in 1997 that wage discrimination did indeed exist within the federal state. Furthermore, it was found that the appropriate methodology for determining the degree of discrimination would produce a wage cost and back pay far greater than that proposed by Treasury Board. This decision was appealed to the Federal Court, with Justice John Munroe Evans finding in favour of the methodology adopted by the tribunal, which, while varying slightly from that of the PSAC, resulted in a settlement for the PSAC members that was significantly higher than that proposed by Treasury Board.[48] Despite the Canadian federal state's adoption of the principle of equal pay for work of equal value, Treasury Board had been determined to keep the wages of women at a lower rate than men's. The decision of Justice Evans, however, clearly revealed the contradiction between the principle of equality and the wage practices of the Liberal government. Faced with this contradiction, Treasury Board reluctantly accepted the settlement.

Conclusion

The case of the PSAC and its growing militancy, especially around the issue of pay equity, reveals a different perspective on the 'civil service' from that of a unified, loyal bureaucracy hierarchically ordered with commands and authority operating from the top down. Conflict over issues emanating from the division of the bureaucracy based on gender and class relations came strongly to the fore in the 1980s and 1990s. Working in the federal civil service at the beginning of the twenty-first century is now more akin to working in the private sector. The aura of being a 'civil servant' that characterized the liberal democratic state in the first half of the twentieth century no longer exists. In the second half of the last century the regulations governing the behaviour of the employees working for the state were contested and modified, both through militant action of employees working within the state and in the courts. Civil servants at the beginning of the twenty-first century have indeed become 'uncivil', and the discourse of civil service has been transformed into one of public employees who are regulated in much the same way as those in the private sector. In this sense, the argument made here agrees with the findings of Michael Mac Neil in Chapter 8. The discourse of civil service that was enforced by external legal regulation has now been stripped bare. What remains is the dominant regime of governance in which the employer seeks to maintain control over the conditions and organization of work within the state. The social category of 'civil servant', with all its accompanying characteristics, including loyalty to the state and privileged status within society, is a discourse that for the moment has disintegrated.

Notes

1. Claus Offe, *Modernity and the State: East, West* (Cambridge, Mass.: MIT Press, 1996).
2. Max Weber, *Economy and Society: An Outline of Interpretive Sociology*, eds Guenther Roth and Claus Wittich (New York: Bedminister Press, 1968).
3. Ibid., 956–94.
4. Ralph Miliband, *The State in Capitalist Society* (London: Weidenfeld and Nicolson, 1969); Nicos Poulantzas, *Political Power and Social Classes* (London: New Left Books, 1973); Leo Panitch, ed., *The Canadian State: Political Economy and Political Power* (Toronto: University of Toronto Press, 1977).
5. Poulantzas argues that state personnel are distinctive because of the normative role they play in carrying out the functions of the state.
6. Rianne Mahon, *The Politics of Restructuring: Canadian Textiles* (Toronto: University of Toronto Press, 1984).
7. Bob Jessop, *State Theory: Putting the Capitalist State in Its Place* (London: Polity Press, 1990), 262.
8. See Gosta Esping-Andersen, 'The Three Political Economies of the Welfare State', *Canadian Review of Sociology and Anthropology* 26, 1 (1989): 10–36.
9. Offe, *Modernity and the State*, 63.
10. Evidence of militancy among civil servants in subordinate positions in the bureaucracy is apparent in a large number of liberal democratic states. See A. Ferner and R. Hyman, *Industrial Relations in the New Europe* (Oxford: Blackwell, 1992); Mona-Josée Gagnon, 'Reflections on the Quebec Public Sector Contract Negotiations', *Studies in Political Economy* 31 (Spring 1995): 169–79; Jens Hoff, 'The Concept of Class and Public Employees', *Acta Sociologica* 28, 3 (1984): 207–26; M. Kelly, *White Collar Proletariat: The Industrial Behaviour of British Civil Servants* (London:

Routledge & Kegan Paul, 1980); Joseph B. Rose, 'Growth Patterns of Public Sector Unions', in Mark Thompson and Gene Swimmer, eds, *Conflict or Compromise: The Future of Public Sector Industrial Relations* (Montreal: Institute for Research on Public Policy, 1984); Marian Simms, *Militant Public Servants: Politicisation, Feminisation and Selected Public Service Unions* (Melbourne: Macmillan, 1987); Leo Troy, *The New Unionism in the New Society: Public Sector Unions in the Redistributive State* (Fairfax, Va: George Mason University Press, 1994).

11. Antonio Gramsci identified intellectuals as a social category and pointed to their importance in both elaborating and applying a particular conception of the world and social relations. Antonio Gramsci, *Selections from the Prison Notebooks*, trans. and ed. Quintin Hoare (London: Lawrence and Wishart, 1971).
12. Nicos Poulantzas, *State, Power, Socialism* (London: Verso, 1978), 154.
13. The demand by francophones in Quebec for recognition as a nation has meant that French language and culture have played an important role in the composition of the federal bureaucracy. This important aspect of the Canadian administrative state is not pursued in this chapter, but the divisions between francophones and anglophones within the federal bureaucracy take a specific form and are a 'condensation' of these relations within Canadian society.
14. Ken Rasmussen, 'Public Service Inquiries, Administrative Reform and the Quest for Bureaucratic Autonomy: 1867-1919', paper presented at the annual meeting of the Canadian Political Science Association, Charlottetown, 1992.
15. Graham S. Lowe, *Women in the Administrative Revolution: The Feminization of Clerical Work* (Toronto: University of Toronto Press, 1987), 97.
16. J.E. Hodgetts, *The Canadian Public Service: A Physiology of Government* (Toronto: University of Toronto Press, 1973); Reg Whitaker, 'Between Patronage and Bureaucracy: Democratic Politics in Transition', *Journal of Canadian Studies* 22, 2 (1987): 55-71.
17. Rasmussen, 'Public Service Inquiries', 29. Rasmussen argues that those in the senior ranks of the civil service showed a strong interest in reform because they wished to increase the autonomy of the service and protect it from political interference. He examined seven public service inquiries that took place between 1867 and 1919 in which senior civil servants advocated reform.
18. Reg Whitaker, 'Images of the State in Canada', in Panitch, ed., *The Canadian State*, 55.
19. Hodgetts, *The Canadian Public Service*, 315.
20. Lowe, *Women in the Administrative Revolution*, 97.
21. Kathleen Archibald, *Sex and the Public Service: A Report to the Public Service Commission of Canada* (Ottawa: Supply and Services, 1970), 16.
22. *Civil Service Review* 20 (1947): 148-9.
23. Lowe, *Women in the Administrative Revolution*, 47-62.
24. Archibald, *Sex and the Public Service*.
25. Ibid., 14-17; Lowe, *Women in the Administrative Revolution*, 86-112.
26. An important exception to the upward mobility of male clerks was the occupation of postal clerk. Postal employees were classified as a separate clerical category, not part of the recruitment ranks for upper-level positions, although the majority were indeterminate employees. Their close association with mail handlers and letter carriers meant that their work was classified as blue-collar rather than white-collar

clerical work. Postal clerks were required to pass sorting examinations and employees prided themselves on their ability to memorize over 2,000 destinations. The social construction of postal employees as blue-collar employees, separate from the general clerical classification, was an important factor in their separate development as public-sector unionists. Long before collective bargaining rights were enshrined in law in 1966, postal workers had developed a trade union identity that included an identification with the mainstream labour movement. See Julie White, *Mail and Female: Women and the Canadian Union of Postal Workers* (Toronto: Thompson Educational Publishing, 1990).
27. Most of the original research for this section is taken from Rosemary Warskett, *Learning to Be 'Uncivil': Class Formation and Feminisation in the Public Service Alliance of Canada 1966-1996* (Ottawa: Carleton University, 1997).
28. Hodgetts, *The Canadian Public Service*, ch. 14.
29. Archibald, *Sex and the Public Service*, 16.
30. Hugh Armstrong, 'The Labour Force and State Workers in Canada', in Panitch, ed., *The Canadian State*, 297.
31. Archibald, *Sex and the Public Service*, 19-27.
32. Greg Albo, 'The Public Sector Impasse and the Administrative Question', *Studies in Political Economy* 42 (1993): 113-27.
33. Claus Offe, *Disorganized Capitalism: Contemporary Transformations of Work and Politics* (Cambridge Mass.: MIT Press, 1985), 302.
34. *Civil Service Review* (June 1970): 6.
35. The Canadian Postal Employees Association was a notable exception in this regard. A long-time affiliate of the Canadian Trades and Labour Congress, the CPEA pressured for the same rights as other industrial workers. They were blue-collar workers who remained on the margins of the social category of civil servants. Belonging to the category held no great promise or benefits for them. When consultation with the government over a wage increase failed in 1965 they simply took the right to strike, put up the picket lines, and forced a different outcome from that recommended by the Heeney Commission. As a consequence, the strike as a legitimate dispute resolution route was enshrined in the Public Service Staff Relations Act.
36. *Civil Service Review* (1952): 453.
37. Data compiled from the following three sources: Public Service Commission of Canada, *Annual Report* (Ottawa: Minister of Supply and Services, 1973-90); Public Service Staff Relations Board, *Annual Report* (Ottawa: Minister of Supply and Services, 1967-90); Treasury Board, *Occupational Group Job Descriptions* (Ottawa: Minister of Supply and Services, various years).
38. *Civil Service Review* (June 1973): 12.
39. Maurice Lemelin, *The Public Service Alliance of Canada: A Look at a Union in the Public Sector* (Los Angeles: Institute of Industrial Relations, University of California, 1978).
40. Iris Marion Young, *Justice and the Politics of Difference* (Princeton, NJ: Princeton University Press, 1990), 200.
41. Rosalie Abella, *Report of the Canadian Royal Commission on Equality in Employment* (Ottawa: Ministry of Supply and Services, 1984).
42. See debates around this question in Joanna Brenner, 'Feminist Political Discourses: Radical Versus Liberal Approaches to the Feminization of Power and Comparable

Worth', *Gender and Society* 1, 4 (1987): 447–65; Young, *Justice and the Politics of Difference*, 40.
43. Canada Employment and Immigration Union (CEIU), '50,000 Strong. The Federal Clerks' Strike of 1980', special issue of *Paranoia* (Toronto: Our Times and the Canadian Association of Labour Media, 1980).
44. Ibid.
45. The librarians were an early case in point. Both the Treasury Board and the PSAC agreed on a settlement based on comparison of wages with the 'historical researcher' group. The settlement was made after laying a complaint with the Canadian Human Rights Commission and raising the issue at the bargaining table. See Rosemary Warskett, 'Political Power, Technical Disputes, and Unequal Pay: A Federal Case', in Judy Fudge and Patricia McDermott, eds, *Just Wages: A Feminist Assessment of Pay Equity* (Toronto: University of Toronto Press, 1991), 172–92.
46. Ibid, 185–8.
47. Ibid.
48. *Attorney General of Canada v. Public Service Alliance of Canada and the Canadian Human Rights Commission* (1999), F.C.C. 1698.

PART III

Dispute Resolution

Chapter 10

Is There Any Justice in Alternative Justice?

NEIL C. SARGENT

Introduction

One of the most salient features of the contemporary justice landscape is its paradoxical character. Janus-like, the legal system presents two quite different faces to the world. On the one hand, the past two decades have witnessed an unprecedented expansion in the scope of formal legal regulation and in the adjudicative function of the courts. Partly inaugurated by the enactment of the Charter of Rights and Freedoms in 1982 and supplemented by substantive and procedural reforms designed to facilitate access to the courts by individuals and groups alike, the courts are increasingly being called upon to adjudicate a wider range of disputed rights claims than ever before.[1] At the same time, however, there has been no reduction in the traditional workload of the courts in areas such as criminal law, family law, and civil litigation, resulting in what almost amounts to gridlock within the formal justice system.

Perhaps it should be no surprise, then, that in tandem with this expansion in rights consciousness we have also witnessed what some commentators have called a whole new social movement centred on the themes of delegalization and alternative justice.[2] Originating in the form of a neighbourhood justice movement in the 1960s and 1970s and drawing much of its inspiration and philosophy from early American community models of dispute settlement, as well as from models of dispute settlement drawn from predominantly non-Western cultures, the alternative justice movement has promoted the use of informal, non-adversarial forms of dispute resolution as a more humane and satisfying alternative to the courts for resolving interpersonal and neighbourhood conflicts, which the formal justice system has traditionally overlooked.[3]

At the heart of this alternative justice movement is the assertion that consensual, participatory approaches to resolving interpersonal or intergroup conflicts are preferable to formal legal intervention through the courts. To many commentators within the alternative justice movement the adversarial nature of the judicial process operates to prolong conflicts and exacerbate hostilities between the disputants, thus adding to the costs of conflict.[4] Moreover, the technical, inaccessible language of the law and the institutional reliance on legal advocates operate to disenfranchise the disputants and deprive them of the opportunity to participate in the resolution of their own conflicts. Consequently, a central theme or goal within the alternative justice movement, from its inception, has been to increase community resources for resolving conflict and empowering citizens to reclaim ownership of their own disputes from an increasingly alienating and bureaucratized legal system.[5]

More recently, the alternative justice movement has outgrown its neighbourhood justice roots, so to speak, and begun to expand its reach into areas formerly seen as part of the exclusive preserve of the courts and the legal profession. The emergence of the practice of divorce mediation as an adjunct to the litigation process for dealing with financial and custody issues following divorce, victim-offender mediation programs for dealing with minor criminal offences, and use of court-connected mandatory mediation programs in the civil litigation process all represent a significant widening and deepening of the reach of the alternative justice movement into what might appear to be the heartland of the formal justice system.[6]

How are we to understand the social implications of this 'movement' towards delegalization and alternative justice? Does it represent a radical decentring of law and a shift towards a different normative mode of governance of social relations, as some of its more fervent supporters claim? Or should it be seen, as some of its critics have argued, as part of an ongoing process of juridification, in which the formal legal order modifies its message in the name of informal justice while enlarging its regulatory grasp over more areas of conflict that hitherto had been out of its reach?[7]

Not surprisingly, perhaps, there is considerable debate within the alternative justice literature over the goals and aspirations of the movement, which in part reflects these very different ideological visions of the relationship between the formal legal order and its informal justice 'cousin'. On one side of the debate are those proponents of informal justice who see the growth of mediation-based alternatives to the courts as part of a process-oriented reform movement aimed at making the justice system more responsive to the needs of disputants.[8] Much of the emphasis within this process reform wing of the alternative justice movement has been directed towards identifying the characteristics of disputes that make them most appropriate for certain kinds of conflict resolution processes. The adjudicative process is not seen as being inherently flawed but as being unsuited for certain types of disputes in which it is preferable to leave decision-making authority to the disputants rather than to a formal adjudicator.[9] This process reform wing of the alternative justice movement has been most active in promoting the concept of a 'multi-door courthouse' where disputants can obtain access to different types of 'justice', depending on their particular needs and the nature of their dispute.[10]

At the other end of this spectrum is what Laura Nader refers to as the harmony ideology wing of the alternative justice movement, which views the increased use of mediation and other forms of alternative dispute resolution as part of a larger change in how society deals with conflicts.[11] Here we encounter an alternative justice movement that defines its project in much more ambitious terms, not just as a process reform movement primarily concerned with efficiency and cost of justice delivery systems but as a transformative social movement, the goal of which is to change the very language and institutional structure through which we are able to envisage the idea of 'justice' itself.[12] Within this restorative justice literature, mediation is viewed as far more than just an efficient 'tool kit' for resolving disputes in a quicker and more cost-effective way than litigation. Instead, mediation and other non-coercive forms of dispute resolution tend to be identified with more communitarian forms of governance, in which individuals and communities are empowered to create 'just' resolutions to their own conflicts without the pressure of coercive intervention by the state and the formal justice system.

It is apparent that these are two very different alternative justice 'stories',[13] which provide very different accounts of the origins and goals of the alternative justice movement; and perhaps more important still, they conceive of their relationship with the formal justice system in quite different terms. For the process reform story, the 'goal' is not to provide a radical alternative to the formal justice system, but to reduce court congestion and provide 'higher-quality' justice for disputants at lower institutional costs for the justice system. For restorative justice theorists, on the other hand, the 'goal' is to transform the institutional processes and value-sets through which interpersonal and intergroup conflicts are addressed, and thereby to effect a transformation in the individuals and communities engaged in conflicts.

Notwithstanding these ideological differences, however, both of these alternative justice stories tend to make use of similar discursive themes in articulating their respective visions of the alternative justice movement. What we encounter in the literature is not so much a struggle between two mutually incompatible sets of ideas about the goals of the alternative justice movement, but a more complex dialogue that takes place around the discourse of self-determination, empowerment, and reconciliation. Most alternative justice theorists, regardless of their ideological affiliation, articulate their project in terms of these three related themes, although different theorists emphasize some themes over others. Thus, restorative justice theorists tend to place most emphasis on the themes of recognition and reconciliation, since without reconciliation justice cannot be truly restorative. The process reform theorists, while also concerned with reconciliation as a justice goal, are primarily engaged in a debate over private ordering versus judicial ordering, and thus place more emphasis on the process values of self-determination and empowerment. We can suggest, therefore, that these three interrelated themes together comprise a discursive continuum around which much of the contemporary dialogue over the future of alternative justice is taking place.[14]

Self-Determination and the Empowered Subject

Purvis and Swan, in Chapters 2 and 6 in this volume, have both identified the significance of the discourse of self-governance and the empowered subject as critical themes within the literature on decentring the state. No longer viewed merely as a passive object of regulatory intervention by the state, the individual emerges anew as the hero of her or his own story, in Purvis's words a 'reflexive rational actor' whose capacity for self-governing choice should be given voice to and actively facilitated by the neo-liberal state. For Purvis, this discursive reconstitution of the subject as a self-governing actor has been accompanied by a strategic withdrawal of the state, particularly in the areas of the economy and welfare policy (though also in some areas of social regulation, such as the regulation of sexuality and sexual orientation). However, Purvis cautions that the state's withdrawal is by no means a complete abdication of its regulatory function but rather a redefinition of its scope, with the state now playing a more facilitative and less directly intrusive role—as he calls it, governing at a distance.[15]

In many respects, similar discursive themes can be seen at work within much of the contemporary dispute resolution literature. Here we tend to encounter the classic liberal notion of self-determination, which refers to the capacity of individuals or groups to make their own self-determining choices in accordance with their own will. Individuals and groups are seen as ends in themselves, rather than as means to an end or objects of collective regulation by the state.[16] The concept of self-determination is closely connected to the liberal values of autonomy and privacy—the claim that individuals or groups should be accorded maximum freedom to make their own choices in pursuit of their own vision of the social good, with minimal intervention or compulsion on the part of others, including the state.[17] In this sense, what makes a decision or an action self-determining is the fact that the decision was made, or the action undertaken, voluntarily and without external threat or compulsion.

Within the dispute resolution literature, the issue of self-determination is often framed in terms of the degree of autonomy left to the disputants to control the process or outcome of their own dispute. Several authors make use of a dispute resolution continuum to illustrate different approaches to resolving conflict, with those dispute resolution processes that leave more control in the hands of the disputants at one end of the continuum and those that leave more authority to a third party to resolve the dispute at the other end.[18] For convenience, this continuum is often expressed as a choice between consent-based processes such as mediation and more formal adjudicative processes in which the judge has authority to decide the dispute.[19]

For Adler, this focus on consensual decision-making and party control is at the heart of the alternative justice or alternative dispute resolution (ADR) movement. Adler suggests that if the alternative justice movement is based on the search for voluntary, consensual modes of reaching agreement, then mediation should be seen as its core institutional form.[20] This is because in mediation, unlike adjudication or arbitration, or other hybrid dispute resolution processes such as early neutral evaluation or judicial mini-trials, it is the disputants, rather than the third party, who have ultimate authority over the resolution of the dispute. The mediator's authority comes from the disputants, and is limited to assisting them in trying to reach their own negotiated settlement rather than trying to impose a resolution. Any resolution to the dispute, therefore, will be the product of the parties' own will. This is also, according to Bush and Folger, what makes mediation unique as a dispute resolution process, since it empowers disputants to resolve their own conflicts, thereby engendering individual moral growth and transformation.[21]

By contrast, adjudication is seen as occupying the opposite end of the consent-coercion continuum, in regard to both the mode of participation by the disputants and the authoritative decision-making power of the judge. Once a legal dispute enters the formal judicial process the parties are compelled to participate in a rigidly structured, adversarial process that provides little opportunity for the disputants to articulate their own needs or desires, or even to tell their own stories in their own words. As Emond observes, the adjudicative process functions as a closed communication system, which actively discourages the disputants from participating, or even communicating with the other party, except on terms dictated by their legal representatives.[22] More disempowering still is the lack of self-determination accorded to the parties over the outcome of the trial. At its heart, the adjudicative process is based on the assumption that justice is derived from the authority of the judge rather than from the will of the parties. Once both parties have presented their proofs and arguments before the court, the judge must hand down an authoritative ruling that will provide a final binding resolution to all the legal issues in dispute between the parties.[23] Moreover, if the court finds in favour of the plaintiff's argument, any legal sanction imposed against the defendant is also coercive in that it is backed by the full legal authority of the state.

Even the language in which the legal dispute is framed may not correspond with the subjective understandings of the disputants about the nature of their own conflict. What may be the most legally relevant issue in dispute may not be the most contested issue to the disputants. Similarly, what may be most important to one or both disputants may not be easily translatable into a legal claim. Consequently, the process of translating the dispute into a legal claim and submitting to the jurisdiction of the court may have the symbolic effect of disenfranchising the parties from ownership of their own dispute.[24]

By comparison with adjudication, mediation is viewed as more empowering to the disputants since it allows them to retain control over the process and language as well as the outcome of the dispute. In this context, the theme of 'empowerment' is not to be confused with the concept of power as control or 'power over'.[25] Power as control implies the capacity of one person or group to impose its will on another person or group. In the context of disputing behaviour it involves a negation of the idea of self-determination, since one person's ability to act in a self-determining manner is limited by the degree to which another person is able to exercise 'power over' that person. To many in the alternative justice movement, this negative concept of power as control tends to be identified with the formal justice system, in which the state is seen as having the ultimate power to impose its will on the disputants through the coercive force of law.

Power as control also implies that power is a finite, limited resource, which increases in value the more exclusively one is able to possess it. Consequently, conflicts over power tend to be seen as adversarial, zero-sum games, wherein one person's empowerment is seen as another side's loss of power or powerlessness. Within the dispute resolution literature, however, the theme of empowerment tends to be associated with 'voice' rather than with coercion, with the ability to express one's point of view, to tell one's story, to articulate one's goals and interests, and to act and make decisions in accordance with those goals and interests, rather than with the ability to impose one's will on others. Chornenki has expressed this as a sense of 'power with' rather than 'power over'.[26] 'Power with' expresses the possibility that both or all disputants can reach more efficient outcomes, and more fully realize their own goals and interests, by acting collaboratively through a process that enhances communication and creates possibilities for joint or mutual gains.[27]

It is clear from this brief discussion that the consent-coercion dichotomy provides a strong ideological underpinning to the mediation movement in its efforts to roll back the boundaries of direct legal intervention and to expand the scope for private ordering by disputants. Yet the emphasis in the literature on the distinctiveness between consent-based dispute resolution processes such as mediation versus coercive processes such as adjudication tends to obscure the extent to which the formal legal system relies operationally on private ordering in practice.[28] As commentators like Mnookin and Galanter have observed, there has always been significant room for private ordering within the adjudicative process.[29] In a civil trial the litigants are free to reach an out-of-court settlement at any time before final adjudication. Typically, the defendant makes an out-of-court settlement offer to the plaintiff in return for the plaintiff giving up all present and future legal claims against the defendant. If the plaintiff accepts the offer, the effect of the settlement agreement is to end the litigation rather than to resolve all the issues in dispute between the parties.

Mnookin has used the term 'bargaining in the shadow of the law' to describe this out-of-court settlement process. According to Mnookin, unlike the formal picture presented in most dispute resolution literature, the court system does not function as a formalized process for adjudicating disputes in which the parties' roles are subordinated to that of the judge. In reality, parties have considerable scope for negotiating the terms on which they may be prepared to settle the dispute. Mnookin observes that legal rules within the bargaining process create bargaining endowments, or claims, for the parties, which they can then trade off in pursuit of a negotiated settlement. At the same time, however, the legal norms tend to set parameters around the likely range of bargaining outcomes available to the disputants, since neither disputant is likely to accept voluntarily a negotiated outcome that leaves him or her materially worse off than if the litigation process had continued.[30]

For proponents of private ordering such as Mnookin, the capacity of disputants to trade off their legal rights in the negotiation process is not seen as an inherent weakness in the private ordering model. On the contrary, this is seen as a strength of the model, since the parties are presumed to be better informed about their own best interests than even the most well-intentioned judge. To critics of private ordering, however, the permissiveness of the model may conceal the operation of private power in the negotiation process.[31] Where there are significant inequalities of bargaining power between the disputants, for example, the potential range of outcomes of the negotiation process is likely to favour the stronger party. Within a divorce negotiation, for example, an ex-wife facing a custody claim by her ex-husband may feel pressured into negotiating away part of her legal entitlement to support or property division in return for the ex-husband agreeing to give up his custody claim. From a private ordering perspective, both parties appear to have made a voluntary, self-determining choice in reaching the negotiated settlement. Seen from another perspective, however, what appears at first sight to be a consensual negotiated agreement may come to look very much like the product of a coercive private ordering process, in which the 'haves' consistently come out better than the 'have-nots'.[32]

Despite these concerns about the invisible costs of private ordering, the impetus behind recent institutional reforms within the civil litigation process has clearly been to promote more private ordering, not less.[33] In this context, potentially the most far-reaching reform initiative has been the introduction of a new court-connected mediation program for civil trials in Ontario.[34] Originally introduced as a voluntary pilot program in Toronto in 1994 and subsequently extended to cover all civil trials other than family law cases in Ottawa and Toronto since 1997, the Ontario Mandatory Mediation Program (OMMP) was designed to promote earlier settlement of disputes (and thereby reduce overloaded court dockets) by requiring the parties involved in a civil case to participate in a mediation session at an early stage in the court proceedings (within a limited number of days after the statements of claim and defence have been filed, but before discovery). The costs of the mediation session are divided between the plaintiff and defendant, with the mediator being selected from an approved list of mediators. If the litigants are unable to settle their dispute the case will proceed to trial in the normal way. The dominating assumption behind the introduction of this mandatory mediation initiative, however, is that adjudication should be seen as a last resort, to be used only when all efforts to arrive at a consensually negotiated resolution of the conflict have failed.

At first glance, the introduction of mandatory mediation into the court system appears to have resulted in a significant shift in disputing behaviour as compared with the adversarial process. Early evaluations suggest that the mandatory mediation program has been successful in achieving relatively high settlement rates across a wide range of disputes.[35] Even in those cases that did not fully settle, the mediation process seems to have reduced the numbers of outstanding issues in dispute between the parties and therefore indirectly facilitated the adjudication process. Likewise, data from interviews with lawyers, mediators, and litigants who have used the new system indicate a quite high degree of satisfaction with the outcomes and cost savings achieved by litigants.[36]

Before declaring this to be the beginning of a new Jerusalem, however, we should keep in mind that even before the introduction of mandatory mediation, the large majority of legal disputes that entered the court system never went to trial. Various estimates indicate that up to 90 or 95 per cent of cases initiated in the courts either settled or were discontinued by plaintiffs before final adjudication.[37] Consequently, the institutional savings for the legal system stemming from the introduction of mandatory mediation programs may not be as great as first imagined. Hence, it may be more accurate to suggest that the real shift in disputing behaviour we are observing may not be that from judicial ordering to private ordering, but rather a shift in the *form* of private ordering.

Indeed, it is possible to argue that the introduction of mandatory mediation results in an extension in regulatory control over the private ordering process. Mandatory mediation, as its name suggests, is a non-voluntary process that creates significant pressures to settle, and it is incorporated within the physical and ideological confines of the formal justice system. In addition, as Picard and Saunders point out in Chapter 11 of this volume, the selection and training of mediators who are qualified to offer mediation services within the program is likely to favour an evaluative or rights-based style of mediation and to privilege lawyer-mediators over those with a non-legal background. Overall, therefore, it appears that the introduction of court-connected mediation programs does not reflect a wholesale decentring of law, but rather a shift in the parameters of legal regulation of the disputing process, over which the legal system still continues to exert a strong normative influence.[38]

From Private Ordering to Community Empowerment

If the process reform wing of the alternative justice movement has defined its project of delegalization in terms of individual empowerment and private ordering, and thereby locates it within the comfortable confines of the liberal state, the same cannot be said for the more communitarian wing of the movement. Here the focus is not just on individual empowerment and self-determination, but more significantly on community empowerment and a shift in the language and value system associated with the idea of justice itself. Within the neighbourhood justice movement, for example, one of the primary objectives of creating neighbourhood justice programs has been to give justice back to the communities, in the sense of creating the conditions within which communities have the resources for responding to their own conflicts and thus for building stronger communities. Similarly, within the contemporary debate over Aboriginal self-government one of the major themes has been the costs to First Nations communities of the coercive imposition of a Euro-Canadian legal system onto Aboriginal legal cultures that defined their sense of community and their modes of responding to conflicts in very different terms from those of Anglo-Canadian criminal law.[39]

Within this communitarian wing of the alternative justice literature the theme of community empowerment is deployed in such a way that the primary obstacle to realizing the goal of more empowered communities is seen as the state. The state's coercive power to define conflicts as crimes, and to impose its own formalized, adversarial system of justice onto the disputants, is seen as fundamentally disempowering. Thus, commentators such as Nils Christie have argued that the formal apparatus of the state and the legal system have progressively robbed communities of the resources for resolving conflicts by transferring control over both the language and the process through which the dispute must be resolved to formal state agencies.[40]

Christie goes so far as to call lawyers professional thieves because of what he sees as their institutional role in disempowering disputants within the formal justice system. Similarly, Howard Zehr argues that true 'justice' is not to be found in the courts or in the sterile language of legal rights and duties, sanctions, and due process.[41] Such a conception of justice is inevitably impoverished, he argues, since it is articulated in terms of rights rather than responsibilities and ultimately relies on coercive sanctions to achieve justice rather than focusing on repairing harm and restoring relationships. To Zehr, justice needs to be defined in terms of reconciliation rather than punishment, and derives its authority from community norms and shared values, not from the force of law.

Like other restorative justice theorists, Zehr insists that much of what we now think of as offences against the state should be seen as breaches of trust rather than breaches of the peace.[42] The idea of trust is a powerful concept connoting a vision of society in which its members exist not as isolated atoms, each of whom may pursue his or her own self-interest provided only that they obey the law, but one in which all members of society are bound together by mutual obligations of trust and responsibility, any breach of which involves a harm to all, not just the immediate victim. From this perspective, retributive justice, or punishing the lawbreaker, may do little to repair the harm that the offender's breach of trust has caused. Instead, trust can only be restored by the offender taking responsibility for the harms his or her actions have caused to the community as well as to the victim. In this way, justice and peace are restored to the community, and in a manner that is unlikely to be achieved through the state's existing emphasis on coercive sanctions alone.

Restorative justice theorists therefore argue that it is time for a new approach to criminal justice reform that does not start by trying to fix the existing system, but instead seeks to transform it fundamentally. In this restorative justice approach, the crime belongs to the community and to the direct participants in the offence—the offender, the victim, and other community members in close relationships with offender and victim—rather than exclusively to the state. By recognizing and restoring the relationships that are harmed through the offender's wrongful actions, more will be accomplished in mending the conflict between them and in promoting the goal of reintegration of offenders back into their communities than can be achieved through the present justice system.[43]

Peacemaking and Reconciliation as Goals of the Alternative Justice Movement

Within the formal adjudicative process the discourse of rights and due process clearly is accorded primacy over the goal of reconciliation. Indeed, within the adversarial system it is hard to see any place for the idea of reconciliation to take root. If a dispute does go to adjudication, any legal remedy must be based on the legal claim made by the successful litigant. In a civil negligence trial, for example, the plaintiff must satisfy the evidentiary burden that the plaintiff suffered injury through the negligent conduct of the defendant. If the plaintiff fails to satisfy this burden, the plaintiff loses the case. In theory, at least, there is no room for compromise. The focus of the trial therefore tends to be on establishing the defendant's legal fault, rather than on investigating the plaintiff's needs. Even if the defendant is found to be legally responsible to the plaintiff, the defendant is not required to apologize or acknowledge any personal responsibility for the harm done. Instead, the legal sanction is considered a complete remedy for the legal wrong done to the plaintiff. The imposition of a legal sanction thus corrects the legal injury and resolves the legal dispute between the parties, but such a resolution may do little to promote reconciliation or to repair any private sense of loss or grief consequent upon the injury.

The emphasis on the fault of the accused, as opposed to the needs of the victim, is even more pervasive within the criminal trial process. Indeed, Zehr argues that the notion of harm has been distorted out of all recognition by the criminal justice system in its exclusive concern with fixing guilt and calculating punishment.[44] In the criminal trial, the offence is considered to have been committed against the state. Consequently, it is the state—in Canada, the Crown— that prosecutes the offence and is considered the victim. One consequence of this is that the victim's private experience of harm or injury is largely factored out of the criminal trial process, leaving many victims feeling disenfranchised and alienated.[45]

Zehr has suggested that even the technical language of legal 'guilt' does not correspond with the offender's subjective notions of responsibility for the harms his or her actions may have caused.[46] Once again, this is in part due to the erasure from the trial process of the victim's subjective experience of injury, which also has as a consequence the erasure of any opportunity for the offender to 'recognize' the harm done to the victim and to the community. This becomes evident where informal plea bargaining takes place, in the hope of obtaining a guilty plea from the accused in return for a lesser charge or a representation by the Crown for a lighter sentence. These 'negotiations' occur between the Crown and the accused, never between the victim and the accused. The plea bargaining process therefore does nothing to satisfy the goals of victim empowerment and accountability and, consequently, does very little to promote the broader goals of reconciliation and healing.[47]

The need to make the justice system more responsive to the demands of victims has helped to mobilize a whole range of institutional reform initiatives within the criminal justice system, from community policing to the development of victim support services programs. Perhaps most significant to the present discussion has been the symbolic force of this restorative justice discourse within the contemporary debates over Aboriginal justice reform and circle sentencing. For advocates of circle sentencing, such as Judge Barry Stuart, circles represent ways of empowering the community to respond collectively to the social problems caused by interpersonal violence and abuse in ways that the traditional sentencing process is unable to do. Rather than the state imposing a coercive sanction on the offender and taking the offender out of the community, the circle sentencing process encourages the offender to take responsibility for the harm caused by his or her actions, both to the immediate victim of the offence and to the community as a whole. By allowing the whole community to have a voice in the outcome of the dispute, it is hoped that the community will be able to provide more support for both victim and offender, thereby helping not only the individuals directly affected by the violence, but also the community, to heal.[48]

At the heart of the sentencing circle process is the desire to achieve reconciliation between offender and victim and between offender and the community. Likewise, the communitarian goals of reconciliation and peacemaking are central to other restorative justice initiatives such as victim-offender mediation programs[49] and family group conferencing.[50] For some critics of this restorative justice approach, however, the goal of reconciliation is inadequate to protect the rights of the victim, since it forces the victim to participate in a 'reconciliation' process in which the offender's needs are considered to be as important as the victim's.[51] In addition, the risk that a victim of violence will be further intimidated when forced to confront the abuser in a mediation process is a significant concern, raising the question of the physical and psychological safety of the victim in such circumstances.[52]

These critiques raise important questions for restorative justice theorists to consider. There is no doubt that the legal system has historically ignored the issue of male violence against women, often viewing domestic violence complaints as private disputes, rather than as criminal assaults, and thus treating offenders too leniently. For some feminist commentators, the state's recent embrace of the goals of peacemaking and reconciliation may amount to a subtle form of decriminalization, by which the state abdicates from its responsibility to protect women from male violence by placing the responsibility back onto women to prevent their own victimization.[53]

As Laureen Snider has observed, however, the logic of this position leads us towards ever more coercive state intervention in pursuit of retributive justice goals.[54] This has resulted in the implementation of zero-tolerance charging policies and the creation of new family violence courts, part of whose stated aim is to increase the rate of processing of male offenders.[55] Yet there is little evidence to show that increasing the rate of incarceration of offenders will do much to address the root sources of crime and violence in society or to ameliorate the experience of powerlessness and alienation on the part of many women forced to participate as victims in the criminal court process.[56] Indeed, Diane Martin has argued that despite its efforts at transformative social change, feminist victims' advocacy claims in the arena of domestic violence have been most successful where their claims have resulted in a strengthening of the existing coercive powers of the criminal justice system.[57] The irony of the feminist critique of restorative justice is thus that it takes us back full circle to a rejection of informalism and an implicit reliance on the state's formal monopoly over the definition of 'justice'.

Is There an Alternative to Alternative Justice?

Having looked briefly at the main discursive themes in the ideology of alternative justice, we are confronted with the question, is there any alternative to alternative justice? At one level the answer to this question is obvious. Not even the most ardent supporter of community justice programs is proposing that the formal justice system should be abolished. This suggests that there will always be a need for the formal adjudicative process, notwithstanding any inroads made on the formal legal system by the alternative justice movement. Not every dispute can be settled through consensus. Not every legal wrong can be made right again. Not every relationship can be reconciled. Not every fault can be pardoned. For those legal conflicts that cannot be settled through consensus and for those legal relationships that have been irreparably harmed, the formal justice system still provides an avenue of last resort, a place where the disputants can always find 'justice', albeit in a very formalized, adversarial, and arguably coercive form.

The more significant question, however, is not whether the formal justice system will remain, but whether it can resist being transformed by the colonizing project of the alternative justice movement. For the present, at least, it does appear that the advantage lies with the alternative justice movement. In the civil law arena, and with increasing confidence in the field of criminal law, in family law, in public environmental dispute resolution, in small claims disputing, and even in the field of human rights litigation, the onward momentum of the alternative justice movement appears to be carrying all before it, breaking down the barriers erected by a resistant legal profession and entering even into the hallowed halls of justice.

The question, then, is: What does this expansion of the institutional scope of the ADR movement mean in practice? Does it mean that many legal professionals will cast aside their litigious inclinations and don the more temperate garb of the mediator? For many legal practitioners, this may not involve so great a stretch as perhaps it first appears. As the process reform theorists observe, adjudication has never claimed to exercise a monopoly of the disputing process, even within the formal justice system. On the contrary, informal settlement techniques based on negotiation and mediation have always been part of the lawyer's arsenal, and these techniques are frequently used by legal professionals in areas as diverse as labour disputes, family law and divorce, commercial disputes, and insurance claims. The shift towards institutionalizing negotiation and mediation-based settlement processes in many of these areas of law, therefore, is not likely to lead to any fundamental changes in the nature of legal practice. Nor is such a shift likely to make the process of settlement any less rights-based in orientation.

What of the judicial function itself? Here it seems as if more profound changes may be underway. For some commentators, the introduction of case management and pretrial settlement conferences marks a dramatic shift in how courts conduct litigation. Thus, Mark Galanter has argued that the introduction of pretrial settlement conferences into the civil litigation process involves a significant change in the role of the trial judge, with many judges acting as informal mediators between the parties in the hopes of arriving at a settlement.[58]

Similarly, in the field of family law, with the shift away from a fault-based system of divorce, there have been corresponding changes in the role of judges towards a greater emphasis on conciliation and problem-solving.[59] Federal and provincial legislative reforms have increased the scope for private ordering on divorce, in the hope of encouraging speedier and less contentious divorce proceedings. This extends not only to financial matters, such as property division and spousal support entitlements, but also to custody and access arrangements, over which the court retains a residual supervisory function.

Even in exercising this supervisory function, however, the traditional judicial function appears to be changing, with many judges relying on court-appointed assessors to determine the best interests of the child(ren) and the capacities of the parents to respond to those interests, rather than relying only on evidence presented by the parents in a formal adversarial proceeding. This has led at least one commentator to observe that the family law adjudication process is being transformed from within and increasingly incorporating the ideology of the family mediation movement.[60]

In other respects, however, the reverse process seems to be in evidence, with the alternative justice movement increasingly compromised by its institutional proximity to the formal justice system, despite its ideological claims to provide a radically different form of justice. Harrington and Merry, in their study of community mediation, examined a number of American programs. They found that those programs without close connections to the court system or well-developed court referral procedures tended to experience lower caseloads.[61] This suggests that the 'success' of the alternative justice movement in attracting clients may depend, at least in part, on its affiliations with the formal justice system. At the same time, however, the creation of mandatory court-connected mediation programs, by effectively removing the decision from disputants as to whether to select mediation or adjudication, arguably violates the very principles of self-determination and consent around which the ideology of alternative justice has been constructed.[62]

Finally, Sally Engle Merry, in a study comparing different forms of mediation, concluded that not all alternative justice programs were necessarily created equal. Merry shows that some ADR processes operate at what she calls the 'top end' of the market, providing financially well-endowed disputants with access to alternative justice processes, such as mini-trials, as either an alternative to expensive litigation or as an aid to preparing a case for litigation. The main users of these expensive, private ADR services were large public- and private-sector organizations involved in complex, often multi-party disputes, for example over commercial, environmental, patent, or anti-trust issues.[63]

On the other hand, the main users of publicly financed, 'low-end' mediation services were lower-income clients, the working poor, and those without the resources to afford a lawyer.[64] For these clients, the choice of mediation or the courts did not mean the same thing as for the users of the top-end private mediation services. In fact, Merry suggests that the increased focus on diverting disputes from the courts through mandatory mediation programs operates to restrict further the less-privileged client group from obtaining access to the courts, thus reserving the formal adjudication process increasingly only for those who can afford to pay for it.[65] Far from improving access to justice, therefore, Merry's research raises the question that the alternative justice movement may in fact reproduce many of the same kinds of concerns about access to justice that it was designed to address.

In conclusion, then, it seems as if we cannot say with any certainty that the formal justice system will capture the alternative justice movement, or that the alternative justice movement will effect a transformation in legal disputing. Perhaps the only real conclusion we can come to is that despite the apparent differences between the ideology of formal justice and the ideology of alternative justice, these two forms of ideological discourse coexist in a mutually constitutive relationship, with each sustaining the other and yet defining its identity in opposition to the other.[66] If the boundaries between these two forms of justice discourse seem to be shifting at present, in favour of a greater scope for alternative justice and away from an overreliance on formal state law, this does not mean that the dance is over.

Notes

1. See W.A. Bogart, *Courts and Country: The Limits of Litigation and the Social and Political Life of Canada* (Toronto: Oxford University Press, 1994), 73–93.
2. See Peter S. Adler, 'The Future of Alternative Dispute Resolution: Reflections on ADR as a Social Movement', in Sally Engle Merry and Neal Milner, eds, *The Possibility of Popular Justice* (Ann Arbor: University of Michigan Press, 1995), 67–87; see also Richard Abel, ed., *The Politics of Informal Justice*, 2 vols (New York: Academic Press, 1981).
3. See Jerold S. Auerbach, *Justice Without Law?* (New York: Oxford University Press, 1983); Raymond Shonholtz, 'Justice From Another Perspective: The Ideology and Developmental History of the Community Boards Program', in Merry and Milner, eds, *The Possibility of Popular Justice*, 201–39; see also Christine B. Harrington, 'Delegalization Reform Movements: A Historical Analysis', in Abel, ed., *The Politics of Informal Justice*, vol. 1, 35–71.
4. H.H. Irving and M. Benjamin, *Family Mediation: Theory and Practice of Dispute Resolution* (Toronto: Carswell, 1987), 39–41; see also Martha Bailey, 'Unpacking the "Rational Alternative": A Critical Review of Family Mediation Movement Claims', *Canadian Journal of Family Law* 8 (1989): 61–96.
5. See Nils Christie, 'Conflicts As Property', *British Journal of Criminology* 17 (1977): 1–12; Shonholtz, 'Justice from Another Perspective'; John Paul Lederach and Ron

Kraybill, 'The Paradox of Popular Justice', in Merry and Milner, eds, *The Possibility of Popular Justice*, 357–78.
6. See Frank E.A. Sander, 'Alternative Methods of Dispute Resolution: An Overview', *University of Florida Law Review* 37 (1985): 1–18; Allan Stitt, Francis Handy, and Peter A. Simm, 'Alternative Dispute Resolution and the Ontario Civil Justice System', in *Rethinking Civil Justice: Research Studies for the Civil Justice Review*, vol. 2 (Toronto: Ontario Law Reform Commission, 1996), 449–90. See generally, Stephen B. Goldberg, Eric D. Green, and Frank E.A. Sander, *Dispute Resolution* (Boston: Little, Brown, 1985).
7. See Richard Abel, 'The Contradictions of Informal Justice', in Abel, ed., *The Politics of Informal Justice*, vol. 1, 267–320; R. Hofrichter, *Neighbourhood Justice in Capitalist Society: The Expansion of the Informal State* (New York: Greenwood Press, 1987); R. Tomasic and M. Feeley, *Neighborhood Justice: Assessment of an Emerging Idea* (New York: Longman, 1982).
8. Frank E.A. Sander and Stephen B. Goldberg, 'Fitting the Forum to the Fuss: A User-Friendly Guide to Selecting an ADR Procedure', *Negotiation Journal* (1994): 49–64; D. Paul Emond, 'Alternative Dispute Resolution: A Conceptual Overview', in Emond, ed., *Commercial Dispute Resolution* (Aurora, Ont.: Canada Law Book, 1986), 1–25; Julie Macfarlane, 'An Alternative to What?', in Macfarlane, ed., *Rethinking Disputes: The Mediation Alternative* (Toronto: Emond Montgomery, 1997), 1–21. See also Stitt et al., 'Alternative Dispute Resolution'.
9. Cf. Lon L. Fuller, 'The Forms and Limits of Adjudication', *Harvard Law Review* 92 (1979): 353–409; Lon L. Fuller, 'Mediation—Its Forms and Functions', *Southern California Law Review* 44 (1977): 305–39.
10. Sander and Goldberg, 'Fitting the Forum to the Fuss'; Stitt et al., 'Alternative Dispute Resolution'.
11. Laura Nader, 'The ADR Explosion—The Implications of Rhetoric in Legal Reform', *Windsor Yearbook of Access to Justice* 8 (1988): 269–91.
12. See, e.g., Howard Zehr, *Changing Lenses* (Scottdale, Penn.: Herald Press, 1990); Daniel Van Ness and Karen Heetderks Strong, *Restoring Justice* (Cincinnati: Anderson Publishing, 1997). For a good discussion of this restorative justice approach, see Jennifer J. Llewellyn and Robert Howse, *Restorative Justice—A Conceptual Framework* (Ottawa: Law Commission of Canada, 1998).
13. The concept of alternative justice 'stories' is borrowed from Robert A. Baruch Bush and Joseph P. Folger's influential book, *The Promise of Mediation* (San Francisco: Jossey-Bass, 1994), 15.
14. See also Christine B. Harrington and Sally Engle Merry, 'Ideological Production: The Making of Community Mediation', *Law and Society Review* 22 (1988): 717–21.
15. See also the discussion of reflexive regulation and 'thin proceduralism' in Swan, Chapter 6, this volume.
16. Steven Lukes, *Individualism* (Oxford: Blackwell, 1973), 54, 55.
17. Ibid., 55, 56, 59, 63. See also J.S. Mill, 'On Liberty', in Mill, *Utilitarianism*, ed. H.B. Acton (London: Everyman's Library, 1972), 80–1.
18. See, e.g., Christopher W. Moore, *The Mediation Process*, 2nd edn (San Francisco: Jossey-Bass, 1996), 7; Cheryl A. Picard, *Mediating Interpersonal and Small Group Conflict* (Ottawa: Golden Dog Press, 1998), 8; Emond, 'A Conceptual Overview', 21; Goldberg et al., eds., *Dispute Resolution*, 8.

19. See Robert A. Baruch Bush, 'Mediation and Adjudication, Dispute Resolution and Ideology: An Imaginary Conversation', *Journal of Contemporary Legal Issues* 3, 1 (1989): 1–35.
20. Adler, 'The Future of Alternative Dispute Resolution', 68.
21. Bush and Folger, *The Promise of Mediation*, 31–2; see also Joseph P. Folger and Robert A.B. Bush, 'Ideology, Orientations to Conflict, and Mediation Discourse', in Folger and Tricia S. Jones, *New Directions in Mediation* (Thousand Oaks, Calif.: Sage Publications, 1994), 3–25.
22. See Emond, 'Alternative Dispute Resolution', 8.
23. Unless one of the parties appeals the ruling to an appellate court.
24. Emond, 'Alternative Dispute Resolution', 8, 9.
25. For a useful discussion of these different understandings of power relations within the dispute resolution literature, see Genevieve A. Chornenki, 'Mediating Commercial Disputes: Exchanging "Power Over" For "Power With"', in Macfarlane, ed., *Rethinking Disputes*, 163–5.
26. Ibid.
27. Ibid. For a classic statement of the theory of mutual gains in negotiation theory, see Roger Fisher and William Ury, *Getting to Yes: Negotiating Agreement Without Giving In*, 2nd edn (New York: Penguin Books, 1991).
28. Esser has argued that the insistence on the supposedly inherent distinctiveness of these dispute resolution processes within much of the research literature amounts to a 'new formalism', which in turn has a distorting effect on what we know about disputing behaviour and institutions. See John P. Esser, 'Evaluations of Dispute Processing: We Do Not Know What We Think and We Do Not Think What We Know', *Denver University Law Review* 66 (1989): 499–547; see also Susan Silbey and Austin Sarat, 'Dispute Resolution in Law and Legal Scholarship: From Institutional Critique to the Reconstruction of the Juridical Subject', *Denver University Law Review* 66 (1989): 437–98.
29. Robert H. Mnookin, 'Bargaining in the Shadow of Law: The Case of Divorce', *Current Legal Problems* (1979): 65–105; Marc Galanter, 'Reading the Landscape of Disputes: What We Know and Don't Know (And Think We Know) about Our Allegedly Contentious and Litigious Society', *U.C.L.A. Law Review* 31 (1983): 4–71.
30. Mnookin, 'Bargaining in the Shadow', 78, 79.
31. See, e.g., Mark Galanter, 'Why the "Haves" Come Out Ahead: Speculations on the Limits of Legal Change', *Law and Society Review* 9 (1975): 95–160. See also Owen M. Fiss, 'Against Settlement', *Yale Law Journal* 93 (1984): 1073–1190.
32. Fiss, 'Against Settlement'; Galanter, 'Why the "Haves" Come Out Ahead'. See also Trina Grillo, 'The Mediation Alternative: Process Dangers for Women', *Yale Law Journal* 100 (1991): 1545–1610; Martha Shaffer, 'Divorce Mediation: A Feminist Perspective', *University of Toronto Faculty of Law Review* 46 (1988): 162–200.
33. See *The First Report of the Ontario Civil Justice Review* (Toronto: Ontario Civil Justice Review, Mar. 1995), 4–5. See also Stitt et al., 'Alternative Dispute Resolution in the Ontario Civil Justice System', 469–72. For a good general discussion of court-connected mediation initiatives in Canada, see Genevieve A. Chornenki and Christine E. Hart, *Bypass Court*, 2nd edn (Toronto: Butterworths, 2001), 175–84.
34. See *Practice Direction Concerning the Alternative Dispute Resolution Pilot Project in the Ontario Court (General Division)* (1994), 16 O.R. (3d) 481; *Practice Direction*

Concerning the Alternative Dispute Resolution Project in the Ontario Court (General Division) (1994), 24 O.R. (3d) 161; Chornenki and Hart, Bypass Court, 179-81.
35. See Evaluation of the Ontario Mandatory Mediation Program (Rule 24:1): Final Report—The First 23 Months (Toronto: Queen's Printer, 2001), 34-54.
36. Ibid., 91-101; see also Chornenki and Hart, Bypass Court, 179-81.
37. Evaluation of the Ontario Mandatory Mediation Program, 63; G. Watson et al., eds, Dispute Resolution and the Civil Litigation Process (Toronto: Emond Montgomery, 1991), 3; Peter Bowal, 'The New Ontario Judicial Alternative Dispute Resolution Model', Alberta Law Review 34 (1995): 207. See generally Marc Galanter and Mia Cahill, '"Most Cases Settle": Judicial Promotion and Regulation of Settlements', Stanford Law Review 46 (1994): 1339-91.
38. See Purvis, Chapter 2, this volume; see also Silbey and Sarat, 'Dispute Resolution in Law and Legal Scholarship'.
39. See Dickson-Gilmore, Chapter 15, this volume. See also People to People, Nation to Nation: Highlights from the Report of the Royal Commission on Aboriginal Peoples (Ottawa: Minister of Supply and Services, 1996).
40. Christie, 'Conflicts As Property'.
41. Zehr, Changing Lenses, esp. 177-214. See also Van Ness and Strong, Restoring Justice; Llewellyn and Howse, Restorative Justice.
42. Zehr, Changing Lenses, 181.
43. John Braithwaite, Crime, Shame and Reintegration (Cambridge: Cambridge University Press, 1989). See also Satisfying Justice (Ottawa: Church Council on Justice and Corrections, 1996); Burt Galaway and Joe Hudson, eds, Restorative Justice: International Perspectives (Monsey, NY: Criminal Justice Press, 1996); Herman Bianchi, Justice as Sanctuary (Bloomington: Indiana University Press, 1994).
44. Zehr, Changing Lenses, 66-74.
45. Ibid., 81-2.
46. Ibid., 67, 69-70.
47. Ibid., 181. See also Braithwaite, Crime, Shame and Reintegration, whose theory of reintegrative shaming has been very influential in the new restorative justice literature.
48. Judge Barry Stuart, 'Sentencing Circles: Making "Real" Differences', in Macfarlane, ed., Rethinking Disputes, 201-32; also Barry Stuart, 'Sentencing Circles: Turning Swords into Ploughshares', in Galaway and Hudson, eds, Restorative Justice, 193-206.
49. See, e.g., Mark S. Umbreit, Victim Meets Offender: The Impact of Restorative Justice and Mediation (Monsey, NY: Criminal Justice Press, 1994).
50. Joe Hudson, Allison Morris, Gabrielle Maxwell, and Burt Galaway, eds, Family Group Conferences (Monsey, NY: Criminal Justice Press, 1996); also Galaway and Hudson, eds, Restorative Justice.
51. See Lisa G. Lerman, 'Mediation of Wife Abuse Cases: The Adverse Impact of Informal Dispute Resolution on Women', Harvard Women's Law Journal 7 (1984): 57-113; E.D. Laroque, 'Violence in Aboriginal Communities', in M. Valverde, L. MacLeod, and K. Johnson, eds, Wife Assault and the Canadian Criminal Justice System: Issues and Policies (Toronto: University of Toronto Centre of Criminology, 1995). See also Carol LaPrairie, 'Community Justice or Just Communities?: Aboriginal Communities in Search of Justice', Canadian Journal of Criminology (1995): 521-45.

52. See Lerman, 'Mediation of Wife Abuse Cases'; N. Zoe Hilton, 'Mediating Wife Assault: Battered Women and the "New Family"', *Canadian Journal of Family Law* 9 (1991): 29–53. For similar concerns about the risks of divorce mediation for women, see Grillo, 'The Mediation Alternative'; Shaffer, 'Divorce Mediation'.
53. LaPrairie, 'Community Justice or Just Communities?'; also Lerman, 'Mediation of Wife Abuse Cases'; Julie Stubbs, '"Communitarian" Conferencing and Violence Against Women: A Cautionary Note', in Valverde et al., eds, *Wife Assault*.
54. Laureen Snider, 'Feminism, Punishment and the Potential for Empowerment', in Valverde et al., eds, *Wife Assault*, 237. See also Dianne L. Martin, 'Retribution Revisited: A Reconsideration of Feminist Criminal Law Reform Strategies', *Osgoode Hall Law Journal* 36 (1998): 151–88; Stuart A. Scheingold and Toska Olson, 'Sexual Violence, Victim Advocacy, and Republican Criminology: Washington State's Community Protection Act', *Law and Society Review* 28 (1994): 729–63.
55. See E. Jane Ursel, *Winnipeg Family Violence Court Evaluation* (Ottawa: Department of Justice, 1995).
56. See L. MacLeod, 'Policy Decisions and Prosecutorial Dilemmas: The Unanticipated Consequences of Good Intentions', in Valverde et al., *Wife Assault*; see also Michelle Berthiaume, 'The Influence of Victims' Advocacy on Criminal Justice Policy: The Case of Caveat', MA thesis (Carleton University, 2001), 38–53.
57. Martin, 'Retribution Revisited', 155; see also Laureen Snider, 'The Potential of the Criminal Justice System to Promote Feminist Concerns', *Studies in Law, Politics and Society* 10 (1990): 143–72; Snider, 'Feminism, Punishment and the Potential for Empowerment'.
58. Marc Galanter, ' ". . . A Settlement Judge, not a Trial Judge": Judicial Mediation in the United States', *Journal of Law and Society* 12 (1985): 1–18; also Galanter and Cahill, ' "Most Cases Settle" '. For a discussion of case management and its implications for the Ontario civil justice system, see Kent Roach, 'Fundamental Reforms to Civil Litigation', in *Rethinking Civil Justice: Research Studies for the Civil Justice Review*, vol. 2 (Toronto: Ontario Law Reform Commission, 1996), 429–40.
59. See Peter H. Russell, *The Judiciary in Canada: The Third Branch of Government* (Whitby, Ont.: McGraw-Hill Ryerson, 1987), 223–33.
60. See Elizabeth Pickett, 'Familial Ideology, Family Law and Mediation: Law Casts More Than a Shadow', *Journal of Human Justice* 3, 1 (1991): 34–40.
61. See Harrington and Merry, 'Ideological Production', 721.
62. For a discussion of the implications of mandatory versus voluntary court-connected mediation programs, see Stitt et al., 'Alternative Dispute Resolution and the Ontario Civil Justice System', 469–72.
63. Sally Engle Merry, 'Varieties of Mediation Performance: Replicating Differences in Access to Justice', in Allan Hutchinson, ed., *Access to Civil Justice* (Toronto: Carswell, 1990), 261, 264.
64. Ibid., 261.
65. Ibid., 271–3. See also Nader, 'The ADR Explosion'.
66. See Jerold S. Auerbach, *Justice Without Law?*, 14; see also Peter Fitzpatrick, 'The Rise and Rise of Informalism', in Roger Mathews, ed., *Informal Justice* (London: Sage, 1988), 178–98; Fitzpatrick, *The Mythology of Modern Law* (London: Routledge, 1992), 169–70.

Chapter 11

The Regulation of Mediation

CHERYL A. PICARD AND R.P. SAUNDERS

> The central quality of mediation is its capacity to reorient the parties towards each other, not by imposing rules on them, but by helping them to achieve a new and shared perception of their relationship, a perception that will redirect their attitudes and disposition toward one another.[1]

Introduction

Disputes in any society are a given; how they are resolved, however, depends on a number of social, legal, economic, historical, and cultural factors particular to the societal setting in which they arise. One of the salient features of this setting is the regulatory framework within which the various parties, rights, and interests interact and play themselves out through both the established and emerging processes. In the Canadian context, how disputes are resolved is similarly a product of the many domestic factors that make up our dispute resolution landscape. As in other societies, while a multitude of everyday disputes are settled by the parties themselves without recourse to any outside intervention, others require a greater degree of effort and attention from third parties and from processes that play some role in the resolution—or the attempt at resolution—of the conflict. It is those disputes negotiated through some more or less recognized mechanism or process that concern us here. More particularly, our attention is focused on the larger framework within which these attempts at resolution are made, in other words, on the issues surrounding the governance of the agents and processes involved.

The variety of processes and mechanisms is great. They can be classified along a continuum from the formal legal arena of lawyers, legal rules and rights, and adjudication to the more informal and voluntary interventions where the parties are assisted in reaching a resolution to their problems. In between, a range of processes, practices, and techniques reflects variations in the degree of formality, voluntariness, party participation, third-party decision-making, and resort to legal rules. Traditionally in law, this continuum of dispute resolution methods has been thought of as a *progression* from relative arbitrariness to highly principled adjudication. This view has long been recognized as an oversimplified characterization of the nature of the continuum, for it fails to take account of the different types of interests, rights, and norms at work in the various processes. However, both subtle and not-so-subtle claims to superiority remain entrenched in the literature and serve as an important backdrop against which to examine the burgeoning area of dispute resolution. The tenacity and the hold of these claims by traditional legal processes cannot be ignored, especially when looking at what many see as the inevitable arrival of regulatory frameworks surrounding dispute resolution.

In this chapter, our attention is focused specifically on the questions and concerns surrounding the governance of one category of dispute resolution processes: mediation. While other methods of resolving disputes are important—perhaps more important in some contexts—mediation has attracted a great deal of attention and support, and a large number of advocates and practitioners, in the last two decades. There has been a great deal of debate over both the ultimate potential and value of mediation as a method of dispute resolution and, more recently, regarding the techniques and methods of the actual practice. This debate has centred on the basic question of *who should be doing mediation*, and has taken different forms and approaches. For example, a substantial body of work has examined in some detail the varieties of education, training, and experience that might (or might not) produce a good mediator.[2] In addition, a small but growing body of literature considers the question of credentials for mediators and the possibility of imposing standards or entry requirements for 'professional' status on those who would call themselves mediators.[3] Our interest in this chapter is to refocus and expand this debate over who should be doing mediation by shifting attention to the potential regulatory framework and consequent regulators of mediation, and to argue that this critical focus can be neither ignored nor marginalized.

Our argument is that the debates to date surrounding mediation have been too narrowly constructed and that consideration must be given not only to the processes and credentials themselves—something we can see in the current debates—but also to the type of governance that might develop and the consequent dangers and possibilities. How these processes of mediation are controlled and by whom will play a critical role in the fulfillment of what many see as their ultimate potential as mechanisms of conflict resolution. As we shall show, this potential for mediation as a transformative, empowering, and communitarian form of conflict resolution—in contrast to the alienating and individualistic traits of more formal and legalistic forms of dispute resolution—has been advanced for some time now. The role that schemes of governance might play in achieving or thwarting this potential is one of our principal concerns. It is crucial to take account of the issue of who or what will regulate the field of mediation and determine the philosophy and direction of the practice.

Moreover, how disputes are resolved and in what type of state-sanctioned process tells us much about the value of the process itself in society and, in turn, something about the power relationships and arrangements within that particular society. Whether from a conventional viewpoint or from a more critical stance, the dispute-resolving process points to something larger going on—that something may be policy setting; it may be about access and who has that access; it may be about power and the ability to decide fates. In any event, the issue of dispute resolution and its governance (whether by the state, the practitioners themselves, or some select group) provides important signposts both to the potency of such processes as solvers of conflict in the long and short term and also to the distribution of power within societies.

Finally, the role of the state in the governance of mediation processes and its consequent role in the provision of 'justice' is important, if not determining, and engages one of the central issues in this book. Specifically, our conclusion on the activity of the state in Canada in regard to the regulation of mediation resonates well with Purvis's conclusions in Chapter 2 on the continuing and important role of the state in the regulation of civil society, albeit often through new methods and often less direct means. The case of mediation strongly supports these arguments in linking the state's actions in relation to the governance of mediation to the establishment of a model of mediation practice best suited to the state's interests and ends.

The Meanings and Practice of Mediation

In its simplest form, mediation can be defined as a process of assisted or facilitated negotiation. It involves the intervention of a 'non-partisan third party, whose authority rests on the consent of the parties to facilitate their negotiations; a mediator has no independent decision-making power beyond what the parties afford.'[4] Mediation has been predicated upon a number of assumptions:

1. It is voluntary (although this is changing with the advent of mandatory mediation programs).
2. It is non-adversarial.
3. It is facilitated by an impartial and neutral third party.
4. The parties in mediation have relatively equal bargaining power.
5. The main function of the mediator is to empower parties to reach a mutually satisfactory agreement.

As we shall show later, however, different meanings, goals, and possibilities have been ascribed to the practice. But first it is necessary to provide a brief background on the development of mediation.

Mediation is one of the oldest and most common forms of conflict resolution.[5] Like many modern concepts and practices, it is an adaptation of a variety of practices and traditions from other countries and other times. For example, in ancient China, mediation was the principal means of resolving disputes and it is still widely practised today. In Japanese law and customs mediation has a rich history. In parts of Africa the *moot* or neighbourhood meeting has long provided an informal mechanism for resolving interpersonal disputes. In England arbitration and mediation were used as early as the Anglo-Saxon period. In fact, mediation has always been seen as a 'natural' way to deal with conflict. Extended families, elders, clan members, religious leaders, friends, and neighbours all offered their wisdom to assist in the resolution of a wide array of social conflicts.

The transplanting of alternative dispute settlement systems from Europe to North America is believed to have come by way of the Quakers. Their settlement procedures, which handled a variety of disputes ranging from commercial transactions to marital disagreements, coexisted with the English system of law. This provided disputants with a choice about how to deal with their disputes. The use of mediation was not unique to the Quakers. The Dutch and Scandinavian settlers also brought with them a tradition of private dispute resolution. It was not long before mediation and arbitration became commonplace. The social role played by these private systems within their communities was important in 'preserving the harmony of the community'.[6] This desire for harmony continues to dominate contemporary discourse on mediation. Some argue, however, that this harmony is more repressive than liberating.[7]

The history of mediation in this country is entwined with values such as justice, self-determination, and the acceptance of humans as rational and capable of solving problems. Rooted in social activism, proponents of mediation sought to assist individuals and groups to use non-violent and more effective problem-solving strategies. Interest in mediation was also tied to attempts at legal reform based on growing public concern about fair and effective dispute resolution as well as concerns about court costs and delays. Several inquiries into public access to justice encouraged the use of alternative dispute resolution, including the Zuber Inquiry[8] in 1987, which recommended sweeping changes to Ontario courts, the Daubney Report[9] to the House of Commons in 1988, and the BC Justice Reform Committee Report,[10] also in 1988.

Important considerations in the growth of mediation in Canada included the voluntary nature of mediation (parties could leave the table if they chose to), the quality of the outcomes achieved, and the frequent inability of the courts and the adversarial process to produce adequate solutions to interpersonal problems. Many cases before the courts involved people who knew each other or who were involved in an ongoing relationship. By the time disputes between neighbours, friends, relatives, or landlords and tenants reached the courts, there were often long and complex histories of disagreements and interpersonal tension. The courts provided little opportunity for a thorough examination of the problems experienced by the parties and could only rule upon the legal merits of the case; it was the incident and not the relationship that took centre stage. While one side 'won' in court, both parties were often dissatisfied with the outcome and the underlying issues and tensions went unresolved. The hope for better resolutions and more control over and involvement in the *process* of resolution spurred the growth of alternative forms of conflict resolution, including mediation.

The growth in mediation was also driven by the many practitioners drawn to a field offering the potential of bettering society and serving the public good by restoring harmony to communities and by transforming (or supplanting) legal institutions. This general sense of purpose continues to motivate many of those entering and working in the field; motives based on self-interest or monetary gains are often secondary. Many practitioners (including lawyers, social workers, business people, mental health counsellors, and judges) have left behind other professions or jobs in order to provide mediation services and to do what they regard as more satisfying and productive work.

The resulting growth and popularity of mediation as an alternative to more formal methods of dispute resolution pose several challenges for both practitioners and the state. For practitioners, the principal tension in recent years has revolved around the purposes of mediation: some stress the goals of efficient case management, cost savings, and timely settlement (an evaluative model), while others focus on a facilitative and transformative model that stresses client needs, communitarian concerns, and social change. This tension has manifested itself in different ways, but most recently it has arisen in the debates surrounding the 'credentialing' of mediators. For the state, the issue has been the uses to which the technique is put, and this has manifested itself in concerns surrounding the control and regulation of *who* is sanctioned to do the work and of what *type* of work is being done. This latter concern is important in that it addresses, albeit in an indirect manner, the question of what model of mediation will be dominant in practice.

The debate over the potential of mediation is well documented and thoroughly discussed elsewhere.[11] Some writers argue for a focus on the transformative potential of mediation and its challenge to the existing mechanisms of dispute resolution and perhaps even to the power balances represented within those mechanisms. Others concentrate on the technical aspects of mediation as offering superior forms of dispute resolution through cheaper, faster, and more efficient techniques. While the former concentrate on values, the client's process experience, client needs and satisfaction, and community harmony, the latter often stress individual rights, outcomes, quality of advice, and mediation's role as an adjunct mechanism to the formal system.

One of the other factors to note is the question of what the clients want or expect from mediation. Here the answer is clear. The research is consistent in telling us that mediation produces high levels of satisfaction and compliance. In a study by Cook, Roehl, and Sheppard, 80 to 89 per cent of disputants of diverse criminal and civil disputes were satisfied with the mediator, the terms of agreement, and the mediation process.[12] Similar findings have been found in other studies.[13] But one of the most striking and significant findings is that parties' favourable attitudes stem largely from how the process works, not from the outcome of the process. Two features have been identified as being responsible for this: (1) the degree of participation in decision-making that the parties experience, and (2) the fuller opportunity to express themselves and communicate their views to all present.[14]

Regulating Mediation

The debate over the regulation of mediation has arisen in an indirect manner. As noted above, the recent focus has been on the issue of credentials, on what to require when it comes to sanctioning mediators and how to accomplish this. The discussion has focused less directly on the issue and value of regulation per se, given the controversy generated by talk of restricting entry into a field that traditionally has prided itself on 'people skills' and accessibility. What started in the mid-1960s as a move to 'deprofessionalize' legal institutions for problem-solving and dispute settlement now appears to many critics to have come full circle.

Throughout the 1990s and continuing today, there has been a heated debate on 'professionalizing' mediation, a debate that in North America revolved around qualifications, standards of practice, and certification. This debate within mediation reflects the range of interests and practices at work:

> at one extreme are the neighbourhood centres espousing voluntarism, self-help and peer relationships. At the other end are highly trained professionals who want to make a decent livelihood carving out a niche in the professional world of help somewhere between career diplomats, organizational consultants, lawyers and therapists. Clustered along both ends are differing views of charging fees and differing perspectives on credentialing.[15]

Discussions of credentialing for practising mediators induce strong reactions, both positive and negative. While applauding the field's growth in stature, some find the discourse about credentialing troubling and fear that elitism may threaten personal and social empowerment. They argue that mediation demystifies and deinstitutionalizes formal settlement mechanisms. They believe that, through mediation, the resolution of conflict is drawn away from 'professionals' and returned to those most affected by it, thereby empowering participants and stimulating social transformation. The fear is that the apparent, and some would say inevitable, move to professionalize the field may place individual and collective empowerment at risk.

A number of questions arise with the move to professionalization. Should standards be set, and if so, by whom? What qualifies a person to practise as a mediator? How do we assess mediator competency? What initial and ongoing training is required? Who should govern the credentialing of mediators? There is no consensus on the answers to these questions. Depending on the forum, there are guidelines describing mediator qualifications and some standards to govern the process, but usually they are advisory and not mandatory. While some mediators argue that it is premature to focus on questions of credentials and that doing so will hinder the development of the field, others believe that the reality is that legislators, judges, and government agencies are already deciding who may and may not mediate. The issues are contentious, and for good reason. The creation of an organized group or subculture that would govern and limit access to the field warrants intense scrutiny and the fear that mediation may become exclusive and elitist (as has happened in the legal and some health professions) is justified.

Fuelling the call for setting credentials is the prevalence of several myths: mediation is 'easy'; non-mediator trainers can teach mediation skills; and the opportunity for mediation is more important than the quality of service.[16] These myths raise concerns about how to stop practitioners and trainers without 'approved' credentials and with little or no hands-on mediation experience from hanging out mediation shingles. Various commissions in Canada and the United States have been struck to report on these and other questions.[17] As yet, neither their findings nor their recommendations have been endorsed by the field. Those advocating for setting criteria for practising mediators argue that professional accreditation would protect the consumer and the integrity of the mediation, while opponents maintain that inappropriate barriers for entry into the field would be created and that broad dissemination of peacemaking skills in society at large could be hampered.

The fear of creating a monopoly lies at the heart of the credentials debate. Certification and licensing would have significant implications not only for mediators who deal with family, corporate, public policy, and international disputes, but also for those working in neighbourhood centres, schools, and other community-based programs. There is the concern that services now offered by these groups will no longer be recognized as legitimate, causing public favour and funding to be lost. Non-adversarial dispute settlement options, it is argued, would be available only to the elite and the wealthy—one of the problems with formal justice systems that mediation innovators set out to change.

Two points can be made about the debates that have formed to date. The first has to do with the tenor or the nature of the discourse. That tenor has taken on a distinctly narrow and legalistic tone that focuses on agents, techniques, and processes, and the access to them, not on the wider purposes or goals of those processes or techniques. Importantly, it uses legal arenas of dispute resolution as the reference point against which the practices and agents of mediation are assessed. The point is that there are wider and more important questions surrounding the governance of mediation than who controls the entry and exit of certain practitioner agents and the standards of evaluation.

The second point is that in the absence of any consensus on the part of practitioners and without an explicit regulatory scheme, informal credentialing has been occurring through the practices of the state in a number of its programs, notably in mandated mediation programs such as that of Ontario (discussed below). The danger is twofold: (1) that the effective direction of the field will be decided in large part by the dictates and needs of the state, of the legal profession, and of the formal legal institutions that mediation was intended to supplant; and (2) that formal regulatory schemes, when they arrive, will follow the practices and directions already firmly in place. This second point requires some elaboration.

It is clear from the North American experience in recent years that interest (if not a preoccupation) in alternate processes of dispute resolution in general and in mediation in particular is increasing on the part of various governments and of the legal profession. In Canada, legislation providing for the use of mediation is relatively new and most of it is silent on the issue of the qualifications of mediators. The first jurisdiction to pass 'enabling' legislation was Yukon, in 1992. This legislation allowed for mediation in environmental disputes. In June of that same year, the Canadian Environmental Assessment Act was given final reading; it provided for the use of mediation in a variety of situations and outlined the procedure for the appointment of a mediator.[18] Canada's Divorce Act requires lawyers to mention mediation to their clients, and the federal young offenders' legislation encourages the use of alternative measures that, in some provinces, results in referral to mediation programs. Legislation passed in Ontario in 1990 made mediation of no-fault benefit disputes mandatory. Consequently, the Ontario Insurance Commission set up a special division, the Dispute Resolution Group, to be responsible for the delivery of fair, fast, cost-efficient and effective methods of resolving disputes relating to benefits awarded between insured persons and insurers. In January of 1992, the Ontario government proclaimed a new Arbitration Act, which encourages business people to use alternative dispute resolution as a way of settling disagreements without the expense and delay of litigation. Although the legislation does not yet define who can mediate, such may well be the case in the future given the American precedent where legislators are deciding who can mediate, who certifies those eligible to mediate, and the standards for mediating particular types of cases.[19]

The Case of Mandated Mediation in Ontario

The case of Ontario's mandated mediation is an interesting one and is typical of what is happening in other areas. Governments are moving forward (regardless of where the debates within mediation are now) and it is important to look at how schemes are being formulated because the result is affecting the direction in which the practice is going. To date, notably in Ontario, the state is endorsing and promoting a model of mediation that favours a particular approach, i.e., the evaluative approach. This approach, as noted earlier, focuses on entitlements, efficient case management, advice, and links to and from the formal court system.

The regulations in Ontario setting out mandated mediation in civil cases do not establish a regulatory *scheme* for its operation.[20] While there is no one standard as to qualifications, credentials, training, models of mediation, and performance standards,[21] the focus on attaining quick settlements and avoiding costly trials is evident. The view of the final report of the Civil Justice Review, on which mandated civil justice in Ontario was based, was very clear as to the place of mediation: 'it is clearly preferable that ADR be integrated into the [civil courts'] case management system as it expands across the province.'[22] The mediation program was to be an adjunct to the formal court system and was to take its cues and clients from that system. The goals were earlier dispute resolution, lower costs, fewer delays, the most efficient allocation of resources, and better access;[23] the more ambitious goals of a facilitative, more relational focus on needs, experience, and transformation are nowhere to be found.

Under the legislation, mediators are provided free to the parties for a three-hour session; if a settlement is not reached, then the parties themselves either pay for further mediation or go to court. A 'good' mediator in such a situation is one who can get a settlement quickly, and because the parties' lawyers are effectively choosing the mediators from a roster of mediators, those who are favoured (and who will therefore survive in business) are those who opt for an evaluative model of mediation.[24] Facilitative model concerns, which are needs-based and focus on the parties themselves reaching a resolution, are therefore lost in the long run. Direct state regulation of mediation may not be here now, but the framework within which a type of mediation occurs and will evolve is very clear.[25]

Lawyers and Mediation

The direction of mediation will not be determined by the state's interest and intervention alone. A significant interest has also been shown by the legal profession, both at the level of the individual lawyer and by professional bodies. The engagement of legal professionals and their associations is not surprising.

Given the traditional work of lawyers and their monopoly over the formal processes of dispute resolution, it is understandable that they would seek to involve themselves in the practice and direction of mediation. Lawyers have always seen themselves as the primary players in the field of dispute resolution, no matter how poorly they engage in it. As Pirie put it, the 'lawyer's role is one that demands participation in the mediation process'; he quoted Chief Justice Burger in referring to 'lawyers' historical and traditional obligation of being "healers of human conflict",[26] and they are therefore rightfully concerned with and involved in all forms of dispute resolution. It is clear that lawyers 'want in' on the process. The difficulty, if not the irony, is that mediation grew in response to the shortcomings and failings of the profession.

As noted earlier, one of the principal attractions of mediation is the potential to *avoid* what law and legal institutions offer and represent: complexity, expense, delay, inaccessibility, formality, alienation, and elitism. Legal formalism, its processes, and its agents represent a well that many argue has long been poisoned, and thus the ability of law and its actors to exert influence over an emerging competitor is troublesome. The situation can be likened to having a group of oil industry executives oversee the cleanup of the environmental damage done by the industry; while they may be well-intentioned in the pursuit of their goal, they are nevertheless limited collectively by their background and orientation.

But the profession's power and influence are already being felt within mediation. In addition to the fact that many lawyers are directly involved in the practice, law schools in Canada and the US have moved aggressively over the last decade to establish and promote ADR courses. Lawyers now advertise their 'specialty' in mediation services, and the Canadian and American Bar Associations both have subcommittees on ADR as well local committees to oversee the practice.[27] As well, there are established guidelines within the professional bodies on mediation. As early as 1990, all 50 states and the District of Columbia had some type of dispute resolution program, totalling more than 1,200 across the US.[28]

The approaches of the legal profession's representatives in dealing with the growth of mediation are varied, but a common thread runs through what is being said: mediation as a form of dispute resolution is within the profession's 'natural' domain. It is a claim of hierarchy and of expertise. This does not always mean that others outside the profession cannot be 'legitimately' involved, but the underlying assumption is clear: lawyers do, will, and *should* play a primary role in the field. In contrast, critics argue that the dominance of legal professionals in the practice and governance of mediation can only taint those practices and prevent them from achieving the potential that might otherwise be possible. It is not difficult to understand the unease that greets the involvement of the legal profession in mediation, particularly when seen in the context of the power, influence, and attraction of the formal law and legal professionals. That said, the profession is well placed to play a pivotal role in dictating the direction of the practice in the short term at the very least.

The danger for mediation lies in being too closely bound to a profession that lacks the makeup, history, motivation, and education and training to engage in the wider project and goals of mediation, and that in many ways has proven itself antithetical to the larger purposes of mediation. The problem lies also in the assimilating and appropriating tendencies of the profession—and critics argue that the field of mediation must be on guard and resist the claims and interventions of law wherever possible. The imposition of the values and direction of a profession that continues to reflect the advantaged and elite of society, and that has proven itself incapable of breaking out of a moribund insularity and arrogance, threatens to leave mediation trapped within the narrow confines of a rights-based, cost-effective, and settlement-focused framework.

The Example of 'Popular Justice'

An excellent example of the colonizing and destructive nature of state and legal influences can be seen in earlier efforts at popular justice in the United States. These programs originated in the 1960s in the hope of providing informal, effective, and accessible forums and mechanisms of social control, and grew out of a frustration with the failings of the formal systems of justice. They are interesting because of the many similarities they share with mediation schemes—both would seek to alter a method of social control and bring it back to the community in which it operates; both seek informal, effective, and more immediate solutions and resolutions; both complained of the lack of affordable access to the formal systems; and both talk of cost-cutting benefits and better and longer-term solutions. Indeed, popular justice schemes often incorporated techniques of mediation within their overall framework. But the experience of popular justice in the US is that it was not able to escape the 'pull' of the formal justice system, both in the sense of the power of the formal system and the state and in its attraction, and eventually 'popular justice' could not escape the resulting control by the state and the formal legal system.

Less than a decade after informal justice was said to offer more flexible and accountable forms of justice, increasing numbers of practitioners and academics began to criticize informalism, suggesting that it had only shifted the parameters of legal control.[29] Abel's *The Politics of Informal Justice*, published in 1982, helped to shape the argument that informal justice expanded state power through non-state forms. Informalism was accused of providing 'second-class justice', which was supported by the fact that a disproportionate number of clients referred to programs in the US were poor, black, and predominantly female.[30] These criticisms were bolstered by complaints of legal rights violations, exploitation, coercion, and expansion of state control into private lives.[31]

The legal sector contended that only formal procedures, based on rules of evidence, could require obedience to judicial decisions, and only legalism could protect the less powerful. Informal practices were faulted for not living up to their claims of reducing the burden and size of the legal apparatus, and were accused of widening the net of social control.[32] Even though informal justice was agreed to be more humane, responsive, and participatory, opponents argued that it marginalized certain crimes and did not have any long-range impact on the distribution of power or on economic pressures.[33] Other lines of attack saw informalism as augmenting oppression by reinforcing patriarchal and middle-class values and by increasing state power.[34] Informal justice was said to be susceptible to domination by stronger parties, and alternatives were criticized for diverting resources and attention from needed court reforms. With Harrington[35] pointing out that only a few of those referred to mediation programs would ever have gone to court, many critics concluded that informal justice expanded social control. In addition, it was argued that it legitimated the basic approach of the formal system. Informal justice was deemed inappropriate for seeking justice, and despite their flaws, the courts were said to remain the best providers of justice.

In research by Roehl and Cook,[36] the conclusion was that goals of citizen empowerment and social change were not realized because community-based programs were underutilized. By the mid-1980s the various attempts were seen as failures by those who earlier had seen a transformative potential in them. These schemes, whether restitution schemes, victim-offender reconciliation programs, neighbourhood justice centres, or neighbourhood watch programs, came to be seen for the most part as cost-saving add-ons or adjuncts to the state's burgeoning criminal justice system. They were seen simply as part of a new delivery system for old policies and practices. The programs that did survive had close and systemic connections to the formal system and usually received their clients, their direction, and their funding from that system. In the end, informal justice did not result in more popular forms of justice and became the poor cousin of the formal system.

The Future of Mediation

Broadly put, the regulation of mediation can be examined in two contexts. The first is that of mediation as a contested mode of regulation—the uses to which mediation may be put in the resolution of disputes in society, in rule formation, and in the governance of civic society. This is an important context and is the focus of another chapter in this book.[37] The second context is the possible regulation of mediation itself and centres on the opportunities and pressures surrounding the institution of controls; this has been our focus. Within this context there is, of course, the question of whether such schemes will make any difference in the end, or whether the fate of mediation has already been sealed by the ongoing actions of the state and the legal profession.

The governance of mediation has not been treated as unproblematic in the literature, but it is the framing of the nature of the problem that causes concern. There is a need to refocus the debates beyond credentialing and its values or limitations to examine directly the issue of who is and who will be setting the framework and (importantly) the themes for the regulation of mediation. It is precisely here that the current interest in the concepts and practices of governance gives an impetus to the examination of the agents and agencies of governance, and their agendas.

Our argument is that how the field is regulated can determine whether the potential value of mediation, beyond the resolution of a particular dispute, can be realized, and whether its traditional values are sustainable within the schemes that are appearing on the horizon. In many ways, the concept of a regulatory scheme itself appears antithetical to this larger potential or worth of mediation, and runs counter (at the very least) to goals of openness, accessibility, and informality. But the more important aspect of regulatory frameworks here is that they also put in place the tools that allow for, if not encourage, the control, appropriation, and conquest of the field. Our contention in this chapter has been that this is especially true and especially dangerous in the face of a predatory legal profession with a history of establishing such control. The fear is that the burgeoning links both to the formal legal system and its agenda and to the legal profession will doom mediation to the role of handservant to the formal system in the future. Any hope for the larger role espoused for mediation would be lost.

What one wants for mediation in its fullest form and manifestation is the antithesis of what mainstream law and lawyering have been in the past and continue to be. Yet the danger ahead for the regulation of mediation is not only that these groups will occupy and colonize the practice (or at least its profitable sectors), but also that they will do the same on the regulatory front. One of the key features and promises of mediation at its origins was the transformation of civic life by bringing the tools and the power to solve disputes back to individuals as well as into the community. The goal of increasing community input, control, and accountability distinguishes this form of conflict resolution from formal legal measures. A regulatory scheme that serves to counter this goal both through the act of controlling access via standards and procedures, and through its orientation towards the very system that mediation has sought to subvert (or at least circumvent), holds little promise. And given the increasing involvement of the state in mediation through mandated schemes attached to the courts, and the strong presence of lawyers within these schemes, both as practitioners and as consumers (via their clients), the danger of the control of the field falling into the hands of the legal profession is clear.

Goodwill and 'nice' intentions will not ensure the triumph of a greater role for mediation; any governance scheme and philosophy must mandate a practice and practitioners of mediation that fulfill its promise. To regulate the practice from a legal professional's point of view and within a law-oriented framework is to lose the fight for an enhanced form of mediation at the outset. Such control offers little for the transformative or communitarian goals of mediation, in the same way that formal legal dispute resolution mechanisms have had very little success in advancing transformative or restorative practices of justice. Law's ways and philosophies are not good enough for the practice of mediation, and they are not good enough for its regulation.

The practice of mediation is not developing in a vacuum and it will not do so in the future; it will be shaped by many forces, including the regulatory schemes that will govern it. The argument in this chapter has been that the governance of mediation is being occupied by those who would see mediation reduced to its least meaningful functions and who are being aided and abetted by state schemes and legislation. As noted in our introduction, in this sense, in the context of mediation, we would agree with Purvis, who argues more generally in Chapter 2 of this book that the state continues to occupy a central role in regulatory schemes by 'governing from a distance' and by providing support and direction for those groups and activities that subscribe and contribute to its overall interests.

In light of the history of mediation to date, the possibility at this point in time seems slight that mediation will be able to overcome its state and legal orientation in order to realize the practice's traditional ideals and purposes. There is also the argument that, ultimately, the governance of mediation may not matter much in the attainment or thwarting of this greater role. The case can be made that when it fully arrives, the die will have been cast, and the regulatory scheme will only enshrine the failed practices and philosophies that will have already firmly and authoritatively taken hold.

Notes

1. Lon Fuller, 'Mediation, Its Forms and Functions', *Southern California Law Review* 44 (1971): 325.
2. See, generally, J. Brett, R. Driegh, and D. Shapiro, 'Mediator Style and Mediation Effectiveness', *Negotiation Journal* (1986): 277–85; R.B. Bush and J. Folger, *The Promise of Mediation* (San Francisco: Jossey-Bass, 1994); Deborah Kolb, *When Talk Works: Profiles of Mediators* (San Francisco: Jossey-Bass, 1994); Christopher W. Moore, *The Mediation Process: Practical Strategies for Resolving Conflict* (San Francisco: Jossey-Bass, 1986).
3. See, generally, C. Honeyman, 'On Evaluating Mediators', *Negotiation Journal* 6 (1990): 23–36; Society of Professionals in Dispute Resolution (SPIDR), *Qualifying Neutrals: The Basic Principles. Report of the SPIDR Commission on Qualifications*

(Washington: National Institute for Dispute Resolution, 1989); Joel Edelman, 'A Commentary on Family Mediation Standards', *Mediation Quarterly* 3 (1986): 97–102; Cheryl Picard, 'The Emergence of Mediation as a Profession', in Catherine Morris and Andrew Pirie, eds, *Qualifications for Dispute Resolution: Perspectives on the Debate* (Victoria, BC: University of Victoria Institute for Dispute Resolution, 1994), 141–64; Ellen Waldman, 'The Challenge of Certification: How to Ensure Mediator Competence While Preserving Diversity', *University of San Francisco Law Review* 30 (1996): 723–56.

4. Julie Macfarlane, *Rethinking Disputes: The Mediation Alternative* (Toronto: Emond Montgomery, 1997), 2.
5. See, generally, Kenneth Kressel and Dean Pruitt, eds, *Mediation Research* (San Francisco: Jossey-Bass, 1989).
6. See Raymond A. Whiting, *The Use of Mediation as a Dispute Settlement Tool: An Historical Review and Scientific Examination of the Role and Process of Mediator* (Washington: National Institute for Dispute Resolution, 1988).
7. Laura Nader, 'The ADR Explosion—the Implications of Rhetoric in Legal Reform', *Windsor Yearbook of Access to Justice* 8 (1988): 269–91.
8. The Honourable T.G. Zuber, *Report of the Ontario Civil Courts Inquiry* (Toronto: Queen's Printer, 1987).
9. David Daubney, *Taking Responsibility: Report of the Standing Committee on Justice and Solicitor General on its Review of Sentencing, Conditional Release and Related Aspects of Corrections* (Ottawa: Supply and Services, 1988).
10. Justice Reform Committee, Province of British Columbia, *Access to Justice: Report of the Justice Reform Committee* (Victoria, BC: British Columbia Ministry of the Attorney General, 1988).
11. See, generally, R.B. Bush, 'What Do We Need a Mediator For: Mediation's "Value-Added" for Negotiators', *Ohio State Journal on Dispute Resolution* 12 (1996): 1–36; Kressel and Pruitt, eds, *Mediation Research*, chs 1, 2, 10, 16; Julie Macfarlane, *Court-based Mediation for Civil Cases: An Evaluation of the Ontario Court (General Division) ADR Centre* (Toronto: Ministry of the Attorney General, 1995); J. Wall, Jr, and A. Lynn, 'Mediation: A Current Review', *Journal of Conflict Resolution* 37 (1993): 160–94.
12. R.F. Cook, J.A. Roehl, and D. Sheppard, *Neighborhood Justice Center Field Test: Final Evaluation Report* (Washington: US Department of Justice, 1980).
13. See Robert Davis, Martha Tichane, and Deborah Grayson, *Mediation and Arbitration as Alternatives to Prosecution in Felony Arrest Cases: An Evaluation of the Brooklyn Dispute Resolution Center* (New York: Vera Institute of Justice, 1980); Craig McEwen and Richard Maiman, 'Small Claims Mediation in Maine: An Empirical Assessment', *Maine Law Review* 33 (1981): 237–68; Kressel and Pruitt, eds, *Mediation Research*.
14. See Bush, 'What Do We Need a Mediator For'.
15. Ron Kraybill and John Paul Lederach, 'Professionalization and the Mennonite Conciliation Service', *Conciliation Quarterly* 11 (1992): 8–9.
16. Margaret Herrman, 'On Balance: Promoting Integrity Under Conflicted Mandates', *Mediation Quarterly* 11 (1993): 123–38.
17. See Society of Professionals in Dispute Resolution, *Qualifying Neutrals*.
18. Bill Diepeveen, 'The Times They Are a-Changing: Legislative Update', *Interaction* 4 (1992): 7.

19. See Picard, 'The Emergence of Mediation as a Profession', 151-2.
20. R.R.O. 1990, Reg. 194 as amended by O. Reg. 451/98.
21. See L. Deborah Sword, 'Lists, Panels, Rosters: What's Next?', *Interaction* 11, 1 (1999): 7: 'Whereas the Attorney General is on record as specifically saying that the mandatory mediation roster is not intended to set standards and credentials, it is having that effect anyway.'
22. Ontario Civil Justice Review, *Civil Justice Review: Supplemental and Final Report* (Toronto: Civil Justice Review, 1996), 40.
23. Ibid., 39.
24. Sword, 'Lists, Panels, Rosters', 3.
25. For the situation of family mediators in Quebec, see Sylvie Matteau, 'Family Mediation in Quebec: The Impact of Free Services on the Practice of Family Mediators', *Interaction* 11 (1999): 5.
26. Andrew Pirie, 'The Lawyer as Mediator: Professional Responsibility Problems or Profession Problems', *Canadian Bar Review* 63 (1985): 381.
27. See National Institute for Dispute Resolution (NIDR), 'Statement of Madeleine Crohn, Before the U.S. House of Representatives Subcommittee on Intellectual Property and Judicial Administration, May 19, 1992', *Forum* (Summer 1992); Ontario Standing Committee on the Administration of Justice, 'Alternative Dispute Resolution', 2nd Session, 34th Parliament, 39 Elizabeth II, 1990
28. NIDR, 'Statement of Madeleine Crohn'.
29. R. Matthews, *Informal Justice* (Beverly Hills, Calif.: Sage Publications, 1988).
30. See Sanford Jaffe, *The Adversary System: Is There a Better Way?* (Washington: National Institute for Dispute Resolution, 1983); R. Tomasic and M. Freeley, *Neighbourhood Justice: Assessment of an Emerging Idea* (New York: Longman, 1982).
31. See Richard L. Abel, ed., *The Politics of Informal Justice*, 2 vols (New York: Academic Press, 1982); Kressel and Pruitt, eds, *Mediation Research*.
32. See M. McMahon and R. Ericson, 'Reforming the Police and Policing Reform', in R.S. Ratner and J.L. McMullan, eds, *State Control: Criminal Justice Politics in Canada* (Vancouver: University of British Columbia Press, 1987), 38-68.
33. Maureen Cain, 'Beyond Informal Justice', *Contemporary Crisis* 9 (1985): 335-73.
34. Richard L. Abel, 'Conservative Conflict and the Reproduction of Capitalism: The Role of Informal Justice', *International Journal of the Sociology of Law* 9 (1981): 245-67.
35. Christine Harrington, 'Delegalization Reform Movements: A Historical Analysis', in Abel, ed., *The Politics of Informal Justice*, vol. 1, 35.
36. See J. Roehl and R. Cook, 'Mediation in Interpersonal Disputes: Effectiveness and Limitations', in Kressel and Pruitt, eds, *Mediation Research*, 31-52.
37. Neil Sargent in Chapter 10 of this book sees a complex interaction occurring between formal state law and alternative justice systems, and we agree with this. However, we would go further and argue, for the reasons given throughout this chapter, that the balance within this interaction is weighted heavily in favour of the formal system, and ultimately the informal will take its shape and direction from the formal.

PART IV

Governing Self-Regulation

Chapter 12

Government, Private Regulation, and the Role of the Market

KERNAGHAN WEBB

Introduction

When an environmental, consumer, health and safety, or other social-oriented problem arises, the instinctive response of many is, 'There ought to be a law.' In particular, the instrument usually turned to is the command-and-control regulatory approach, consisting of mandatory legislated prescriptions backed up by penalties, compliance monitoring provided by inspectors, enforcement through investigations, and ultimately penal sanctions rendered by the courts. Since legislation represents a respected, high-profile statement of societal concern about a problem—the product of the state's central democratic organ—and is designed to apply to all equally, it is not surprising that the public regulation option is the first choice of many. However, conventional regulatory approaches are not without their share of drawbacks, including jurisdictional constraints, limited incentives for regulated actors to undertake protection activity beyond that required by prescribed standards, as well as protracted and expensive development and implementation processes.

The first major proposition put forward in this chapter is that, in the right set of circumstances, private regulatory approaches can and do usefully supplement public regulation. While functioning largely outside the limelight, private market-oriented regulatory initiatives are nevertheless flourishing across many sectors and activities, operating as an overlay to the labyrinth of public regulations already in place. Businesses, working with governments, consumers, community representatives, and non-governmental organizations, have developed and implemented rule systems to monitor and control misleading and inappropriate advertising,[1] to ensure quality in the manufacture of products and the delivery of services,[2] and to maintain ethical business-consumer relations.[3] At a sectoral level, a wide range of private regulatory regimes are in place, including those pertaining to the protection of personal information,[4] sustainable harvesting practices of forestry companies,[5] environmental practices of chemical producers,[6] customer service standards of cable television companies,[7] and protections against sweatshop and child labour in overseas apparel factories.[8] In addition, individual firms have established worker health and safety codes that their suppliers must meet.[9] Also, non-profit consumer organizations have successfully initiated and operate private regulatory systems with a market orientation, most notably in the area of customer service standards,[10] and environmental non-governmental organizations have spearheaded the development and application of environmental certification systems that have then been applied by industry.[11]

As even this brief listing suggests, within the broad category of 'private regulation' there is tremendous variety. Indeed, there is considerable value in viewing private regulatory systems not simply as a single, monolithic category, but rather as a continuum, from those where government departments or officials and legislation have played a key role in directly initiating, approving, or framing the private regulatory structure (what might be called 'co-regulation'[12]) to those with no direct government involvement (what might be called 'pure' self-regulatory models[13]). Within the private regulation category, this chapter focuses on those that government has not directly initiated, approved, or framed in law. Yet, as is discussed in greater detail below, even private regulatory approaches developed without direct government prompting are indirectly stimulated, sustained, and recognized by the legal system. While law remains an important 'driver' of private regulation, one of the distinctive features of non-governmental self-regulation is that it also often attempts to harness the energies of a different set of societal forces, players, and institutions from those typically involved in the development and use of public regulation. To be more precise, private regulation frequently relies to a significant extent on market (e.g., customers, suppliers, insurers, investors), peer, and community pressure to stimulate publicly beneficial improvements, and private regulation makes extensive use of non-governmental 'intermediaries' such as industry associations, environmental, consumer, and other non-governmental organizations (NGOs), and standards organizations in program development and implementation.

In so doing, private regulatory approaches can help to address some of the aforementioned weaknesses of public regulation; specifically, voluntary approaches are largely unaffected by many of the jurisdictional obstacles that plague legislated command-and-control regimes, may more easily stimulate behaviour that goes beyond baseline prescribed standards, and can be easier to develop and implement than public regulatory approaches. Moreover, private regulations can refine and clarify ambiguous legal concepts such as 'reasonable care'. But while private market-oriented regulatory approaches can potentially have positive impacts for consumers, workers, the environment, the broader community, and the legal system, this is not to suggest that they are without problems. Common weaknesses include generally lower visibility and credibility than legislative regimes, less likelihood of rigorous standards being developed, difficulty in applying the rules to free riders, variable public accountability, and a more limited array of enforcement options.

Not surprisingly, then, there is considerable skepticism about the efficacy of private regulatory systems and the motivations behind those who champion them. Some might claim that it is unreasonable to expect industry to regulate itself effectively, and that private regulation represents a classic case of foxes guarding the henhouse.[14] This line of argument seems to rest on the premise that public and private regulatory approaches represent diametrically opposed and competing approaches, that society (consisting of, among others, governments, businesses, NGOs, and the greater community) must choose one or the other and cannot have both. While there may be rare occasions when the two represent stark and incompatible 'either/or' alternatives, in the vast majority of circumstances, public and private regulatory approaches can and do operate concurrently, usually in a mutually supporting fashion. Returning to the foxes and henhouses metaphor, the foxes (who know a thing or two about the weaknesses of henhouses) are but one set of actors guarding the henhouse—and they themselves are regulated, by both governmental and non-governmental actors.[15]

This leads to the chapter's second major proposition, that while private regulatory systems are voluntarily engaged in by industry and others—voluntary in the sense that they are not directly, legislatively required—they nevertheless operate within an overarching framework of law, are influenced by law, and influence law, and many have significant legal implications:

- Frequently, industry will adopt private regulatory systems in an attempt to stave off pressures for new regulations.[16] The perceived failure of a particular private regulatory approach to address a problem may hasten the introduction of a new governmental regulation.[17] Private standards may be incorporated into legislation[18] or imposed by courts in sentencing.[19]
- Adherence by firms to the terms of a private regulatory system may decrease the likelihood of regulatory non-compliance, government enforcement actions, and regulatory liability.[20]
- Many private regulatory systems are consent-based in nature, and as such they may be enforced through contracts.[21]
- Businesses entering into private regulatory approaches that have the effect of maintaining prices, decreasing choice, or preventing new firms from entering into an activity may attract the attention of antitrust or competition law agency officials.[22]
- Private regulatory systems may also be taken by courts as evidence of

acceptable industry standards of care—standards ultimately enforceable through regulatory prosecutions and tort actions in negligence.[23]

In short, a dynamic if largely ad hoc 'conversation' is continuously taking place between private regulatory systems (i.e., the players, the institutions, and the rules developed) and the overarching legal system within which private regulatory regimes operate. However, until recently, there has been little attempt to think systematically about the relation between the two, from a public policy standpoint, or to consider how to better integrate them in order to take advantage of the resulting synergies and minimize the negative consequences. In effect, public and private regulatory approaches represent a potentially fruitful partnership, when the two are used in the right set of circumstances and in the right way. A key challenge facing public policy decision-makers is to determine the right set of circumstances, what possible alterations can be made to laws, policies, and activities to encourage effective private approaches, and generally how to better use the two approaches in a manner that maximizes their potentialities. Recent governmental,[24] industry,[25] NGO,[26] and scholarly activities[27] suggest that many parties have realized the importance of addressing this challenge.

This chapter is an exploration of some of the issues associated with this challenge. Following a brief historical discussion, the body of the chapter is devoted to a critical description of one existing private regulatory system, the Responsible Care program, as it operates within a broader legal, political, economic, and social environment. A concluding section brings together critical observations regarding the interrelationship of public and private regulatory systems, and speculates on future prospects for this potentially important partnership.

Private Regulation: An Old Story

Private regulation is not a new phenomenon. If regulation is defined as a system of rules designed to structure or control behaviour and relations, and private regulation is considered to be a system of rules devised and applied by entities other than the state, then perhaps the earliest and most common form of private regulation known to humankind is organized religion. In fact, at many points in history organized religion represented the dominant form of regulation in society. However, with the exception of religious-oriented food preparation certification systems such as those associated with kosher (Jewish) and halal (Muslim) foods,[28] there is rarely an overt market-oriented element to religious forms of private regulation, and so they rest outside the focus of this chapter, which is devoted to an examination of market-oriented private regulatory regimes.[29]

Like organized religion, predominantly market-oriented private regulatory systems also have a long history. An early example of a pervasive private commercial regulatory model was the organization of merchant and craft guilds throughout Europe and Asia. While most readers are probably familiar with the guilds that emerged in the Middle Ages in Europe, guilds have an even more ancient lineage in non-Western society. In *The Rise and Decline of Nations*, Mancur Olson writes that '[t]here have been guilds, for example, in Moslem countries (and even in Mecca), in Byzantium, in China, in Hellenistic times, and even in Babylonia.'[30] In Western society, the growth of the guilds represented one of the first signs of a commercial revival in Europe, a revival that seems to have started in ninth-century Venice and then spread throughout Continental Europe and England. Organizations of merchants and craftspeople formed guilds in each town and addressed virtually every aspect of the commercial activity:

> when a merchant lost his stock in a wreck or at the hands of a robber baron, his fellows aided him to start again. If a man in another town refused to pay his debts to a guildsman, the guild would seize the next inhabitant of the same town who came its way. When a member died, his fellows would bury him and care for his widow and children. . . . Often the guild maintained a school to train the member's sons . . . on the economic side, the guild secured a monopoly of the town's business for its members. No one not a member could sell at retail in the town. If a foreign merchant brought goods to the town, he had to sell them to a member of the guild or at least pay a very heavy sales tax. Often the guild was the agent for its members in dealing with the lord.[31]

The regulation of commercial activity by the guilds was a comprehensive and intensive effort, frequently involving control of the means of production (no use of new technologies without agreement of guild members), regulation of working and production conditions (no member was allowed to make goods cheaper or faster than the others, and limits were maintained on the number of hours of work), control of entry into the trade and apprenticeships, regulation of price and quality, and dispute resolution.[32] Although guilds clearly possessed an important social welfare component for their members, they were essentially commercially oriented. In effect, the exigencies of the market—consumers looking for the lowest price and some assurance of quality, and merchants/manufacturers looking for some way of maintaining price and quality and controlling entry into the trade—stimulated those in the business to organize themselves and create rule structures. Needless to say, in their objectives of maintaining prices and production and restricting entry of newcomers, the guilds represented classic examples of anti-competitive behaviour.

Gradually, the guilds lost their power in the face of the closure of the commons and with the shift to factory-based production brought about by the Industrial Revolution. The state and its regulatory agencies assumed primary control over many issues formerly the purview of the guilds, from consumer protection to worker health and safety standards and anti-trust (competition) laws. But concomitant with the rise of industrialization and regulation, new forms of private regulation also emerged. The Better Business Bureau has its origins in the late 1800s.[33] The Automobile Association of America and its Canadian counterpart began at the turn of the century.[34] Industry associations were created, and many entered into agreements encompassing both price and labour issues.[35] What distinguishes the modern proliferation of private regulatory systems from the earlier models, such as the guilds, is the fact that the modern versions tend to be supplemental to an extensive backdrop of public regulation.

The Canadian Responsible Care Program[36]

Program Description

The major impetus for the Responsible Care initiative of the Canadian Chemical Producers' Association (CCPA) was a series of chemical disasters in the late 1970s and early 1980s (most notably, Love Canal, Seveso, and Bhopal) that eroded public confidence in the industry and raised the spectre of tighter government regulation and controls. In an attempt to address these perceived problems, the CCPA developed the Responsible Care initiative, which has evolved into an elaborate system of principles and rules designed to improve the safe and environmentally sound management of chemicals over their life cycle. The key objectives of the program are improved environmental performance, improved government-industry relations, and increased public trust. Agreement to adhere to the Statement of Responsible Care and Guiding Principles, and to report on emissions, is now a condition of membership in the CCPA. Between 90 and 95 per cent of the chemicals manufactured in Canada are produced by members of the CCPA, although some of the smaller manufacturers are not represented in the CCPA. Following the Canadian lead, there are now Responsible Care programs operating in 40 countries.

As it operates today in Canada, the Responsible Care program includes a statement of policy, guiding principles, a national advisory panel, a chemical referral centre, a verification process, and six codes of practice covering community awareness and emergency response, research and development, manufacturing, transportation, distribution, and hazardous waste management. The program has evolved over time, becoming more rigorous and transparent in the process. Initially, in 1979, the Statement of Policy on Responsible Care was simply a one-page description of 'good intentions' and adhering to it was voluntary. In 1983, 12 CCPA members agreed to sign the Statement of Policy. In 1984, commitment to the Statement became a condition of CCPA membership. A National Advisory Panel (NAP), consisting of 12–16 external experts and advocates (including academics and environmental group representatives), was established in 1986 to provide ongoing advice on the program's development and implementation, including six codes of practice developed, with advice from the NAP, to 'flesh out' the Statement. In 1993, the collection and publication of emissions and waste data became mandatory. Member companies were also required to engage in ongoing community advisory processes and to conduct internal audits of compliance with the codes of practice. In 1994, the CCPA introduced a system of external verification of performance. More changes are being considered for the future, part of a guiding philosophy of 'continuous improvement' that is said to underlie the program.

Today, when a company joins the CCPA, it commits to implementing fully the 152 elements of the codes of practice within three years. Compliance verification involves site visits by a four-person team, consisting of two industry representatives who are independent of the company (e.g., competitors) and two non-industry members, one of whom is a community representative. The costs of the compliance verification process (estimated at $10,000–$14,000, on average) are borne by the firm being inspected. The verification teams review a plant's management systems, based on interviews with plant officials, suppliers, customers, and community residents, and document reviews and site visits. At the time of writing, all CCPA members, except for the newest ones, have completed or initiated external evaluations.[37] The verification system focuses on the presence of a management system but does not measure actual performance.

Peer pressure plays a key role in creating an atmosphere of mutual accountability and encouraging members to comply. Regional Leadership Groups, consisting of the chief executive officers of member companies, meet quarterly to discuss compliance issues and assist each other in developing solutions to problems. According to one account, the consequences of a chief executive having to stand up in front of his peers and say 'We didn't make it' are severe enough that only once has a member withdrawn after failing to meet even the minimum standards.[38] An additional incentive for participation in Responsible Care can be the prospect of lower insurance premiums.[39]

By 2000, CCPA members had achieved a 62 per cent reduction in their total emissions of substances (excluding CO_2) compared to 1992.[40] These cuts have been made at the same time as the industry has grown in output. It is not suggested here that the Responsible Care program was solely responsible for these reductions. Undoubtedly there were a host of other factors as well, including legislation, changes in the market, and changes in technology. By 2005, member companies are anticipating that emissions per unit of production will decline by 74 per cent based on 1992 volumes.[41] Members of the CCPA are reported as saying the Responsible Care program has a wide variety of financial benefits, including reduced waste management, cleanup, and disposal costs, improved emergency response capability, improved community relations enabling companies to avoid protracted disputes based on distrust, faster government permitting, better ability to obtain financing and insurance at reasonable costs, and potentially less legal liability.[42]

The Responsible Care Program and the Law

The Free-Rider Problem

A standard criticism of private regulatory initiatives is that, since the initiatives are consent-based, those who do not wish to participate (free riders) are not subject to the terms of the program. This is in stark contrast to legislated programs where regulations apply equally to all regardless of the desires of individual industry members. Thus, for example, in the case of the Responsible Care program, some of the smaller chemical manufacturers are not members of the CCPA and are non-participants in the program. Non-participants are likely to be of considerable concern to firms that subscribe to a voluntary program, since the activities of non-subscribers could 'taint' those who do adhere to the program. This, in turn, creates a disincentive for firms to join the program: why bother if their efforts can be sabotaged by those of non-joiners? Free riders are also a matter of public concern in the sense that those who most need to comply with the terms of a voluntary program may be the ones not joining, with attendant bad effects on the environment, workers, and the general community.

The law can help to address this type of free-rider problem in a number of ways.[43] First, industry associations can welcome regulations where they feel that voluntary initiatives are not working. Thus, for example, the vice-president of the CCPA has been reported as saying, 'We don't have any problem with having regulations when voluntary programs don't work. . . . When they don't work, they have to be backed up by a government willing to actually regulate, or else you will have free riders—companies that don't take care of their problems.'[44] In effect, the CCPA seems to be using the threat of regulation, and its approval of its use, as a stimulus for greater adherence to the voluntary program.

But this is not the only way that law can help to address the free-rider problem associated with voluntary programs, which, after all, operate within a framework of law whether or not individual industrial firms may wish for a regulatory regime. An important type of private tort action (negligence) and the central type of regulatory offence (the strict liability offence) hinge on a concept referred to as either 'reasonable care' or 'due diligence'.[45] A negligence action can be brought by a harmed community member who is the subject of a spill, while a regulatory action is normally brought by a government agency. Whenever any firm is the subject of a private negligence suit or a strict liability offence prosecution, once the plaintiff or the Crown has proven to the satisfaction of the court that a duty was owed and not met, and foreseeable harm took place (in the case of a negligence lawsuit), or that the factual elements of an offence took place, the accused or defendant can in most circumstances only escape liability by demonstrating that the activity took place despite all reasonable care or due diligence having been exercised.

In considering whether reasonable care or due diligence has been exercised, the court will typically look to the industry practice. In so doing, the court is essentially attempting to determine whether or not the firm in question is meeting the accepted standards of industry. If, as in the Responsible Care example, a leading industry association has put in place a program requiring that members meet certain commitments, this can be taken into consideration by the court as evidence of accepted industry practice.[46] As a result, if a particular company has not participated in the voluntary program or does not meet the terms of the program, this can play an important role in establishing liability. In this way, even a firm that has not participated in a voluntary initiative can be made subject to its terms and the free-rider problem can be defeated. Adherence or non-adherence to a voluntary program can also play an important role in determining an appropriate sentence.

Two interesting points emerge here. First, the law can act to reward those who participate in voluntary programs and reinforce the value of voluntary programs through recognition of the programs in due diligence defences, at the same time as it acts to punish those who do not participate or meet the terms of a voluntary program. Second, a failure on the part of non-industry participants (e.g., environmental groups, community representatives, government officials, etc.) to become involved in the development of a voluntary program may detract from the rigour of those standards and thereby make it easier for firms to meet the industry standard in negligence or regulatory legal actions. Thus, the fact that a voluntary standard can play an important role in legal actions should create an incentive for non-industry participants to become involved in the development of those standards so that they can ensure that the standards are as rigorous as possible.

The legal system can also reinforce voluntary programs and isolate free riders in another way. In light of the fact that limited resources are available to enforce regulatory programs, it would be possible to develop an enforcement strategy that took into consideration adherence with voluntary programs as one criterion in deciding which firms will be the subject of inspections (or how regularly those inspections will take place). This is not to suggest that inspectors would cease to conduct inspections of firms that are members of voluntary programs. However, regulatory forbearance programs involving expedited permitting, less frequent inspections, and public recognition can be designed to reward firms for participating in 'beyond compliance' voluntary programs.[47] Environmental agencies are also indicating in their public enforcement programs that they will not seek access to voluntary audits as part of normal inspection activities.[48] This suggests awareness on the part of some governments that firms taking the initiative to monitor their own facilities should not be discouraged from so doing by having the fruits of their self-regulation (e.g., a voluntary audit that revealed a problem) used against them in normal inspections. Finally, in apparent recognition of the value of voluntary environmental protection programs such as that of Responsible Care and ISO 14000, the Canadian Environmental Protection Act has recently been amended to require courts to take into account whether an alleged offender has in place an environmental management system that meets a recognized Canadian or international standard. The Act also authorizes courts to make orders as part of sentencing that include use of environmental management standards.[49] In short, the free-rider problem associated with voluntary programs such as Responsible Care can to some extent be offset by a number of different aspects of regulatory law.

Other Collective Action Problems

A second common criticism of voluntary programs such as Responsible Care is that they have 'no teeth'—that is, they provide no effective penalties or disincentives when incidents of non-compliance occur, or there is no effective use of such penalties or disincentives. The law can be of considerable assistance in addressing contraventions of voluntary programs. Among participants, voluntary programs are consent-based, and as such, in the normal course of events are structured through contracts.[50] Thus, for example, members of an association that pay dues, have obligations, and receive services are in a contractual relationship with that association. Similarly, firms wishing to use an industry standards trademark or logo such as Responsible Care on their products must agree to adhere to the terms of that program in a licensing agreement (a form of contract). In all cases, failure to meet the terms of the contract can result in sanctions such as fines, negative publicity, withdrawal of benefits, or termination of relations, all ultimately backed up by legal actions if necessary. There are many cases of disciplinary action being taken by associations against non-complying members.[51] And as mentioned above, the fact that non-adherence to a voluntary program can be a factor taken into account in a negligence or strict liability offence action can create a strong incentive for compliance for participants in a program. In addition, positive incentives for compliance can play an important role: for example, being a member of the CCPA means having access to the latest information on up-and-coming issues and being able to participate in their development, and being able to display the Responsible Care logo on company promotional material.

Legal Constraints on the Responsible Care Program

While the discussion above has focused on how the law can be used to reinforce the terms of voluntary programs and encourage their use, it is also true that constraints on what can be done through voluntary action are imposed through law. In particular, a key concern for government and, indeed, society is the potential for anti-competitive action through multi-firm arrangements. A central focus of the federal Competition Act is a robust and competitive marketplace, in part achieved through the prevention of barriers to entry into the market or restrictions on the ability of a firm or firms to expand within a market.[52] The Competition Bureau has prepared an information bulletin, 'Strategic Alliances Under the Competition Act', that provides guidance on the circumstances in which inter-firm alliances are permissible or restricted. Early consultations with the Competition Bureau can help to avoid any problems. Regular and meaningful involvement of third parties other than the competitors (e.g., members of the public, interest groups) in program development and implementation can also lessen competitiveness concerns. In the case of the Responsible Care program, CCPA officials consulted regularly with the Competition Bureau, and third-party involvement in the program has become an integral component of the program.

The Role of Industry Self-Regulation in Addressing Regulatory Weaknesses

A central and distinguishing feature of command-and-control regulatory systems, as opposed to voluntary programs, is their coercive base—the fact that the state and only the state has the authority to imprison and/or fine persons for contravening a law. That this punishment is available for many environmental offences is undoubtedly a powerful 'stick' that is provided to legislators and not to industry associations and non-governmental organizations which are administering voluntary programs. However, many of the weaknesses of regulatory programs also stem from their coercive base. Because imprisonment is one of the most serious punishments that can be levied against an individual, it is not surprising that there are many constraints on the ability of the state to use it. To reflect its democratic foundation, and in an effort to prevent its abuse, the legislative process and implementation of penal legislation are formal (i.e., subject to many substantive and procedural constraints), open, slow, and expensive to operate. Thus, for example, the process of identifying and regulating toxic substances pursuant to the Canadian Environmental Protection Act has been extremely rigorous and painfully slow,[53] which provides an opening for voluntary programs such as Responsible Care that can operate in a more proactive and less formal fashion.⁵⁴ The process of regulatory enforcement must be done in a manner compatible with the Charter of Rights and Freedoms, but disciplinary actions of industry associations need not comply with the Charter.⁵⁵

Jurisdictional/Sovereignty Limitations
A legislative body can only promulgate statutes and regulations within its jurisdictional competence. Because much of environmental and economic activity crosses provincial/state and national borders, or has cross-jurisdictional implications, regulatory solutions may be very difficult to negotiate and implement. There may be considerable dispute among legislative bodies about the extent of the problem, its priority, which legislative body should address it, and the best approach to correcting it. In a federation such as Canada, where both levels of government have powers over the environment, disputes over the extent of their powers are common and can interfere with the development of effective solutions. At the international level, differences of perspectives between developed and less-developed countries and even within developed countries can delay or stymie formulation and effective implementation of international conventions.

Voluntary measures such as the Responsible Care program, which are developed by non-governmental sources, can to some extent avoid these types of difficulties. Within Canada, the Responsible Care program applies to all chemical facilities owned by CCPA members, regardless of uneven federal and provincial laws or enforcement. And at the international level, as we have seen, Responsible Care programs pursuing similar outcomes to those in Canada are operating in over 40 countries.

Inadequate Enforcement Capacity/Integrity
To work as intended, command-and-control regulatory measures require a dedicated, expert administration and enforcement capability with sufficient resources. Cutbacks and changes in government priorities can severely undermine effective regulatory enforcement. In a federation such as Canada, there is considerable potential for uneven enforcement because of differences in legislation, resources, expertise, and commitment from region to region. In less-developed countries, there may have never been an adequate enforcement capacity.⁵⁶ To some extent, voluntary approaches of international scope financed and driven by non-governmental actors avoid problems of governmental implementation, underfinancing, and inadequacies.⁵⁷ In some situations, it may be possible for local community associations to contract directly with industries concerning such matters as emission controls and compensation. These types of community-industry pacts may be particularly attractive in jurisdictions where the regulatory system is substandard.⁵⁸ As was discussed earlier, the Responsible Care program, as a national program to be applied consistently by members across Canada regardless of location, can stimulate com-

pliance through its peer pressure reviews and compliance verification activities, thus complementing government enforcement efforts.

Limited Incentives for Improvement beyond Baseline Standards
A difficulty with conventional command-and-control approaches is the lack of incentives for firms to improve once the law-based standard has been reached. The only rewards a firm has for exceeding standards set in regulations are possible competitive gains received from anticipating new regulations or, possibly, from developing new technologies that can then be sold to others when and if the standard is ratcheted up at some later point. On the other hand, by relying on market or peer pressures to stimulate compliance, voluntary approaches can catalyze firms to go further than baseline standards. As discussed earlier, the Responsible Care program is premised on the concept of 'continuous improvement' and, according to members, has resulted in numerous financial benefits.

Expense and Time To Develop and Amend
The process of developing and amending legislation is expensive, lengthy, and uncertain. There is no doubt that developing and amending voluntary approaches can also share these characteristics. However, there are some significant differences. First, because a comparatively smaller number of players are involved (e.g., not a nation's entire legislative assembly) and because the processes for developing and amending are tailored to a particular initiative, there is a greater likelihood that the processes for formulating and amending voluntary measures will be more streamlined and thus less expensive and lengthy than conventional regulatory processes.[59] Second, the expense of developing and amending a voluntary measure is typically incurred by the particular players involved in the initiative themselves rather than by all taxpayers. Thus, for example, the cost of developing the Responsible Care initiative is borne by the chemical industry, with costs typically being passed on to the ultimate consumers of their products. As even the brief description provided above suggests, the Responsible Care program has been amended regularly and significantly since its introduction, becoming more rigorous and comprehensive. The formality, scale, and expense of the regulatory development and amendment process militate against such frequent and substantive changes to a regulatory regime.

Conflicts between Legal and Voluntary Approaches
Although discussion to this point has revolved around how the two approaches can and do work together in a positive, reinforcing manner, this is not to suggest that there are not areas of serious potential conflict. First, voluntary programs can be used to engage in anti-regulatory lobbying when in fact new legislation and regulations are needed.[60] Second, there is the potential for government involvement or support of voluntary programs to undermine govern-

ment's ability to enforce laws in an even-handed, rigorous, and effective manner.⁶¹ The challenge is for government to operate in a scrupulously open, accessible, and even-handed manner in both its legislative and enforcement activities so that the existence of voluntary programs and government activities in relation to those voluntary programs does not affect its neutrality and the fundamental obligation of the state to protect and promote the public interest both rigorously and consistently.

Conclusions

The main proposition put forward in this chapter is that the relation between private market-oriented regulatory systems and the law is currently and potentially a very constructive one. While not denying the potential conflicts, the two approaches can, in the right set of circumstances, reinforce to some extent each other's strengths and make up for each other's weaknesses. The coercive base and universal application of regulatory legislation provide it with necessary attributes for addressing many of today's social problems. But command-and-control regulations are subject to significant jurisdictional constraints, are developed and amended only after protracted formal processes, do not encourage improvements beyond stipulated standards, and involve resource-intensive administration.

Private regulatory approaches, which are often driven by commercial and market factors and tend to harness the energies of intermediaries, may be able to overcome many of the jurisdictional obstacles facing regulatory programs, can encourage improvements beyond minimum standards, are generally easier to put in place, and usually cost less to implement than coercive-based regulations. The weaknesses of private regulatory systems in terms of contending with non-joiners and effectively addressing non-compliant behaviour of participants can be countered to some extent through exploitation of the contractual underpinnings of such programs, as well as by creative use of the potential linkages between such programs and due diligence defences in both tort and penal actions.

Although private regulatory systems and the law can have a constructive, reinforcing effect on each other, such mutually beneficial relationships appear to be more by happenstance than design.⁶² That is, there is little evidence of a conscious awareness (at least until recently) on the part of public policy-makers of the value in establishing a legal framework to stimulate development and implementation of private regulatory approaches that supplement regulatory programs. Perhaps this is as it should be, since it means that the private initiatives that do emerge do so in an ad hoc, organic fashion (i.e., because industry, NGOs, or standards organizations have reached the conclusion on their own that they want to do it), and not because a government bureaucrat thought it would be a good idea. Gunther Teubner has suggested that it is more realistic to 'replace the over-optimistic model of "incentives through legal norms" by the more modest "social order from legal noise"'. He states that the corporate world 'perceives legal norms highly selectively and reconstructs them in a wholly different meaning context. Legal signals are reinterpreted anew, according to the inner logic of the concrete market and the concrete organization.'[63]

While there is much to commend the modest, organic approach, perhaps more can be done at a macro level to construct a legal environment that recognizes the limitations of law and is openly conducive to use of private regulatory programs as supplements to legal regimes: on a modest scale, legislators could statutorily recognize and encourage use of private regulatory programs,[64] regulators in their enforcement and compliance policies could encourage the development of voluntary programs,[65] and courts could more rigorously examine voluntary programs and encourage those that are rigorous and involve meaningful third-party input.

A more ambitious, conscious, and strategic effort to 'build' a voluntary environmental program on a legislative base is currently being undertaken by the European Union through the Eco-Management and Audit Scheme (EMAS).[66] Essentially, a legal framework has been put in place to encourage industry to adopt explicit and comprehensive environmental management procedures, as verified and audited by independent third parties. In the future EMAS may become a mandatory system, but for now it is not. Whether this legislative framework will successfully encourage voluntary adoption of EMAS, and whether it is transferable to a North American context, is not clear at this point. The ISO 14000 environmental management system represents a similar initiative to EMAS, yet it was developed with Canadian participation in the absence of the elaborate EMAS legal structure.

Whether a modest or a more ambitious approach is adopted, the analysis undertaken here suggests that legislators, regulators, and courts may need to develop a more sophisticated approach to voluntary initiatives as supplements to the laws. Similarly, intermediaries may need to adopt a more ambitious and sophisticated approach to the use of voluntary initiatives. Industry associations, following the lead of the Canadian Chemical Producers' Association, should consider moving beyond lobbying activity to undertake standard-setting and implementation activities on behalf of their members. Standards organizations may have to revisit the processes they use to develop standards, and the 'products' they produce, if they wish to have the support of NGOs in their initiatives.

More non-governmental organizations, following the lead of the World Wildlife Fund, Greenpeace, and Friends of the Earth in their Forest Stewardship Council initiative, and the UK Consumers' Association in its Web Trader initiative, may wish to explore the feasibility of developing and implementing voluntary programs in other areas, rather than simply participating in industry- or government-initiated programs. When NGOs actually seize the initiative and spearhead private regulatory regimes on their own, there is the opportunity for these groups to move from being 'invited guests' in governmental or industry projects to becoming 'hands-on' experts in implementation. The experience and understanding gained from seizing the reins in this manner, and working with others in the process, move NGOs from being lobbyists to being empowered direct agents of change. In so doing, NGOs are arguably capitalizing on their credibility in the eyes of the general populace, and the decreasing credibility of government and industry.

While it might once have been seen as sacrilege (i.e., the question could be asked, 'What self-respecting environmental NGO would risk the likelihood of losing its moral cachet, and its base of funding, to enter into arrangements with the private sector?'), we now have several examples of one-on-one partnerships between respected environmental organizations and major corporations striving for the reduction of pollutants[67] or more sustainable management of resources.[68] Perhaps even more surprising, governments are now lining up to obtain certifications for their resource management programs from NGO-led voluntary programs.[69] And in another reversal of roles, at least one high-profile consumer-oriented American multinational has agreed to require that its suppliers located in developing countries comply with US safety and health standards, with compliance monitored by an external panel of advisers.[70] Here we see an example of a firm in an industrialized country incorporating domestic legislative requirements into its contractual arrangements with its 800 supplier factories in 58 developing countries.

In short, it would appear that the potential for voluntary initiatives to work in conjunction with legal systems to achieve public policy objectives is only just beginning to be tapped, with the governments, industry associations, individual firms, and NGOs all beginning to take on new roles and responsibilities as never before. While not without weaknesses, and potentials for conflicts, together the two approaches seem capable of achieving more than either could working independently. At this point, it remains to be seen whether this flurry of experimentation with innovative modes of governance is just a passing phase, and whether legal systems, governments, industry, standards associations, and NGOs will fully embrace the opportunities and appropriately guard against the possible problems. Interesting days lie ahead.

Notes

This chapter represents the opinions of the author only. It draws substantially on the following: K. Webb, 'Voluntary Initiatives and the Law', in R. Gibson, ed., *Voluntary Initiatives: The New Politics of Corporate Greening* (Peterborough, Ont.: Broadview Press, 1999), 32–50; K. Webb and A. Morrison,

'Voluntary Approaches, the Environment, and the Law: A Canadian Perspective', in C. Carraro and F. Lévêque, eds, *Voluntary Approaches in Environmental Policy* (Boston: Kluwer Academic Publishers, 1999), 229–50; K. Webb, 'The Legal Framework for Voluntary Arrangements: Present and Future,' paper presented at the Canadian Law and Society Association annual meeting, St John's, May 1997; K. Webb, ed., *Voluntary Codes: Private Governance, the Public Interest and Innovation* (Ottawa: Carleton University Research Unit for Innovation, Science and the Environment, 2002).

1. See, e.g., Office of Consumer Affairs (Canada), *Market-Driven Consumer Redress Case Studies: Advertising Standards Canada* (2001), at: <http://strategis.ic.gc.ca/SSG/ca01652e.html>. For more information on the ASC, and the Canadian Code of Advertising Standards, see: <www.adstandards.com>. In the United Kingdom, a similar organization is the Advertising Standards Authority (ASA). For more information on the ASA, see: <www.asa.org.uk>. See also the European Advertising Standards Alliance, at: <www.easa-alliance.org>. For the United States, see the National Advertising Division of the Better Business Bureau Web site at: <www.bbb.org/advertising/nad>. See also the International Advertising Association, at: <www.iaaglobal.org>.
2. For example, the ISO 9000 quality management series. ISO 9000 is concerned with 'quality management'. It consists of 'plan-do-check-act' systems pertaining to what the organization does to enhance customer satisfaction by meeting customer and applicable regulatory requirements and continually improving its performance in this regard. ISO 14000 is primarily concerned with 'environmental management', that is, what the organization does to minimize harmful effects on the environment caused by its activities, and with continually improving its environmental performance. For more information concerning ISO 9000, see: <http://www.iso.ch/iso/en/iso9000-14000/index.html>.
3. The mission of the Better Business Bureau (BBB) is 'to promote and foster the highest ethical relationship between businesses and the public through voluntary self-regulation, consumer and business education, and service excellence'. See <http://www.bbb.org/about/aboutCouncil.asp>. The BBB has established the 'Uniform Standards of Membership' of the Better Business Bureau, and authorizes the use of its logo to businesses that comply with its standards, as discussed in Office of Consumer Affairs (Canada), *Market-Driven Consumer Redress Case Studies: Better Business Bureau* (2001), at: <http://strategis.ic.gc.ca/SSG/ca01654e.html>.
4. For example, see discussion of TRUSTe, a non-profit organization that licenses the use of its privacy seal ('trustmark') to companies that have in place appropriate privacy safeguards. See <www.truste.org>. This and other privacy programs are discussed in C. Bennett, 'Privacy Self-Regulation in a Global Economy: A Race to the Top, the Bottom, or Somewhere Else?', in K. Webb, ed., *Voluntary Codes: Private Governance, the Public Interest and Innovation* (Ottawa: Carleton University Research Unit for Innovation, Science and the Environment, 2002).
5. Programs include the Canadian Standards Association's Sustainable Forestry Management Standard, the Forest Stewardship Council's Principles and Criteria for Forest Management, the American Forest and Paper Association's Sustainable

Forestry Initiative, and the Pan European Forestry Certification Council as discussed in G. Rhone, D. Clarke, and K. Webb, 'Two Voluntary Approaches to Sustainable Forestry Practices', and K. Webb and D. Clarke, 'Voluntary Codes in the United States, the European Union and Developing Countries: A Preliminary Survey', in Webb, ed., *Voluntary Codes*.

6. See, e.g., the Responsible Care program of the Canadian Chemical Producers' Association, as discussed in J. Moffet, F. Bregha and M. Middelkoop, 'Responsible Care: A Case Study of a Voluntary Environmental Initiative', in Webb, ed., *Voluntary Codes*.

7. See, e.g., Office of Consumer Affairs (Canada), *Market-Driven Consumer Redress Case Studies: The Cable Television Standards Council* (2001), at: <http://strategis.ic.gc.ca/SSG/ca01656e.html>.

8. See discussion of the US Apparel Industry Partnership (AIP), consisting of labour, human rights, consumer groups, and several major apparel makers, in R. Liubicic, 'Corporate Codes of Conduct and Product Labeling Schemes: The Limits and Possibilities of Promoting International Labor Rights Through Private Initiatives', *Law and Policy in International Business* 30 (1998): 111–58; see also G. Rhone, J. Stroud, and K. Webb, 'The Gap's Code of Conduct for Treatment of Overseas Workers', and K. Webb and D. Clarke, 'Voluntary Codes in the United States, the European Union and Developing Countries', in Webb, ed., *Voluntary Codes*.

9. For example, see discussion of The Gap's apparel worker code in Rhone, Stroud, and Webb, 'The Gap's Code of Conduct'.

10. For example, see the UK Consumers' Association's Web Trader program, which bestows the Web Trader trustmark to on-line merchants who agree to adhere to a set of criteria for merchant reliability set by the organization. See <http://whichwebtrader.which.net/webtrader/>. See also the non-profit Canadian Automobile Association's Approved Automobile Repair Service (AARS), which grants use of the CAA AARS logo to garages that adhere to standards and submit to random inspections. Garages that fail to comply with the standards can lose their CAA status. See Office of Consumer Affairs (Canada), *Market-Driven Consumer Redress Case Studies: Canadian Automobile Association's Approved Automobile Repair Service*, at: <http://strategis.ic.gc.ca/SSG/ca01655e.html>.

11. For example, the Forestry Stewardship Council's Sustainable Forestry initiative was spearheaded by the World Wide Fund for Nature (also known as World Wildlife Fund), with the support of Greenpeace and Friends of the Earth, as discussed in Rhone, Clarke, and Webb, 'Two Voluntary Approaches to Sustainable Forestry Practices'.

12. I. Ayres and J. Braithwaite, in *Responsive Regulation: Transcending the Deregulation Debate* (New York: Oxford University Press, 1992), 102, describe co-regulation as 'industry-association self-regulation with some oversight and/or ratification by government'.

13. T. Wotruba, 'Industry Self-Regulation: A Review and Extension to a Global Setting', *Journal of Public Policy and Marketing* (1997): 38–54.

14. See, e.g., M. Baker, 'Private Codes of Corporate Conduct: Should the Fox Guard the Henhouse?', *University of Miami International-American Law Review* 24 (1993): 399.

15. The Chairman of the United States Federal Trade Commission, Robert Pitofsky, has

stated that 'The presence of government regulation on a given subject does not necessarily "preempt" private efforts to advance similar goals.' He also stated that 'Self-regulation can bring the accumulated judgment and experience of an industry to bear on issues that are sometimes difficult for the government to define with bright lines. Finally, government resources are limited and unlikely to grow in the future. Thus, many government agencies, like the FTC, have sought to leverage their limited resources by promoting and encouraging self-regulation.' R. Pitofsky, Chairman, FTC, 'Self Regulation and Antitrust', prepared remarks for DC Bar Association Symposium, 19 Feb. 1998, Washington. Available at: < www.ftc.gov/speeches/pitofsky/self4.htm > .

16. The Canadian cable television industry assumed self-regulation responsibilities over customer service activities following suggestions from the Canadian Radio-television and Telecommunications Commission (CRTC) that if the industry did not take on these responsibilities, the CRTC would. See Office of Consumer Affairs (Canada), *Market-Driven Consumer Redress Case Studies: The Cable Television Standards Council*. In other cases, the spectre of regulations being imposed is not directly mentioned by government officials, but is nevertheless understood by industry. For example, the vice-president of one US supermarket chain that adopted a voluntary third-party audited food safety program for its suppliers was quoted as saying, 'If we don't give them internally generated voluntary programs, we're going to get it from regulatory agencies.' R. Vosburgh, 'Produce Safety Audits are Consumer Driven', *Supermarket News*, 6 Mar. 2001, at: < www.primuslabs.com/ap/SN_0300.htm > . The desire to stave off new regulations was also clearly a motivation of the Canadian Chemical Producers' Association in its development of the Responsible Care program, as discussed in greater detail later in the chapter.

17. The Canadian Personal Information Protection and Electronic Documents Act, SC 2000, c. 5, began life as a voluntary 'Model Code for the Protection of Personal Information', developed by governments, the private sector, and consumer groups through the Canadian Standards Association (CSA). When the CSA Model Code was finalized, the Canadian Marketing Association (which had participated in the drafting of the CSA Code) urged the federal government and the provinces to develop legislation based on the Code. Canadian Direct Marketing Association (now the Canadian Marketing Association), 'Direct Marketing Industry Welcomes Federal Privacy Bill', 1 Oct. 1998, at: < www.the-cma.org/main.html > . This encouragement for government action was in apparent recognition of the fact that a purely voluntary approach would not fully address free riders who were not members of their association, but who could nevertheless sully the Association's good efforts.

18. For example, the CSA Model Code for the Protection of Personal Information, now enshrined in a schedule to the Personal Information Protection and Electronic Documents Act, as discussed above. In addition, the CSA hockey helmet standard is referentially incorporated into a schedule to the Hazardous Products Act, RSC 1985, c. H-3. See also A. Morrison and K. Webb, 'Bicycle Helmet Standards and Hockey Helmet Regulations: Two Approaches to Safety Protection', in Webb, *Voluntary Initiatives*.

19. For example, in *R. v. Prospec Chemicals* (1996), Alberta Judgment No. 174, Alberta Provincial Court, 25 Jan. 1996, counsel for Prospec, which was convicted of contravening Alberta sulphur emission limits, proposed that Prospec be permitted to

seek ISO 14000 certification as part of a court-ordered sentence. The judge agreed, ordering Prospec to complete the ISO program and post a letter of credit for $40,000 subject to forfeiture if the company failed to comply with the certification order. See K. Webb, 'Voluntary Initiatives and the Law', in R. Gibson, ed., *Voluntary Initiatives: The New Politics of Corporate Greening* (Peterborough, Ont.: Broadview Press, 1999), 33–5.

20. Both judges and legal commentators have suggested that compliance with environmental management systems such as ISO 14000 can reduce legal liability. See, e.g., comments to this effect by an American judge and a New Zealand lawyer in 'Reducing legal liability with an ISO 14001 EMS', in *Standards New Zealand Environmental Newsletter* 1, 1 (Feb. 1996): 1. See also J. Melnitzer, 'Fix environmental snags before seeking ISO 14000 certification', *Law Times*, 16–22 June 1997, 14–15, as discussed in Webb, 'Voluntary Initiatives and the Law'.

21. See, e.g., *Ripley v. Investment Dealers Association (Business Conduct Committee)*, [1991] 108 Nova Scotia Reports (2d) 38 (NS Court of Appeal).

22. As discussed by Pitofsky, 'Self Regulation and Antitrust'; see also K. Webb and A. Morrison, 'Voluntary Approaches, the Environment, and the Law: A Canadian Perspective', in C. Carraro and F. Lévêque, eds, *Voluntary Approaches in Environmental Policy* (Boston: Kluwer Academic Publishers, 1999), 229–50.

23. In a recent Ontario case, the court held that non-compliance with a recognized industry standard constituted evidence of lack of due diligence on the part of the accused. See *R. v. Domtar*, [1993] Ontario Judgments, No. 3415 (Ont. Court, General Division). In Italy, the Supreme Court recently held that the rules of the Italian Code of Advertising Self-Regulation were legitimate criteria for defining 'professional correctness'. European Advertising Standards Alliance, *Alliance Update* no. 14 (May 1999): 2.

24. For example, Industry Canada and Treasury Board, *Voluntary Codes: A Guide to Their Development and Use* (Ottawa, 1998). Available at: <www.strategis.ic.gc.ca/volcodes>.

25. For example, see quote of European Advertising Standards Association on the value of the two systems working together, but in harmony. See also the 'Criteria and Principles for the Use of Voluntary or Non-regulatory Initiatives to Achieve Environmental Policy Objectives', developed by the New Directions Group (NDG), a joint industry-environmental organization coalition, which has established principles for effective voluntary approaches.

26. See, e.g., earlier discussion of the UK Consumers' Association's Web Trader, CAA's Approved Automobile Repair Service, and FSC's Sustainable Forestry Management initiatives for examples of NGO leadership in use of private regulatory approaches. See also Pollution Probe's publication 'Voluntary Initiatives: Policy Framework and Roles' (Conference Proceedings, 14–15 June 1999): <http://www.pollutionprobe.org/Publications/Policy.htm>.

27. See, e.g., J. Braithwaite and P. Drahos, *Global Business Regulation* (Cambridge: Cambridge University Press, 2000); V. Haufler, *Public Role for the Private Sector* (Washington: Carnegie Endowment, 2001); K. Webb, *Voluntary Codes*.

28. See discussion of legal aspects of kosher and halal food certification schemes in K. Webb and A. Morrison, 'The Law and Voluntary Codes: Examining the "Tangled Web"', in Webb, ed., *Voluntary Codes*.

29. Interestingly, Braithwaite and Drahos, in *Global Business Regulation*, 497, n. 6, maintain that '[s]trictly speaking, following Durkheim, the Christian church was the first great NGO to supply a set of norms which regulated business practice.'
30. M. Olson, *The Rise and Decline of Nations: Economic Growth, Stagflation, and Social Rigidities* (New Haven: Yale University Press, 1983), 147.
31. S. Painter, *A History of the Middle Ages: 284–1500* (New York: Alfred A. Knopf), 229.
32. Ibid., 230–1; R. Larmour, 'A Merchant Guild of Sixteenth-Century France: The Grocers of Paris', *Economic History Review* 20 (1967): 469–79; J. Kellet, 'The Breakdown of the Guild and Corporation Control over the Handicraft and Retail Trade in London', *Economic History Review* 10 (1957–8): 381–94; A. Kiralfy, *Potter's Outlines of English Legal History*, 5th edn (London: Sweet and Maxwell, 1958), 103 ff.
33. History discussed in Office of Consumer Affairs (Canada), *Market-Driven Consumer Redress Case Studies: the Better Business Bureau*.
34. History discussed in Office of Consumer Affairs (Canada), *Market-Driven Consumer Redress Case Studies: the Canadian Automobile Association's Approved Automobile Repair Services*.
35. As discussed in Olson, *The Rise and Decline of Nations*.
36. The following discussion draws substantially on Moffet, Bregha, and Middelkoop, 'Responsible Care: A Case Study of a Voluntary Environmental Initiative'.
37. Ibid.
38. Jean Belanger, the CCPA president, as reported ibid.
39. The author has been told by CCPA officials that members of Responsible Care have been offered lower insurance rates.
40. CCPA, *Reducing Emissions 9: A Responsible Care Initiative. 2000 Emissions Inventory and Five Year Projections* (Ottawa: CCPA, 2000): < http://www.ccpa.ca/english/library/RepDocsEN/NERM00EN.pdf > .
41. Ibid.
42. Ibid.
43. Discussed in greater detail in Webb, 'Voluntary Initiatives and the Law', 32–50.
44. A. Duffy, 'Industry told its "free ride" about to end', *Ottawa Citizen*, 20 Mar. 1999.
45. For a more detailed discussion of the due diligence defence used in regulatory offences, see K. Webb, 'Regulatory Offences, the Mental Element, and the Charter: Rough Road Ahead', *Ottawa Law Review* 21 (1989): 419–78.
46. For example, in *Visp Construction v. Scepter Manufacturing Co.*, [1991] O.J. No. 356 (Ont. C.J., General Div.), the court found that a burst pipe that met a CSA standard met the due diligence standard. See also 'Reducing Legal Liability with an ISO 14001 EMS', *Standards New Zealand Environmental Newsletter* (Feb. 1996): 1.
47. Regulators in the United Kingdom have offered the prospect of reduced inspections to firms putting in place ISO 14001 environmental management systems. See K. Kollman and A. Prakash, 'Green by Choice? Cross-National, Variations in Firms' Responses to EMS-Based Environmental Regimes', 53 *World Politics* (2001): 422. The United States Environmental Protection Agency has announced a Performance Track program that, among other things, will include 'a lower priority for inspection targetting purposes' for firms with strong compliance records, an environmental management system of some form (not necessarily ISO 14001 or one audited by a third party), appropriate public reporting and outreach, and a commitment to pollution reduction and compliance. See National Academy of Public Administration (NAPA),

Environmental Government: Transforming Environmental Protection for the 21st Century (Washington, 2000), 47. The state of Connecticut is providing benefits such as expedited permit reviews, reduced fees, less frequent reporting, facility-wide permits, and public recognition for firms that are registered to ISO 14001, have adopted approved principles of sustainability, and have good compliance records. See State of Connecticut, *An Act Concerning Exemplary Environmental Management Systems*, Public Act No. 99–226 (1999).

48. For example, see Environment Canada's *Enforcement and Compliance Policy for the Canadian Environmental Protection Act, 1999*, at: <www.ec.gc.ca/enforce/homepage/cepa/CEPA99_final_eng.pdf.pdf>. However, this policy also indicates that, where there is a suspected violation, audit information may be used as part of enforcement activities.
49. Canadian Environmental Protection Act, 1999, SC 1999, c. 33, ss. 287(c) and ss. 291(1)(e).
50. As discussed in greater detail in Webb and Morrison, 'Voluntary Approaches, the Environment, and the Law'.
51. See, e.g., *Ripley v. Investment Dealers Association (Business Conduct Committee)*, [1991] 108 NSR (2d) 38 (NSCA).
52. The Competition Act implications of voluntary arrangements are discussed in greater detail in Webb and Morrison, 'Voluntary Approaches, the Environment, and the Law'. For the American perspective, see Pitofsky, 'Self Regulation and Antitrust'.
53. See Auditor General of Canada, *Managing the Risks of Toxic Substances: Obstacles to Progress* (Ottawa: Auditor General of Canada, 1999), at: <www.oag-bvg.gc.ca>.
54. Responsible Care members have voluntarily agreed to reduce use of a wide number of chemicals that are not yet regulated pursuant to the Canadian Environmental Protection Act, 1999. See CCPA, *Reducing Emissions 9: A Responsible Care Initiative. 2000 Emissions Inventory and Five Year Projections* (Ottawa: CCPA, 2000).
55. See, e.g., *Ripley v. Investment Dealers Association (Business Conduct Committee)*, in which the court held that disciplinary matters between associations and members were a matter of contract not subject to the Charter of Rights and Freedoms.
56. See, e.g., discussion of lax enforcement in Mexico in E. Brown Weiss, 'Environmentally Sustainable Competitiveness: A Comment', *Yale University Law Journal* (1993): 2132.
57. Of course, private voluntary programs may run into their own problems of underfinancing. According to Bregha and Moffet, 'Canadian Chemical Producers' Responsible Care Program', CCPA membership fees have doubled in recent years, in part to finance the Responsible Care program. The fact that the on-site independent compliance verifications are paid for by the companies concerned is one way of keeping the costs of the central administrative agency (e.g., the industry association) to a minimum.
58. See, e.g., S. Pargal and D. Wheeler, 'Informal Regulation of Industrial Pollution in Developing Countries: Evidence from Indonesia', *Journal of Political Economy* 104 (1996): 1314.
59. The formal, inclusive, highly visible, and accountable processes of public policy formulation can be viewed as one of its main features, a point discussed in Webb and Morrison, 'Voluntary Approaches, the Environment, and the Law'.
60. This prospect is discussed in Moffet, Bregha, and Middelkoop, 'Responsible Care: A

Case Study of a Voluntary Environmental Initiative'.
61. This is discussed in some detail in Webb, 'Voluntary Initiatives and the Law,' 33-5.
62. The following discussion introduces the concept of what is being referred to by commentators as 'reflexive law'. This concept and its implications in the environmental area are the subject of an excellent article by E. Orts, 'Reflexive Environmental Law', *Northwestern University Law Review* 89 (1995): 1227-1340.
63. G. Teubner, 'The Invisible Cupola', in Teubner, L. Farmer, and D. Murphy, eds, *Environmental Law and Ecological Responsibility* (West Sussex, England, 1994), 33.
64. The statutory recognition of use of environmental management programs in the Canadian Environmental Protection Act, 1999 represents a small step in this direction.
65. Through sentencing guidelines and enforcement policies, this is beginning to take place in some jurisdictions.
66. See discussion of EMAS in Orts, 'Reflexive Environmental Law'.
67. For example, see discussion of the collaboration between Dow Chemical and the Natural Resources Defense Council to identify and implement pollution prevention strategies at Dow's plant in Midland, Michigan, in National Academy of Public Administration, *Environmental Governance* (2000), 58. According to the report, teams found pollution prevention strategies to reduce nearly seven million pounds of emissions and wastes per year at an annual savings of some $5 million.
68. For example, the Marine Stewardship Council, devoted to sustainable management of fish and other marine animals, began in 1996-7 as a legally ratified partnership between Unilever (one of the world's largest buyer's of frozen fish) and the World Wide Fund for Nature. Later, it became fully independent from Unilever and the WWF. See *MSC Background*, at: < www.msc.org/html/content_5.htm >.
69. For example, the state of Alaska's commercial salmon fisheries management program has been certified as sustainable by the non-governmental Marine Stewardship Council, pursuant to its sustainable fishery standards. See < www.state.ak.us/adfg/geninfo/special/sustain/susthome.htm >. And agencies responsible for the management of state-owned lands in Minnesota, New York, and Pennsylvania have either achieved Forest Stewardship Council certification or announced that they intend to do so. Per E. Meidinger, 'Environmental Certification Programs and U.S. Environmental Law: Closer Than You May Think', *Environmental Law Reporter* 31 (2001): 101-69.
70. For example, as part of its apparel workers' code, Nike has committed that all Nike supplier factories would meet certain US Occupational Safety and Health Administration indoor air quality standards. See < http://www.nikebiz.com/labor/code.shtml >. According to a 21 September 2001 article in the *Seattle Times*, Nike has agreed to create a committee to oversee the company's labour, environmental, and diversity policies. The committee will include a Nike director and visiting scholar at the Massachusetts Institute of Technology, a former Nike president, and a former dean of Stanford University's Graduate School of Business. According to the same article, Nike has 500,000 contract workers in about 800 factories in 58 countries. 'New Nike panel to tackle company's factory issues', *Seattle Times*, 21 September 2001, at: < www.globalexchange.org/economy/corporations/nike/seatimes091101.html >. Watchdog NGOs such as Global Exchange admit some progress has been made, but remain skeptical. See, e.g., T. Connor, 'Still Waiting for

Nike to Do It' (San Francisco: Global Exchange, 2001), available at: < www.globalexchange.org >. According to Connor, 'Health and safety is one area where some improvement has occurred. But even here the company is not willing to put in place a transparent monitoring system involving unannounced factory visits' (p. 5).

Chapter 13

Cyberspace Governance: Canadian Reflections

MICHAEL MAC NEIL

Introduction

The Internet provides new tools and new perils for individuals as they engage in the complex interactions of their everyday lives. Whether for entering into consumer and commercial transactions, gathering information, participating in political debate and action, doing work, or communicating as part of their personal and intimate relations, Canadians are finding that the Internet is quickly becoming ubiquitous. It is one manifestation of the groundswell of a host of digital technologies that force us constantly to re-evaluate how we govern ourselves. Debates abound about the role of territorial-based governments, the efficacy of traditional legal techniques, the desirability for new forms of self-regulation, and the need for international co-operation. Cyberspace poses many thorny questions about how best to facilitate relationships and transactions, protect people from harm, encourage democratic practices, and ensure fundamental rights and freedoms. Cyberspace also raises fears among some about the ungovernability of harmful conduct, including such activities as illegal interception of telecommunications, electronic vandalism and terrorism, stealing telecommunications services, pornography and other offensive content, telemarketing fraud, electronic money-laundering, and telecommunications in furtherance of criminal conspiracies.[1]

A fundamental theme for this chapter is the complex interaction between state-centred law and the governance of cyberspace activity. Three key concepts underpin the following analyis: law, governance, and cyberspace. Law can be thought of in many different ways, from positivistic command-and-control regulation emanating from territorially based sovereign states to extremely informal customs and practices developed by the organizational structures and informal communities within which activity in cyberspace is conducted. This chapter adopts Peter Huber's position that the fundamental question is not whether there will be rules of law to regulate human interaction in cyberspace, but rather, where will the rules of law come from.² A central question in the debate is whether special rules should be developed to regulate cyberspace activity, or whether the traditional forms of regulation, especially the background rules of statute-based criminal law and private law developed primarily through the common law, provide a sufficient basis for ordering cyberspace interactions.

Governance as a concept is concerned with the processes by which people are able to regulate their interactions with each other, whether social, political, or economic. Governance concerns range from issues of how a sovereign governs its subjects to how communities and institutions govern themselves and how individuals govern their daily lives. Cyberspace, a relatively new 'place' in which people interact, includes the Internet and other forms of telecommunications.

Technological changes enable new forms of conduct to arise that were not previously possible and allow other forms of conduct to be engaged in more efficiently, extensively, and quickly.³ The technology may enable forms of conduct regarded as offensive, but equally, technology may provide the means by which that and other offensive conduct may be controlled. Because of the shift in the nature of the activities that are facilitated by the technology, one can also expect a shift in the social importance of competing values. For example, where more information is easily collected about individuals, their preferences, and their habits, one can expect that privacy will emerge as a much more important value to be cultivated. How privacy is cultivated—through technological protections, development of norms, or centralized regulatory efforts—is a central issue to be explored in this chapter.

This chapter also takes up the theme, set out in the introductory chapter of this volume, of the decentring of the state as it relates to regulation and governance of cyberspace. The chapter provides an overview of the current literature on the governance of the Internet, highlighting the plurality of perspectives on the future of law and the state in defining regulatory specifics and framing governance techniques. It then reviews a number of Canadian initiatives in regulating privacy, encryption, and Internet broadcasting. The chapter demonstrates the continuing importance of the Canadian state in governance of a new technological and social phenomenon, while also indicating the complex transformation in regulatory practices in a late-modern liberal democracy.

Theories of Governance

A variety of governance approaches have been advocated in recent scholarly literature about the Internet. These approaches reflect a diversity of viewpoints about the challenges posed by cyberspace to issues of legal techniques, regulatory schemes, the purpose of government, and the adaptation of law, regulation, and governance mechanisms to technological change. The approaches to governance can be divided into a range of categories. The anarchists emphasize supposedly unique features of the Internet, such as its global, decentralized, and non-hierarchical architecture, that make it extremely difficult and probably illegitimate for territorially based governments to enforce either their traditional legal rules or Internet-specific rules. The incapacity of nation-states to regulate is often lauded as a positive feature of the Internet, for, in the words of Wisebrod, '[a] policy of freedom will allow the Internet to evolve naturally and beneficially, becoming the paradoxical sum of its users' involvement: a self-controlling, yet uncontrollable, functional anarchy.'[4]

The anarchist position melds into one that emphasizes forms of self-regulation or transnational regulation as either the only possible or perhaps the best means of governing cyberspace. A central tenet of the advocates of self-regulation is that state-based law is not able or is not best-suited to the regulation of cyberspace activity. They argue that the preconditions for successful governance of physical space, on which traditional state sovereignty rests, no longer apply to cyberspace. Hence, geographically based state governance techniques cannot be effective in the governance of cyberspace, and indeed may not even be legitimately used.[5] These state skeptics tend to turn either to self-governance or to supranational institutions to provide the legal and co-ordinating bases under which cyberspace can flourish.[6]

What exactly might self-regulation or self-governance mean in this context? It could clearly take on a number of dimensions, ranging from self-help initiatives, including the use of both exit strategies[7] and technological or similar constraints for controlling behaviour,[8] to the development and enforcement of social norms,[9] to governance mechanisms that draw on the background rules of property and contract to give power to individuals to structure the terms and conditions on which they will engage in on-line interactions with others.

Self-governance may be desirable because it is more efficient. Cheaper and quicker cyberspace communication can contribute to more effective dissemination and enforcement of norms. As well, 'real-world' norms may be impractical to apply to cyberspace, and a self-regulating cyberspace community may make and enforce rules for itself that are more appropriate. Cybercitizens may be more willing to comply with these rules if they have helped to refine them through suitable self-governance regimes.[10] Hence the self-governance position is not solely dependent on the claim that nation-states cannot effectively regulate cyberspace activity, but actually demands a carefully nuanced evaluation of the appropriate locus of control for different law-making, norm-making, and enforcement activities. We already have many examples of self-governing communities and institutions in the material world, and hence it is not surprising that similar communities and institutions would arise in cyberspace as well.

Advocates of self-regulation do not, however, necessarily exclude any role for territorially based governing institutions. The limits on the capacity of self-regulating institutions are recognized, especially where the actions of members of sub-communities have an impact on non-members of those communities. These external effects are worthy of regulation by more traditional means. Hence, according to Reidenberg, a vision of governance is needed that acknowledges the complex mix of state, business, technical, and citizen forces. While the private sector is a driving force in the development of the information society, governments must be involved to protect public interests.[11]

The development and implementation of technical and network standards that define many facets of Internet access and use are important for governance. Several commentators emphasize the importance of such standards or 'code' in regulating choices for Internet users as they engage in their daily communicative interactions. Reidenberg, for example, describes the set of rules for information flows imposed by technology and communication networks as forming 'a "Lex Informatica", that policy makers must understand, consciously recognize, and encourage'.[12] Such a Lex Informatica provides means for resolving the difficulties of multi-jurisdictional application of local rules to such problems as content restrictions, treatment of personal information, and protection of intellectual property.[13] Reidenberg, Lessig,[14] and Boyle[15] all emphasize, however, the dangers that private development of technical and network standards pose to the achievement of public purposes. For example, a technical solution for the protection of copyright that makes it impossible to copy undermines the legal framework for copyright, which encourages the development of a public domain and allows copying for fair use or fair dealing. The technology may not only undermine the capacity of the state to engage in regulation in pursuit of public purposes, it may also provide a means by which governments can enlist private parties to accomplish indirectly that which they may be forbidden from doing directly.[16] They may aid in the proliferation of a surveillance society that will enable both public and private actors to track closely the actions of citizens, consumers, employees, and others in a fashion that seriously undermines respect for privacy and autonomy. The irony is that the same technologies can also be used to enhance the ability of individuals to take control over their own actions, whether through the use of encryption to protect their privacy or the use of filtering tools to determine what information gets through the software fences that they create for themselves.

Many commentators have emphasized the importance of contractual paradigms to cyberspace governance. Gibbons,[17] for example, argues that contract has the advantage of flexibility, adaptability, and respect for libertarian values where there is a true meeting of minds through negotiated terms. The danger of contract is that it may lead to the imposition of standard forms that jeopardize fundamental liberties or fair treatment.[18] Hence, Gibbons advocates the adoption of a contractual governance paradigm, subject to the caveat that contracts of adhesion be avoided. For those cyberspace activities with external impacts, Gibbons argues that existing legal rules should apply, without the need to develop special regulatory rules for cyberspace.

Yet another perspective on cyberspace governance emphasizes either the likelihood that international regimes will emerge to govern cyberspace activity or the desirability of such a result. The global interconnectedness of Internet users, combined with concerns about the feasibility of regulation by nation-states, leads some to advocate the development of international agreements on cyberspace standards and enforcement processes. As Purvis states in Chapter 2, 'the forces of "globalization" have impinged on the capacity of states to act autonomously in a world that has been described as simultaneously shrinking and fragmenting.' Even if nation-states are able to regulate cyberspace activity, there may be good reasons for them to act in co-operative fashion to establish international governance structures, or to grant a minimal level of autonomy to independent cyberspace governance mechanisms.[19] One possibility is to treat cyberspace as an international space in the same fashion that the high seas, Antarctica, and outer space are treated as international spaces, the legal rules for which are rooted in principles of international law.[20]

Other commentators contest the claim that cyberspace marks a radical challenge to the nation-state. These critics question the anarchical assertion that the nation-state cannot regulate content and activity on the Internet, noting both the potential of technical solutions to law enforcement and recounting law enforcement 'success' stories.[21] Shapiro argues that the anarchists are mistaken when they conceptualize cyberspace as a separate 'place'.[22] He claims that the most important point to remember is that on-line actions by human beings have real impacts on the lives of other human beings. The origins of the Internet, with its difficult technical interfaces and specialist participants, may have encouraged the idea that getting into cyberspace was like going to a different 'place', but increasingly the Internet is seen in instrumental terms as a system that facilitates new forms of economic exchange. '[T]he almost supernatural notion of cyberspace as a place "to go" is losing currency. The mystery of cyberspace is fading.'[23] Instead, Shapiro advocates that we think of cyberspace as a locus of control, 'an interface that allows us to control other things— the information we are exposed to, the people we socialize with, the resources of the physical world.'[24] The consequence of perceiving cyberspace as non-autonomous does not lead to de-emphasizing its importance. Rather, the consequence is that cyberspace activity will be seen as having profound impacts on everyday activity. In terms of governance, Shapiro argues that there is no need to develop a separate law of cyberspace. Rather, traditional legal categories continue to apply, albeit with necessary adaptations for on-line activities. The real danger, though, is that the regulatory power of cyberspace itself, through the architectural constraints that it places on individual choice, has the potential to undermine existing values and rights. The governance challenge is not merely to regulate on-line behaviour, but to place limits on the regulative impact of cyberspace in everyday life.

Trachtman, too, rejects the argument that because cyberspace cannot be neatly cabined within the territorial jurisdiction of any one state, no state can legitimately or effectively regulate cyberspace.[25] In particular, he rejects the idea that domestic governing institutions must inevitably defer either to private market forms of governance or to international governance institutions. He contends that some things are likely to be governed by the private market, some by international institutions, and some by the state. Ultimately, he argues that the choice of which institution is best suited to deal with which problems should be guided by a search for economic efficiency, but inevitably will be influenced as well by existing institutional contexts. For Trachtman, cyberspace marks 'a technical change that modifies the transaction costs and benefits profile of various social and private arrangements.'[26] The rise of cyberspace may be cause for revising how sovereignty and jurisdiction work in governance structures, but should not be seen as a reason for abandoning those concepts altogether. Hardy[27] echoes these themes, concluding that the best approach is to apply a presumption in favour of decentralization of solutions, as such solutions are likely to be most flexible and least intrusive, criteria that are especially important in the face of the rapidly changing nature of communication technologies. Hence, problems should be approached with a presumption that centralized law-making authorities should initially respond by doing nothing, allowing individuals to further their interests either by withdrawing from activities they do not like or by acting in a co-operative fashion, through contracts and associative activity, to regulate activity. Only where contractual and associative activity create or allow significant external (and presumably harmful) effects on third parties is intervention by statute or judicial decision-making justified.

Goldsmith also argues that the regulation skeptics have incorrectly assessed both the capacity and the legitimacy of state-based institutions to regulate cyberspace activity,[28] at least from the perspective of jurisdiction and choice of law. In particular, Goldsmith argues that it is legitimate for nations to engage in regulatory activity that attempts to control the local effects of extraterritorial conduct, and that efforts at local regulation with an extraterritorial effect on transnational activity are not inherently illegitimate. It may be desirable to subject many of the default rules governing cyberspace to mechanisms of private ordering, in which those who control access to the network can use their property and contractual powers to structure the conditions of use. However, many may be affected by cyberspace activity who do not have the opportunity to consent to a cyberspace regime, there may be significant costs to establishing a consensual-based regime, and mandatory laws may be designed to protect citizens from being unfairly taken advantage of through private ordering mechanisms. In all these situations, a nation-state can and should be expected to engage in regulatory activity even though it may have extraterritorial effects. It may be true that the ability of a nation-state to enforce its regulations could be limited by various exigencies, and therefore regulatory impact is likely to be greatest on those users and service providers who are physically present or who have assets within the territory over which the state exerts jurisdiction.

Even if one accepts the ability of nation-states to regulate Internet activity, one must still ask whether such regulation would be a good thing, and whether it is necessary to develop new rules that are sensitive to the unusual features of cyberspace. Some warn against undue haste by courts in accepting jurisdiction over Internet activity where the contact with the political jurisdiction is minimal.[29] The concern is that the risk of multi-jurisdictional liability, especially for small businesses seeking to take advantage of the marketplace opportunities created by the Internet, may raise the price of participation beyond the price such entrepreneurs are willing to pay.

Finally, some commentators have noted that an important function that nation-states or international institutions may be expected to fulfill is to provide a legal basis for the organizational structures that carry out many of the governing tasks in cyberspace, including the assignment of domain names, the resolution of domain name disputes, and the development of technical protocols.[30] Similarly, government participation in the development of standards may be an important contribution to the orderly and efficient development of those standards.[31]

Canadian Governments and Governance

Over the past five years, the Canadian government has been struggling to define its role in relation to governance of the information highway, cyberspace, and the Internet. While the Canadian state is more likely to rely on self-help, private, and market initiatives to respond to many complex problems, it nevertheless has not abandoned a regulatory function and does play a vital role in shaping the space in which 'non-state' initiatives take place. It has created advisory committees, task forces, and other consultative mechanisms to obtain advice on what should be done. A number of themes have begun to emerge as governments sort out their role.

In its 1995 report, the Information Highway Advisory Council stated that the Canadian government had no choice but to become involved, its primary responsibility being to establish the ground rules and to ensure that there are effective enforcement mechanisms.[32] A major theme of the recommendations of the report is that the technical and economic infrastructure must be built on market-driven, competitive forces, but with a sensitivity to traditional Canadian policies, which seek to promote equality and maintain our cultural identity. As well, the increased availability of information presents a threat to security and personal privacy that governments have a role to play in protecting. In addition to calling on the government to set the ground rules, update the regulatory environment, and move to sustainable competition and marketplace rules, the advisory report suggests that the government should play a crucial role by being a co-ordinator, a catalyst, and a model user.

The broad range of issues facing the government in setting down the ground rules requires careful reflection on what to do on many legal fronts. The application of broadcasting and telecommunications policy, intellectual property rules, the facilitation of electronic commerce, the appropriateness of traditional common-law rules, the need for new criminal-law rules, and the establishment of appropriate governing mechanisms all must be confronted by the government. It is clear that the government has rejected the arguments of the anarchists that cyberspace is ungovernable by territorial-based governments. Rather, the challenge has been to articulate the values that Canadians want to see preserved and defended in the face of the difficult-to-predict social impact of cyberspace on social behaviour and moral attitudes.

Subsequent reports and studies have laid out or are seeking to determine some of the frameworks for further government action. For instance, the Canadian Radio-television and Telecommunications Commission (CRTC) initiated a series of hearings on the regulation of the new medium, with the goal of articulating how the broadcasting and telecommunications regulatory framework should (or should not) apply to Internet activity. Another report commissioned by Industry Canada served to demonstrate the feasibility of applying existing legal regimes to cyberspace activity.[33] This chapter will focus on three specific issues with which the government has concerned itself in the past two years: protection of privacy, regulation of encryption, and regulation of the Internet as a form of broadcasting.

Privacy

Privacy is emerging as a central issue in the governance of cyberspace. The ubiquitousness of digital information creates new, cheap, and insidious challenges to the protection of privacy. Concerns about privacy are a quintessential twentieth-century phenomenon, with the earliest focus on the fear of surveillance and monitoring by Big Brother state apparatuses. Whether for purposes of tax collection, rule enforcement, or policing, surveillance is a crucial tool of the state.[34] However, capitalist enterprise is also a primary site for the exercise of surveillance and for technological and organizational innovations in carrying out such surveillance.[35] Ranging from monitoring workers to tracking consumers, private enterprise has a keen interest in collecting and using personal data in pursuit of profit-making activities.

Limited forms of legal protection of privacy emerged through the development of common-law tort protections, the reading in of privacy as a protected constitutional value, followed by legislated standards designed to control the collection and use of information by governments. With the increasing pervasiveness of a digital economy, the concerns have expanded to the security of information as it is transmitted through the labyrinth of telecommunications systems and as it is collected, stored, and used by private actors for their own commercial, political, or other ends. The fears about privacy are seen as a potential barrier to the full development of electronic commerce, and therefore as an issue that must be addressed in order for the digital economy to achieve its full potential. How it should be addressed is a key question in the context of cyberspace governance. Choices range from enhancing individual autonomy and self-help by the development of privacy-enhancing technologies, to self-regulation through the development of standard practices and privacy guidelines, to the extension of legislated solutions to private activity.

The federal government's approach was revealed in a 1998 discussion paper indicating that it would introduce legislation to balance 'the business need to gather, store, and use personal information and the consumer need to be informed about how that information will be used and assured that the information will be protected'.[36] Note the implicit assumptions. First is the recognition that private businesses have become major gatherers, users, and archivists of information and that this is a legitimate private-sector activity. Second, it is assumed that consumers, on their own, will not be able to take sufficient self-help measures to ensure the needed confidence level for them to engage in electronic transactions. Third, it is assumed that legislative intervention is needed because the use of standard codes developed by industry or standards associations are insufficient to ensure an adequate level of privacy protection. Fourth, although the quotation itself does not say so, another guiding principle behind the government actions is the recognition that national policies must co-ordinate with the privacy policies of other national governments. Each of these assumptions deserves further elaboration.

The first assumption behind the development of Canadian privacy policy is a key to unlocking what is going on. Privacy protection is being promoted not as an end in itself, but as a means of inducing the requisite degree of consumer confidence to ensure that information is made available to maximize the efficiency of information use by business. Thus the development of more codes, conventions, and laws to protect privacy proceeds apace, without in any significant way dampening the trend towards the collection of 'more data on more people by more powerful systems and for more purposes than at any time in history.'[37]

The second assumption highlights the inadequacy of self-help mechanisms and individual responsibility as a means of protecting privacy. Although some observers may insist, consistent with classical liberal views, that self-help, rather than government intervention, is the only appropriate means of privacy protection,[38] regulation of privacy over the past 20 years demonstrates an increasing willingness of legislatures to extend controls to ensure the protection of data in both private and public settings. Privacy is not necessarily protected for the sake of privacy itself, but to create a space within which individuals as consumers will be active participants in market activity. Individual responsibility will not work on its own because of inadequate consumer knowledge and insufficiently effective techniques. This view of the need for state intervention is consistent with the claim made by Purvis in Chapter 2 that neo-liberalism assumes a role for the state in shaping, regulating, and governing in order to facilitate individuals as actors in a world infused with enterprise culture.

The third assumption is that self-regulating mechanisms are insufficient as well. Although industries such as banking and direct marketing have sought collaboratively to develop codes of practice defining appropriate privacy measures, these initiatives are regarded as inadequate for several reasons. First, not all businesses subscribe to the codes or regard themselves as stakeholders in their promulgation. Second, the codes do not provide sufficiently robust enforcement mechanisms. Third, there is a concern that the codes do not sufficiently balance the interests of citizens and consumers against the needs of business. While the federal government in introducing the Personal Information Protection and Electronic Documents Act[39] demonstrated considerable willingness to rely on the privacy framework developed by the Canadian Standards Association in its Model Code for the Protection of Personal Information, the key to the government's actions was the introduction of mandatory provisions and the creation of an enforcement mechanism. Governmental action in this instance reflects what Webb, in Chapter 12, describes as common weaknesses of private, market-oriented regulation: lower credibility and visibility than legislative regimes, less likelihood of rigorous standards, difficulty in applying the rules to free riders, variable public accountability, and a more limited array of enforcement options.

Fourth, the government has been very explicit in acknowledging that a primary incentive to the introduction of legislated privacy protection for the private sphere is a response to global developments. In particular, the European Directive on the Protection of Personal Data[40] makes it a condition of its database regulation initiatives that information can be exported only to those countries that have introduced similar protections for the data. This reflects what Colin Bennett has described as the strong pressures for policy convergence that have forced many states to legislate a broadly similar set of statutory principles protecting private information.[41] It is also consistent with what Purvis, in Chapter 2, describes as the dramatic internationalization of economic relations that contributes to the extension of political action beyond the boundaries of the nation-state.

As a result of these assumptions, and following consultations pursuant to the 1998 discussion paper, titled *The Protection of Personal Information*, the federal government has enacted the Personal Information Protection and Electronic Documents Act.[42] The Act purports to protect personal information and to set the ground rules for facilitating increased use of electronic documents. Although background documents explicitly link the legislative initiative to the promotion of electronic commerce, the Act in fact applies to all organizations that collect, use, or store personal information in the course of commercial activities.[43] This reflects a recurrent theme throughout the debate about cyberspace governance: the need to develop legal regimes that are neutral with respect to technology.

All organizations covered by the Act are required to comply with the standards set out in Schedule 1 of the Act. These standards are the ones contained in the National Standards of Canada Model Code for the Protection of Personal Information.[44] The Act goes beyond the Model Code, both by imposing some more rigorous obligations and by providing an enforcement mechanism under the aegis of the federal Privacy Commissioner. Whether the Act is likely to achieve the objective of protecting privacy and inducing a sufficient degree of consumer confidence to enable electronic commerce to blossom remains to be determined. It is also evident that the protection of privacy may quickly be eroded in the interests of fighting terrorism following the 11 September 2001 terrorist attacks on the World Trade Center in New York.[45] Nevertheless, the initiative is important because it demonstrates an ongoing commitment to develop regulatory frameworks where there is considerable evidence to demonstrate the inadequacy of individual self-help and self-regulation models.

Encryption

Prescriptive governmental regulation is not the only means by which privacy can be protected. A recurring theme in cyberspace governance debates is that governance is best effected through private means. The adoption of the Model Code for the Protection of Personal Information by the Canadian Standards Association is one form of such private action that could be extended to cyberspace activity through the co-operative norm-building processes of cyberspace stakeholders. Another is through more direct self-help means, empowering individuals to take effective action to protect their own privacy. This empowerment is made possible by the development of technical tools that allow individuals to control what, if any, information about themselves is made available. If an individual is able to engage in cyberspace relations and transactions on an anonymous basis, then there may not be any pressing need for governmental intervention. Similarly, if the individual is assured that private information cannot be monitored by organizations, governments, or others to whom the individual has not granted access, the psychological assurance of privacy and security needed to facilitate cyberspace activity might be achieved without government participation. The development of encryption techniques may well facilitate such individual empowerment.

As I was gathering information for this chapter, I did a search of the Web pages of IntellectualCapital.com,[46] an on-line site dedicated to discussion of current public policy issues. After I had entered the word 'encryption' in the site's search form, the following message was flashed on my computer screen, as a result of the machinations of the browser I was using[47]: 'Any information you submit is insecure and could be observed by a third party while in transit. If you are submitting passwords, credit card numbers, or other information you would like to keep private, it would be safer for you to cancel the submission.' The message highlighted a number of things: that I was actually submitting information over the Internet, information that I had chosen to submit; that such information could be observed by a third party while in transit; that in some situations I, as a transmitter of information, may have an interest in keeping information private, at least from third-party snoopers who may be monitoring my communications; and, finally, that I can exert some control in order to look after my own privacy interests.

That control can be exerted in several ways. One would be not to submit the information at all. The problem with that is that I would not be able to carry out the search for information that I was wanting to acquire. A second choice would be to transmit the search term, knowing that it could be read by somebody else. In making that decision, I might carry out a cost-benefit analysis. I could consider what use might be made by others of the information in a way that could damage my interests. I would consider the likelihood that somebody was actually monitoring my communications. If I was engaged in research with potentially lucrative returns, and if I was fearful that a competitor was keeping track of what I was up to, that may influence my decision about what to do. Alternatively, if I knew that the government of the jurisdiction where I live had imposed repressive measures on the use of encryption and might be monitoring citizens to determine if they were attempting to evade those restrictions, I would think twice about submitting the search term. If the costs were too high, I would forgo the opportunity of gathering the information in this way. Finally, however, another choice might allow me to have the benefits of my search without the corresponding costs. If I could send the information in encrypted form, so that third-party snoopers would only see unintelligible data, and if the Web site to which I was directing the communication could unencrypt the information to carry out the search I had intended, then I might feel secure that the dangers of sending the information are minimized. This option, of course, depends on the ability to encrypt the information and have it unencrypted only by the intended recipient.

Encryption is a process whereby plain text can be transformed into cyphertext—unintelligible data—to ensure its confidentiality. In turn, a person with the appropriate key can take what appear to be unintelligible data and transform the scrambled text to its original intelligible form. Modern encryption techniques rely on the science of cryptology to produce encrypted text using mathematical algorithms. An essential element in encryption is the use of a key to encrypt the text and the use of a related key to decrypt the text. The encoding or decoding keys can be the same, or they may be different. Prior to the 1970s, the efficacy of encryption techniques depended on the encrypter and the decrypter sharing a common key. The security of encoded material required a means of ensuring that the key was shared between users without having it revealed to non-users. This entailed some secure channel for sharing the key, separate from the channel used to transmit the coded text. For persons who were not in face-to-face contact, this significantly limited the availability of encryption to those who could devote the resources to sharing keys by secure means. The science of cryptology has now progressed to the point where a dual-key system can be used—one a public key available to all, the other a related private key available only to the encrypter or decrypter. This makes encryption a relatively easy and inexpensive process that can be available to almost anybody. In a digital world where many are concerned about issues of privacy, confidentiality, security, and authenticity, effective and efficient encryption processes are a crucial prerequisite to making the on-line world a desirable place in which to carry out transactions and to communicate with others.

It is clear that security of on-line communications and the privacy of transmitted information are major concerns for the average citizen.[48] Those visionaries who dream that cyberspace will transform the marketplace realize that these concerns must be addressed. Encryption offers an essential component to the architecture of cyberspace that would enable privacy and security concerns to be dealt with in ways that induce the needed trust, which is a precondition for cyberspace to become favoured as a place for economic transactions.

Encryption does have a darker side, however. The fear has been expressed that encryption will also facilitate communications by organized crime and terrorists in ways that will allow them to evade the monitoring and detection efforts of law enforcement and national security officials. This concern was shown in a 1997 resolution of the Canadian Association of Chiefs of Police urging the government to implement a public key infrastructure requiring licensing of certification authorities and a mandatory key recovery regime that would provide for lawful access to cryptographic keys.[49] Encryption schemes have become sufficiently robust that it is highly unlikely that encoded material can be decoded unless one has access to the key.[50] A key recovery regime would require the establishment of a system whereby encryption keys would be maintained by trusted third parties, with the use of encryption not providing for key recovery being prohibited. In passing the motion, police chiefs were reflecting the view that cyberspace was just one more place where criminal activities could take place and that governments have a legitimate role to play in structuring relations and activities in cyberspace to maximize the potential effectiveness of law enforcement procedures.

On the other side of the debate are the proponents of limited government interference with cyberspace activity, who are greatly concerned not only about establishing the conditions for robust commerce in cyberspace, but also about maintaining civil liberties. The Electronic Frontier Foundation's policy statement on encryption best exemplifies this approach.[51] It argues that there should be no regulation on the use of encryption, either for stored data or for real-time communications, and that there should be no limits on the export of encrypted material. The EFF's arguments are based on feasibility of government regulation, the potential harms to privacy and confidentiality, and the desirability of promoting a robust, indigenous, Canadian encryption industry. In other words, members of this Foundation are looking, among other things, to their own potential for economic gains.

Until recently, Canada's public policy on encryption was entirely concerned with the prevention of export of encryption products that might jeopardize its strategic interests or those of its allies. It is a signatory, along with 32 other countries, to the Wassenaar Arrangement, which requires export controls to be placed on a long list of dual-use products, including cryptography products. In 1998, the government published a discussion paper outlining policy options in the regulation of encryption.[52] The paper indicated a range of goals that should be addressed in such a policy, which must:

- help realize the economic and social benefits that can be derived through the use of cryptography in secure global electronic commerce;
- ensure business and public confidence in the use of certification authorities, other cryptographic service providers, and cryptography product suppliers in Canada;
- respond to the challenges when lawful access to encrypted real-time communications or encrypted stored data is mandated;
- and respond to the challenges posed to national security information-gathering capabilities by the international spread of strong cryptographic products.[53]

The consultations following the release of the discussion paper demonstrated a strong commitment to minimizing governmental regulation of encryption.[54] Despite the calls of law enforcement officials and national security interests for establishing a key escrow system to facilitate the ability to gain access to encrypted information, the government announced in October 1998 that it would not engage in extensive regulation.[55] While it would continue to fulfill its obligations under the Wassenaar Arrangement, it committed itself to streamlining the processes by which export permits could be obtained. It refused to introduce legislation requiring the use of key escrow systems, although it did commit itself to promoting 'best practices' by certification authorities. The only concession to the law enforcement concerns was a commitment to introduce legislation making it a criminal offence to use encryption in committing a crime.

This new encryption policy has several notable features. It reflects the recurring theme throughout much of the federal government's attempt to develop a regulatory strategy, namely, that such strategy is primarily aimed at encouraging private initiative and the facilitation of marketplace transactions. Second, it also reflects a recognition that individual self-help mechanisms, through the use of technological or 'code' solutions, may be the most efficient means of addressing individual concerns about privacy and the security of online transactions. Third, the continued commitment to the Wassenaar Arrangement reflects yet again on the commitment to international co-ordination in the development of cyberspace policy. Finally, the failure to address substantively the law enforcement demands for assured access to encrypted communications may be explained on a number of grounds. It may reflect concerns about the effectiveness of such intervention strategies as an aid to law enforcement, especially given the chilling effect it could have on the use of encryption in electronic commerce. Alternatively, it may also reflect to some extent the claims by civil libertarians and social activists regarding rights of free expression and the potential importance of encryption in international human rights struggles. Finally, it may merely be an admission that the government is powerless to prevent the use of encryption in the face of easy distribution of encryption software.

Broadcasting, Telecommunications, and Cyberspace: Governing Convergence

Canadian governments have a well-established history of regulating broadcasting and telecommunications. Broadcasting regulation has been justified on the ground that broadcasting spectrum was scarce, and therefore it was desirable to ensure that its use fulfilled the public interest. This has been accomplished both by establishing a government-owned broadcasting system, the Canadian Broadcasting Corporation, and by licensing broadcasters. Licensing enables the control of broadcasters by setting terms and conditions under which the licensee must operate. Two major themes have informed the licensing process: the desire to encourage Canadian content and to provide universal service.

Similarly, the regulation of telecommunications has a venerable history in Canada, along with the regulation of utilities and transportation. A major justification for regulatory intervention has been the fear that market competition is likely to be undermined by a natural tendency towards monopolization. Hence, regulatory intervention in the past has been designed not to ensure that competition occurs, but to ensure that the natural monopolies that do develop are not able to exploit unduly their market position and extract exorbitant rents. In addition, telecommunications regulation has been designed to ensure universal access to services at relatively equal cost to all citizens, a goal that has often been achieved only by cross-subsidization of some services by others.

The assumptions on which the regulatory framework were built have begun to disintegrate. The development of new technologies such as cable television, satellite television, fibre optics, and digitization have all served to alter significantly the scarcity problem, so that there may be room for all entrants to the broadcasting field. Similarly, significant deregulation of telecommunications services has occurred as a result of the growth of competition in areas where it was only possible or efficient in the past for one organization to deliver the services. Furthermore, the distinction between broadcasting and telecommunications has begun to break down, with telephone companies now able to deliver broadcasting services and cable television companies able to provide phone services. Equally significant, from our perspective, is that the growth of the Internet is challenging traditional means of delivering these services. The convergence of broadcasting, telecommunications, and cyberspace thrusts into the public policy forum the question of what role, if any, the government should play in regulating the Internet, especially in terms of protecting Canadian culture, promoting Canadian content, or ensuring universal access to cyberspace.

This debate has most recently played out through consultations conducted by the CRTC. In a 1995 report[56] the CRTC reported on the trend towards convergence, recommending that regulatory initiatives be directed at the promotion of fair and sustainable competition, thereby providing consumers with access to choice of distributors of telecommunications and broadcast services. In effect, the traditional divide between telecommunications and broadcasting was recognized as disintegrating, and competition was seen to provide a key mechanism for promoting the public interest. There was, however, a continuing recognition of the role of governmental regulation in protecting and promoting a separate Canadian culture and identity, to be achieved by continuing to impose rules to ensure the availability of Canadian content. The report did not, however, fully address how the Internet fit within the larger regulatory picture. This became the focus of a subsequent round of consultation, leading to a May 1999 report of the CRTC on the regulation of what it called the 'new media'.[57]

During the extensive consultations leading to the report, there was a strong consensus that the CRTC should not, pursuant to its powers under the Broadcasting Act, engage in the regulation of the Internet. This recommendation was accepted by the CRTC, which in its report concluded that the majority of Internet activity does not fall within the definition of broadcasting under the Act, either because it is primarily alphanumeric text or because the individualized, one-on-one nature of the interaction means that the material being delivered is not a program being transmitted to the public.[58] For that portion of Internet content that does come within the definition of broadcasting, namely, audiovisual signals and digital audio services, the CRTC decided that it would exercise its jurisdiction to grant an unconditional exemption from the licensing requirements of the Broadcasting Act.[59] It concluded that licensing was inappropriate because unregulated private industry was already producing a significant amount of Canadian content and because licensing of 'new media would not contribute in any way to its development or to the benefits that it has brought to Canadian users, consumers and businesses.'[60] The CRTC did not, it claimed, want to impede the energy and creativity of private actors by imposing unneeded regulation.[61]

The CRTC also addressed the issue of regulating objectionable content such as hate propaganda and obscenity on the Internet. The CRTC plays a role in regulating such objectionable content within its broadcasting jurisdiction, and hence it was natural to consider whether such forms of regulation are appropriate to the Internet. The CRTC rejected the broadcasting model, first noting that it had no jurisdiction to regulate Internet material that was primarily alphanumeric text, but more importantly arguing that laws of general application, combined with self-regulation models, were more suitable for the task. In particular, the CRTC pointed to both industry-developed codes of conduct and the availability of filtering software that would allow individual users greater control over access to Internet content within their homes or places of business.

The CRTC position reflects the ongoing tensions about how to govern cyberspace. The unwillingness to regulate content, in terms of either promoting and protecting Canadian culture or controlling objectionable content, confirms a shift in regulatory discourse that increasingly emphasizes the desirability of self-regulation and privatized, market solutions. Although not fully explicated in the CRTC's reasoning, there also appears to be considerable doubt about the efficacy of command-and-control regulation, especially given the global dimensions of the Internet. The reluctance to intervene at this stage of cyberspace growth is attributed to the fact that very little 'real' broadcasting activity is taking place on the Internet. This position, of course, can be challenged, as there are already a large number of radio stations broadcasting over the Internet, and many organizations offer significant amounts of video, including the Canadian Broadcasting Corporation. The question, then, is whether it will be possible to maintain a separate regulation of traditional broadcasting content in the face of Internet competition. Convergence demands a uniform regulatory regime, and the option is either to abandon attempts to protect Canadian content through licensing mechanisms and rely solely on incentive-based approaches or to extend broadcasting-style protection to Internet-based broadcasting. The CRTC has left open the possibility of revisiting the question in several years when the industry is more mature. However, the interim hands-off regime may have a significant impact on exactly how the maturation of the industry proceeds.

In refusing to engage in the regulation of offensive content, the CRTC has repeated the mantra of the desirability of technologically neutral rules, the desirability of providing individualized control to parents and end-users, and the importance of industry taking responsibility for developing standards or codes. These are the same themes raised in the debates about the regulation of privacy and encryption, and reflected in the many government reports that have been attempting to define the government's role in the governance of cyberspace.

Conclusion

This brief survey of several Canadian initiatives demonstrates several points. First, it is quite apparent that the nation-state is not irrelevant to the question of how cyberspace will be regulated. The range of initiatives being considered or already undertaken by the Canadian federal government shows that cyberspace is not regarded by lawmakers and governmental actors as a separate place beyond the jurisdictional reach of the nation-state. Shapiro's claim[62] that the Internet is increasingly seen in instrumental terms as facilitating new forms of economic exchange seems to best characterize the Canadian government's initiatives. This is consistent with the point made in Chapter 2 by Purvis that neo-liberal influences on the state have brought a market rationality into virtually all spheres of life. The challenge, then, becomes to put in place a basic framework that enhances the potential of the Internet, but includes taking into account the need to control some of the potential dangers of cyberspace interactions, in order to provide the sense of security and confidence needed to maximize the potential of the new economic spaces. We are quickly getting beyond the vision of cyberspace as a wild frontier in which justice is meted out by informal means free from state-based authority.

Nevertheless, there are clear constraints on the extent to which traditional forms of command-and-control regulation are likely to be used and the extent to which Canada-only solutions to the problems are likely to be adopted. For example, the move towards the enactment of privacy legislation demonstrates considerable reliance on private initiatives, using standards that are drawn from an industry-established code of practice. Although the proposed legislation would modify some of those standards, and while an enforcement mechanism has been added, the protection offered still is premised on individuals being able to take responsibility for themselves. This approach is further reinforced by the government's position on the use of encryption, which effectively accepts the legitimacy of technological solutions that empower the individual, even if some costs are imposed on the effectiveness of law enforcement processes.

The emphasis on private initiative is further emphasized by the decision of the CRTC to not involve itself, for the present, in the regulation of broadcasting over the Internet. A strong faith in the power of competition to accomplish the goals of ensuring the availability of Canadian content, combined with some confusion about the extent to which the Internet is likely to become a dominant means of delivering audio and video materials, caused the CRTC to exercise its discretion to exempt such activity as may be in its jurisdiction from any licensing requirements.

The government is clearly feeling its way towards a compromise that accepts much of the discourse of privatization and self-regulation but also claims a space within which governmental regulation is still legitimate and necessary. It is not likely that a separate cyberspace governance will emerge. Rather, a mixture of private initiatives and self-help mechanisms, aided by technological tools, standard-setting processes, international co-operative efforts, and, where needed or where politic, governmental regulatory structures, will form the structures and processes within which denizens of cyberspace will create their relationships and resolve their disputes.

Notes

1. P.N. Grabosky and Russell G. Smith, *Crime in the Digital Age: Controlling Telecommunications and Cyberspace Illegalities* (New Brunswick, NJ: Transaction Publishers, 1998).
2. Peter Huber, *Law and Disorder in Cyberspace* (Oxford: Oxford University Press, 1997).
3. Grabosky and Smith, *Crime in the Digital Age*, 210.
4. Dov Wisebrod, 'Controlling the Uncontrollable: Regulating the Internet', *Media & Communications Law Review* 4 (1995): 331–63. Updated on-line: <http://www.CataLaw.com/dov/docs/dw-inet.htm>. Accessed 22 June 2001. (Unless otherwise indicated, all Web sites listed in the notes to this chapter were accessed on this date.)
5. David R. Johnson and David G. Post, 'Law and Borders: The Rise of Law in Cyberspace', *Stanford Law Review* 48 (1996): 1367–1402; Joel R. Reidenberg, 'Governing Networks and Rule-making in Cyberspace', *Emory Law Journal* 45 (1996): 911–30.
6. Proponents of self-regulation include Mark Gould, 'Governance of the Internet: A UK Perspective', in Brian Kahin and James H. Keller, eds, *Coordinating the Internet* (Cambridge, Mass.: MIT Press, 1997), 39–61; David R. Johnson and David G. Post,

'And How Shall the Net be Governed? A Meditation on the Relative Virtues of Decentralized, Emergent Law', in Kahin and Keller, eds, *Coordinating the Internet*, 62-91; John T. Delacourt, 'The International Impact of Internet Regulation', *Harvard International Law Journal* 38 (1997): 207-35.

7. David G. Post, 'Anarchy, State, and the Internet: An Essay on Law-Making in Cyberspace', *Journal of Online Law* (1995): article 3. Available at: < http://www.wm.edu/law/publications/jol/95_96/post.html >.
8. See notes 12-14, below.
9. See Edward J. Valauskas, 'Lex Networkia: Understanding the Internet Community', *First Monday* 1 (7 Oct 1996): 5. Available at: < http://www.firstmonday.dk/issues/issue4/valauskas/index.html >. Valauskas claims that the Internet community has created an uppermost layer of social or cyber-etiquette protocol that 'occurs on top of all of the technical standards and protocols that keep the Internet humming.'
10. Henry Perritt, Jr, 'Cyberspace Self-government: Town Hall Democracy or Rediscovered Royalism?', *Berkeley Technology Law Journal* 12 (1997): 413-76.
11. Reidenberg, 'Governing Networks and Rule-Making in Cyberspace', 926.
12. Joel R. Reidenberg, 'Lex Informatica: The Formulation of Information Policy Rules Through Technology', *Texas Law Review* 76 (1998): 555-93. The analogy is to the *lex mercatoria* or mercantile law that developed to regulate commercial transactions in Europe in the fourteenth century.
13. Kenneth W. Dam, 'Self-help in the Digital Jungle', *Journal of Legal Studies* 13 (1999). Available at: University of Chicago Law School, John M. Olin Program in Law & Economics Working Papers (2nd Series), No. 59 < http://papers.ssrn.com/sol3/papers.cfm?abstract_id = 157448 >.
14. Lawrence Lessig, 'The Constitution of Code: Limitations on Choice-based Critiques of Cyberspace Regulation', *Commlaw Conspectus: Journal of Communications Law & Policy* 5 (1997): 181-91; Lawrence Lessig, 'The Zones of Cyberspace', *Stanford Law Review* 48 (1996): 1403-11.
15. James Boyle, 'Foucault in Cyberspace: Surveillance, Sovereignty, and Hardwired Censors', *University of Cincinnati Law Review* 66 (1997): 177-205.
16. Ibid.
17. Llewellyn Joseph Gibbons, 'No Regulation, Government Regulation, or Self-regulation: Social Enforcement or Social Contracting for Governance in Cyberspace', *Cornell University Journal of Law and Public Policy* 6 (1997): 475-551.
18. David E. Sorkin, 'Revocation of an Internet Domain Name for Violations of "Netiquette": Contractual and Constitutional Implications', *Marshall Journal of Computer and Information Law* 15 (1997): 587-607, provides an extended analysis of a hypothetical, but realistic, example of how a standard form contract can be used to abet the governance function (enforcement of netiquette) of a major Internet institution (the company with sole authority to register and enable the use of second-level domain names in the .com domain space), and the counter-arguments to the enforceability of such contract.
19. Timothy S. Wu, 'Cyberspace Sovereignty?—The Internet and the International System', *Harvard Journal of Law and Technology* 10 (1998): 647-66.
20. Darrel Menthe, 'Jurisdiction in Cyberspace: A Theory of International Spaces', *Michigan Telecommunications Technology Law Review* 4 (1998): 69-103. Available

at: <http://www.law.umich.edu/mttlr/volfour/menthe.html>.
21. Ibid.
22. Andrew L. Shapiro, 'The Disappearance of Cyberspace and the Rise of Code', *Seton Hall Constitutional Law Journal* 8 (1998): 703-23.
23. Ibid., 710.
24. Ibid.
25. Joel P. Trachtman, 'Cyberspace, Sovereignty, Jurisdiction, and Modernism', *Indiana Journal of Global Legal Studies* 5 (1998): 561-81.
26. Ibid., 580.
27. Trotter Hardy, 'The Proper Legal Regime for "Cyberspace" ', *University of Pittsburgh Law Review* 55 (1995): 993-1055.
28. Jack L. Goldsmith, 'Against Cyberanarchy', *University of Chicago Law Review* 65 (1998): 1199-1250.
29. Dan L. Burk, 'Jurisdiction in a World without Borders', *Virginia Journal of Law and Technology* 1 (1997): 3. Available at: <http://vjolt.student.virginia.edu/graphics/vol1/home_art3.html>.
30. Alexander Gigante, 'Blackhole in Cyberspace: The Legal Void in the Internet', *Marshall Journal of Computer and Information Law* 15 (1997): 413-36.
31. Marcus Maher, 'An Analysis of Internet Standardization', *Virginia Journal of Law and Technology* 3 (Spring 1998): 5. Available at: <http://vjolt.student.virginia.edu/graphics/vol3/home_art5.html>.
32. Industry Canada, *Connection, Community, Content: The Challenge of the Information Highway*, Final Report of the Information Highway Advisory Council (Ottawa, 1995). Available at: <http://strategis.ic.gc.ca/SSG/ih01070e.html>.
33. Michel Racicot, Mark S. Hayes, Alec R. Szibbo, and Pierre Trudel, *The Cyberspace Is Not a 'No-Law' Land: Internet Content-Related Liability Study* (Ottawa: Industry Canada, 1997). Available at: <http://strategis.ic.gc.ca/SSG/sf03117e.html>.
34. Reg Whitaker, *The End of Privacy: How Total Surveillance Is Becoming a Reality* (New York: New Press, 1999), 41. See also the discussion by Barry Wright in Chapter 4, this volume.
35. Ibid., 40.
36. Task Force on Electronic Commerce, *The Protection of Personal Information: Building Canada's Information Economy and Society* (Ottawa: Industry Canada and Justice Canada, Jan. 1998), 2. Available at: <http://e-com.ic.gc.ca/english/privacy/632d2.html>.
37. Simon G. Davies, 'Re-engineering the Right to Privacy: How Privacy Has Been Transformed from a Right to a Commodity', in Philip E. Agre and Marc Rotenberg, eds, *Technology and Privacy: The New Landscape* (Cambridge, Mass.: MIT Press, 1998), 144.
38. 'Relying on the government to protect your privacy is like asking a peeping tom to install your window blinds.' John Parry Barlow quoted in Robert B. Gelman et al., *Protecting Yourself Online: The Definitive Guide on Safety, Freedom and Privacy in Cyberspace* (New York: HarperEdge, 1998), 39.
39. S.C. 2000, c. 5.
40. European Union, *Directive 95/46/EC of the European Parliament and of the Council on the Protection of Individuals with regard to the Processing of Personal Data and on the Free Movement of Such Data*, OJ No. L281 (Brussels, 24 Oct. 1995). (Referred

to as *EU Data Protection Directive*). Available at: < http://europa.eu.int/eur-lex/en/lif/dat/1995/en_395L0046.html >.
41. Colin J. Bennett, 'Convergence Revisited: Toward a Global Policy for the Protection of Personal Data?', in Philip E. Agre and Marc Rotenberg, eds, *Technology and Privacy: The New Landscape* (Cambridge, Mass.: MIT Press, 1998), 99.
42. S.C. 2000, c. 5.
43. Some privacy advocates, however, decry the linking of privacy to the facilitation of electronic commerce as an unnecessary and undesirable feature of the legislation, because it tends to undermine the commitment to privacy as a fundamental value. See, for example, the presentations of the Consumers' Association of Canada and the Public Interest Advocacy Group to the House of Commons Standing Committee on Industry, 8 Dec. 1998. Available at: < http://www.parl.gc.ca/InfoComDoc/36/1/INDY/Meetings/Evidence/INDYEV81-E.HTM >.
44. CAN/CSA-Q830-96.
45. Bill C-36, the Anti-terrorism Act, introduced by the Canadian government in the wake of the terrorist attacks in the US, amends both the Privacy Act and the Personal Information Protection and Electronic Documents Act, potentially barring access by individuals to information that is being collected about them.
46. < http://www.intellectualcapital.com >.
47. Netscape, version 4.04. For release notes, see < http://home.netscape.com/eng/mozilla/4.0/relnotes/windows-4.04.html >.
48. Task Force on Electronic Commerce, Industry Canada, *The Protection of Personal Information: Building Canada's Information Economy and Society* (Ottawa: Industry Canada, 1998), 6. Available at: < http://e-com.ic.gc.ca/english/privacy/632d2.html >. The document cites a survey by Ekos Research Associates showing that over half of Canadians agree that the information highway is reducing the level of privacy in Canada. See Ekos Research Associates, Press Release: 'The Electronic Market Place: The Information Highway and Canadian Communications Household Study' (Ottawa-Hull, 2 Oct. 1998), which indicates concerns about security and privacy continue to be major factors affecting consumers' decisions to engage in on-line commercial transactions. Available at: < http://www.ekos.com/ecom.htm >. Accessed 30 Jan. 2000.
49. Annex A in *Cryptography Policy Discussion Paper: Analysis of Submissions* prepared by AEPOS Technologies Corporation for Industry Canada, 15 Sept. 1998. Available at: < http://e-com.ic.gc.ca/english/crypto/631d3.html >.
50. This is an oversimplification. While it appears possible to develop highly robust schemes, various forces have slowed their widespread adoption. Several recent initiatives have demonstrated that it is possible to 'break' some encryption techniques commonly used in communicating digital data. For instance, in July 1998 a group of collaborators was successful in decoding a message encoded using Data Encryption Standard (DES), the most widely used and government-approved commercial data encryption algorithm. One purpose of the collaborators was to demonstrate the need for strong encryption, a development being hampered, according to some, by US government restrictions on the export of strong encryption products. See John Markoff, 'U.S. Data-Scrambling Code Cracked With Homemade Equipment', *New York Times* 17 July 1998. Available at: < http://www.nytimes.com >.

51. Electronic Frontier Canada, 'Submission to Task Force on Electronic Commerce in response to Call for Comments on *A Cryptography Policy Framework for Electronic Commerce—Building Canada's Information Economy and Society*, 21 Apr. 1998. Available at: < http://www.efc.ca/pages/crypto/efc-letter.21apr98.html >. Accessed 30 Jan. 2000.
52. Task Force on Electronic Commerce, *A Cryptography Policy Framework for Electronic Commerce—Building Canada's Information Economy and Society* (Ottawa: Industry Canada, 1998). Available at: < http://e-com.ic.gc.ca/english/crypto/631d11.html >.
53. Ibid., Part 2.
54. AEPOS Techologies, *Cryptography Policy Discussion Paper: Analysis of Submissions* (Ottawa: Industry Canada, 1998). Available at: < http://e-com.ic.gc.ca/english/crypto/631d3.html >.
55. John Manley, 'Presentation to the National Press Club: Canada's Cryptography Policy', 1 Oct. 1998. Available at: < http://e-com.ic.gc.ca/english/speeches/42d3.html >. Perhaps surprisingly, Bill C-36, the Anti-terrorism Act, introduced after 11 September 2001, does not seek to regulate access to encrypted communications.
56. CRTC, *Competition and Culture on Canada's Information Highway: Managing the Realities of Transition* (*The Convergence Report*) (Ottawa: Government Services Canada, 1995). Available at: < http://www.crtc.gc.ca/ENG/HIGHWAY/HWY9505E.HTM >.
57. CRTC, Broadcasting Public Notice CRTC 1999–84/Telecom Public Notice CRTC 99–14, Ottawa, 17 May 1999. Available at: < http://www.crtc.gc.ca/ENG/BCASTING/NOTICE/1999/P9984_0.txt >.
58. 'Broadcasting' is defined in section 2 of the Broadcasting Act as 'transmission of programs . . . for reception by the public'; 'program' is defined to exclude visual images that are primarily alphanumeric text.
59. Section 9(4) of the Broadcasting Act requires the CRTC to exempt broadcasting undertakings from the licensing requirements if it is satisfied that such licensing will not contribute significantly to the implementation of the Act's policy objectives.
60. CRTC, Broadcasting Public Notice CRTC 1999–84/Telecom Public Notice CRTC 99–14.
61. Ibid., para. 90.
62. Shapiro, 'The Disappearance of Cyberspace'.

PART V

Historical and Cultural Perspectives on Regulation and Governance

Chapter 14

Extending *La Longue Durée*: Commercial Impact in the Reform and Use of the Law of Quarantine

LOGAN ATKINSON

Introduction

During December of 1998, Canadian public health officials faced a dilemma of particularly intense proportions. They were advised of the possible contamination of a very large quantity of blood products imported from the United States, contamination occurring as a result of the donation of blood by a man subsequently diagnosed with a variety of spongiform encephalopathy. The most notorious forms of this group of neurological wasting diseases are bovine spongiform encephalopathy (BSE, popularly referred to as 'mad cow disease') and, in the human population, Creutzfeldt-Jakob Disease (CJD). The blood donor's condition had been communicated by his family to Bayer Inc., the American corporation responsible for fractionating his blood, and Canadian authorities were advised as a result.

The Canadian regulator immediately responded by issuing an order for the quarantine of the suspected blood products until more information could be gathered on the donor's condition. The difficulty faced by the regulator was compounded, however, by some disturbing factors. First, while the donor was obviously suffering from some variety of spongiform encephalopathy, the exact form of his disease remained undiagnosed. He might have been suffering from classical CJD, or perhaps he had contracted a new variant of the disorder (known as 'new variant CJD' or 'nvCJD') that emerged when BSE apparently leapt the species barrier and infected the human population in Great Britain. Or, maybe, he had contracted yet another form of the disease, possibly by consuming game hunted in the western United States. Second, the means by which the family of spongiform encephalopathies is transmitted from victim to victim is simply unclear, and it is not known how it transfers from one species to another, if it does so at all. Third, the quarantined blood products were in high demand among Canadian consumers, such that the regulator was immediately pressured to resolve the question of the safety of the product before users ran out of supplies.

The quarantine was lifted in less than a week, reversing a Canadian regulatory policy that dictated the quarantine of all products known to have been contributed by donors with CJD. The explanation offered by Canadian officials was simply that no studies had been done to suggest that CJD could be transmitted through blood products and that the consumers of these products desperately needed a constant supply. As it turned out, the United States had similarly adjusted its quarantine policy the summer before, allegedly as a result of 'pressure from the blood industry'. In fact, scientists at Bayer Inc. were quoted as saying that '[a] scientist can never say never with absolute certainty. But that doesn't mean you have to paralyze yourself into inaction.' And further, Bayer claimed its manufacturing processes were such as to reduce the traces of the disease agent to 'negligible levels' before the products were released to the distributors.[1]

The threat to public health posed by the global distribution of agricultural and other products is one that speaks directly to many of the observations about the reduced effectiveness of the interventionist state summarized by Peter Swan in Chapter 1 of this volume. The distribution of many products that people require, including blood and pharmaceuticals, but also including countless agricultural products and their derivatives, tests the sovereignty of the regulator through the pressures of both commercial expediency and consumer demand. As late as June 2001, Dr Samuel Jutzi of the Food and Agriculture Organization of the United Nations was quoted as saying that 'free market forces and deregulation have contributed to the heightened risks of mad cow disease crossing borders through trade in goods and animals. Market liberalization and globalization may indeed, in the absence of strict enforcement of biologically and ecologically justified regulation, contribute to worldwide spread of such risks.'[2] In effect, the decision to control the movement of products that arguably pose a threat to the public health has been removed from responsible domestic authorities and is now the subject of nebulous pressuring by a vague combination of commercial interests, consumers, and civil libertarians.

But does this development simply function as a further illustration of the fracturing of the locus of regulatory authority described by Swan as the concern of theorists of the 'crisis of the state' in the early twenty-first century? Are the structural constraints experienced by state regulatory authorities concerned with the protection of public health in the face of international risk unique to our particular moment in history? Employees of Canada's Department of Health seemed to recognize the urgency of resolving the tensions between public health protection and commercial efficacy in particular when they presented a petition to Health Minister Alan Rock in late October 1999, calling for sweeping reforms to the regulatory regime around food safety. According to this petition, Canada's Food Inspection Agency was caught between its mandate to protect public health, on the one hand, and its mandate to promote healthy production of and trade in agricultural products, on the other.[3] Arguably, Canada's regulator is torn between satisfying the idea that a healthy economy means a healthy society and protecting the physical well-being of the citizenry. Does this splintering of regulatory authority in the arena of public health, this apparent conflict between the obligation to regulate in the area of public health protection and the need to promote a healthy global trade in products of all sorts, simply justify the observations made by theorists of the 'crisis of the state' in the post-interventionist era?

In this chapter, I want to advance the argument that, in the area of public health protection at least, the preoccupation with the shift from an interventionist state to one whose regulatory effectiveness is challenged through globalization, new technologies, and increased ethnic diversity neglects the historical tension between two essentially modern versions of the public good. Since at least the early seventeenth century the protection of public health has been recognized by many as a matter of state regulatory responsibility. In England (and then the United Kingdom), state authorities tried to meet this responsibility by enacting a series of trade, immigration, and domestic quarantine laws to secure public health as an aspect of the public good. Yet, the efficacy of that regulatory regime has always been jeopardized by the tensions between the strict enforcement of public health regulation and the demands of the commercial community. From the perspective of the civil libertarian and the entrepreneur, a healthy economy, encumbered by few restrictions and supported by the state as facilitator, was essential to overall community well-being. In other words, much of the crisis of the state identified by contemporary theorists through the fracturing of state authority has been suffered in the area of public health protection throughout modernity.

Quarantine is a public health measure that emerged in England coincident with the development of other early public health measures to combat outbreaks of disease. Since the first quarantine statute in England in 1604,[4] there has been remarkably little change in the law, despite medical advances in prevention, diagnosis, and treatment, and despite regulatory innovation in the area of public health generally. The reason for this, I will suggest, is that the vision of the public good inherent in the substantial public health reforms of the 1800s—a vision that had its roots in public health legislation as early as the beginning of the seventeenth century—has been undermined at least since the early eighteenth century by a competing vision of the public good advanced by powerful commercial interests. Essentially, this competition has left the law of quarantine in its early modern, pre-scientific form. This inertia in development of the law of quarantine underscores the idea that policy development in the area of public health protection, which in some interpretations of the interventionist state leads to innovation in regulatory intervention in the market and the community generally, functions as an exception to the more general theorization of the regulatory state outlined by Swan in Chapter 1.

While ordinary sanitation regulation might be understood as conducive to successful commercial enterprise, quarantine has always been considered to be a threat to commercial life because of its interference in the ordinary movement of people and goods, both domestically and in the context of immigration and the import/export trade. Because it has posed this threat to apparently healthy commercial activity, reform of the law of quarantine has lagged behind other innovative regulation in the area of public health. The law of quarantine has remained unchanged in most substantive aspects down to the present day. In essence, *la longue durée*[5] of pre-Enlightenment, pre-scientific medical legal history is extended into modern life through the unchanging law of quarantine, a regulatory framework that remains embedded in very early modern ideas through the pressures of commercial expediency.

In some respects, the argument advanced here is not dissimilar to that in much of the literature critical of the position that regulation is ordinarily designed to advance the public good. Gabriel Kolko, for example, in his 1965 study of American railroads and regulation,[6] takes issue with the idea that early regulatory intervention in the business of transportation, which took the form of restraint on the swashbuckling railroad entrepreneurs who, if left unchecked, would plunder the public wealth, was promoted purely or even primarily for a broadly defined public good. To the contrary, suggests Kolko, the regulation of the railroads was welcomed, even encouraged, by the owners of the railroads as a way to control competition and reduce predatory invasions of a nicely profitable market. In other words, even the nineteenth-century state was constrained in its ability to design policy and regulation free of the pressures of particular constituencies that ultimately had as much to do with the design of regulation as the regulator itself.

Kolko advances the position that regulation is often contrary to the public good as conceived by those responsible for its development, largely because that adverse effect is manifested in a regulatory framework supportive of conditions conducive to the interests of a select group. In the case of the law of quarantine we find something of the same concerns for the disengagement between the ultimate form of regulation and the public interest intended to be served. That separation evidences the constraints that operate on the policy and regulatory processes. But in the way in which the law of quarantine has remained relatively unchanged in substantive respects since the seventeenth century, we find a resistance to regulatory innovation rather than a reformed regime that falls short of its goal to enhance the public interest. In this respect, interests that may be seen to be contrary to public well-being have exerted sufficient influence on the process of regulatory reform to stand successfully in the way of innovation by advancing a particular interpretation of the public interest that, perhaps coincidentally, corresponded nicely with their own commercial success.

Two separate historical incidents are described here to underscore the influence that commercial pressures have brought to bear on attempts to reform or put in practice particular quarantine measures. In the first example, we will consider the attempt at reform of the law of quarantine undertaken by Robert Walpole's Whig administration in England during the early 1720s.[7] As he contemplated the possibility of a serious epidemic in England, Walpole was working with twin disabilities: a regulatory regime in the area of public health and disease prevention already 120 years old and the pressure of a public alarmed at the approach of bubonic plague from Continental Europe. Despite the perceived urgent need for a newly energized regulatory initiative, the attempt at reform failed, in large part due to pressures from a well-organized, well-connected commercial lobby.

Our second example involves some of the events in Upper Canada during the cholera outbreak of 1832. Despite the fact that most neighbouring jurisdictions had either enacted quarantine and other public health statutes some decades earlier or responded to the threat of the importation of cholera with relatively precise legislative and administrative initiatives, Upper Canada was virtually completely unprepared for the emergence of cholera among its settlers in 1832. There was no relevant legislation in Upper Canada until 1833,[8] after the first epidemic had run its terrible course, and the official administrative response to the disease as it made its way up the St Lawrence River during June of 1832 was half-hearted at best. While this apparent lack of foresight and action might be explained by various factors, the preponderance of evidence suggests that the executive government at York (Toronto) deliberately failed to react aggressively to the threat of the epidemic, primarily because the conventional quarantine response was considered anathema to the continued commercial development of the colony.

Both of these examples illustrate the influence that extra-legal considerations often have on how the formal law develops, or fails to develop, and both underscore how public health and safety can be defined in different ways to achieve different legal and administrative results. In both cases, we see governments confronted with a choice between competing visions of the public good—one that speaks directly to the preservation of health through an aggressive legal and administrative intervention in private life, the other that suggests that an improved public health can only be accomplished when the public economy is healthy and vibrant. This latter vision, of course, can be realized most completely when private lives are left to be enjoyed free of regulatory interference, and it is arguably as a result of the success of this vision that the law of quarantine remains to a large extent entrenched in its early modern form.

Commerce and Quarantine in England to the 1720s

Despite the periodic devastation of the population by plague,[9] the English Parliament failed to pass anything like a law of quarantine until 1604 (the first year of the reign of King James I), and this law probably resulted from the severity of the plague of 1603. This experience led to many changes in administrative practices, but the change in which we are most interested was a wholesale change in government's regulatory responsibility for public health protection. For the first time in English legislative history, a statute was passed by which some provision was made for dealing with the consequences of epidemic disease. King James's Act, passed in the first session of the first Parliament of his reign, had two principal thrusts. First, it established a regime by which local authorities might raise revenues for the relief of the poor from the effects of the epidemic. Second, it enacted a series of measures by which domestic quarantine might be instituted in given conditions. King James's Act was unique in both of these respects.

While the taxing regime established under King James's Act was significant, the second major thrust of King James's Act is most relevant for our purposes, for here we find the establishment of the first system of domestic quarantine with legislative sanction in English history. The power to establish and enforce the quarantine was vested personally in local officials, in the 'Mayor, Bailiffs, Constable or other Head Officer of any City, Borough, Town Corporate, privileged Place or Market-Town' where the plague was found, and likewise in the head officer of any infected county (Section VII). The powers granted to these officials were the first legislative endorsement of what would become the notorious practice of the 'shutting-up of houses', by which any person infected with the plague, residing in an infected dwelling, or merely 'being in' an infected dwelling might be ordered by the local official to remain in that dwelling for the purpose of preventing the spread of infection. Nothing in King James's Act provides for exceptions, limits the length of time within which the quarantine might continue, or directs that the certificate of a physician would be enough to allow the quarantine of a particular person to be lifted. In effect, Section VII established a system of discretionary house arrest, with no possibility of appeal for either the healthy or the infected. As a result, the plague raised the prospect of the complete loss of individual liberty at the whim of a local official, all in furtherance of a vision of the public good that subordinated individual freedom to the prospect of an enhanced communal welfare.

To assist in the enforcement of quarantine, the power was granted to the local official by Section IX to appoint 'Searchers, Watchmen, Examiners, Keepers, and Buriers', all of whom were to be employed in establishing the need for quarantine and in protecting against its violation. The powers of such appointees were provided in Section VII. First, they were to ensure that those ordered to remain inside an infected house actually complied with that order. And second, they were empowered to use force ('violence', according to Section VII) in the event that attempts were made to break the quarantine. The watchmen and others appointed by the local official were given absolute immunity from both prosecution and civil liability for injuries arising from the exercise of their duties, at least where those injuries occurred to 'such disobedient Persons', and this further underscores the official commitment to advancing the perceived public good at the expense of individual freedoms.

The offence of breaching quarantine and escaping the efforts of the watchmen was a felony, and penalty in such a case was death. The severity of this measure was tempered, however, by subsection VII(2), where it was provided that the felony would arise and the death penalty apply only where the escapee 'shall converse in Company, having any infectious Sore upon him uncured.' Sick persons at large but not in communication with others, then, were spared conviction for the felony, as were healthy escapees and those who might have recovered from their illness prior to their recapture and examination. These offenders were to be 'punished as a Vagabond' and 'bound by his or their good Behaviour for one whole Year' (subsection VII[3]).

This, then, is the essence of King James's Act, the first legislation of its kind in England.[10] It established a number of measures best explained in the context of emergency powers during times of plague, the severity of which had never been experienced before. Subsection IX(2) provided that the statute would continue only 'until the End of the first Session of the next Parliament'. King James's Act, therefore, was intended to be temporary, a special creation and extension of legal powers, and an exceptional imposition on citizens and their freedoms, both in the taxing powers and the quarantine sections, in circumstances of particular threat and anxiety. But probably because the plague continued to threaten the English during the first five decades of the seventeenth century,[11] King James's Act remained in force, first temporarily by virtue of 3 Car. I c. 4 (passed in 1628), and then perpetually by virtue of 16 Car. I c. 4 (passed in 1641).

King James's Act, as it was enforced from time to time throughout the seventeenth century, remained complete and unaltered until 1711, when, with the passage of 9 Ann. c. 2 ('Queen Anne's Act'), Parliament again directed its attention to measures designed to prevent the spread of plague. Strictly speaking, Queen Anne's Act did not amend King James's Act; questions of domestic quarantine and the special taxing powers were not mentioned in the new statute. Rather, Queen Anne's Act introduced another important aspect of the law of quarantine, that is, the quarantine of ships arriving in England, having sailed from foreign ports infected with the plague.

According to recitals in Queen Anne's Act, the Queen had been issuing orders on a random basis for about two years, requiring this ship or that to be quarantined for a period of time before off-loading cargo or crew. Apparently, however, it was felt that an act of Parliament would make administration of the law more expeditious. There were three broad aspects to the new statute. First, and most obviously, ships arriving from infected ports were to be quarantined at the direction of the Queen. Second, movement of people back and forth between ship and shore was prohibited; anyone leaving the ship would be compelled to return, and anyone visiting the ship was required to stay. The responsibility for policing this part of Queen Anne's Act was left to the master of the vessel, and the failure to carry out these duties properly meant forfeiture of the ship to the Queen. Officials on shore were directed to seize any small boats belonging to the ship, ensuring that the ship would remain isolated during the period of quarantine. Third, Queen Anne's Act required that, after completion of quarantine, the ship's cargo was to be 'open'd and air'd'.

Over the course of just more than 100 years, the English Parliament had developed a relatively comprehensive law of quarantine designed to address what was one of the most serious social issues of the seventeenth century. It is not clear why, prior to 1603, officials had responded to the continual threat of plague in an apparently ad hoc, unsystematic way, but it seems relatively certain that, after that date, Parliament and local authorities began to employ their law-making powers to prevent the spread of epidemic disease, with a view to the collective good, social control, and public responsibility. Despite the harsh terms of much of the regulatory regime established in response to the plague, and despite the fact that the disease seemed to spread across the country irrespective of regulatory intervention, the law remained unchanged in any material respect until the innovations captured in the provisions of Queen Anne's Act.

Viewed retrospectively, it seems that lawmakers had little choice but to enact a series of harsh, restrictive measures, especially given what we now know to be the errors in contemporary medical opinion on the cause, diagnosis, and treatment of the plague. But opinion on the wisdom of the law of quarantine was sharply divided among intellectuals of the late seventeenth century. Many observers understood the law as the mundane product of the activity of legislators, rather than as a mysterious, divinely ordained guide to a good life, and realized that the law so constructed could be remade in a form more humane, more effective, and less invasive. For these observers, the ineffectiveness of law simply heightened the indignities attendant upon the plague itself. Implicitly at least, those critical of the law[12] suggested that the disruption inevitably accompanying plague and the law of quarantine might be softened, and might be justified more easily, if positive results could be expected.

On taking control in 1720, Robert Walpole's administration inherited this regulatory regime and these debates over the effectiveness of the law. To help us understand the quandary of the Walpole administration at the time it attempted to overhaul the regulatory regime in the early 1720s, we must consider in some detail the criticisms of the prevailing law and the fear that accompanied the news of the spread of plague across Europe during the early years of the reign of George I. Critics focused on the law's effectiveness on the one hand, and on the law's humanity on the other, concentrating on the essentially modern difficulties inherent in resistance to reform of the law of quarantine, particularly in connection with the impact of reform on England's growing foreign trade and the struggle to define a fair and consistent set of civil liberties.

Daniel Defoe, for example, took great pains as early as 1712 to point out the extent of suffering from the plague experienced at the time by people in Denmark and Sweden. He highlighted the weaknesses of legal attempts to control the spread of the disease, and warned England to enact measures to prevent the spread of plague from across the sea. Even at this early date, Defoe appears to doubt the ability of Queen Anne's Act to control effectively imports from infected ports. He said that, in Denmark, 'all possible measures' had been taken, to no avail, to protect the Danes from an importation of the plague from Sweden, including the isolation of returning soldiers and the encirclement of infected towns.[13] And, in speaking particularly of the difficulty in getting foodstuffs through the lines of guards and into the villages, Defoe lamented that the inhabitants 'suffered great hardships by this severe restraint, having not been allowed to flee for the safety of their lives'.[14]

In these publications and many others, Defoe focused on the suffering of plague victims and the weakness of the regulatory regime. He was alarmed at English indifference to the spread of plague and he warned of the terrible suffering the English might expect should the plague return. His call for enhanced legal protection intensified with commentary such as this:

> as none but a stupid Generation, that scoff at every Judgment of Heaven, and value not their Maker till they feel his Hand, can be so blind, as not to see the visible approaches of such a Thing to us; so nothing, but a dreadful indifferency in the Event, can make the Hints I have given, Unseasonable.[15]

Very soon after the passage of Queen Anne's Act, Defoe, at least, did not hesitate in expressing a forceful opinion in favour of reform measures designed to protect the English from plague, and his preoccupation with the plague heightened considerably in the period from 1719 to 1722. He published some 16 letters on the issue during that time,[16] most of which made reform of quarantine law and domestic restrictions their focus. In many of these letters, Defoe took great pains to alert his readers to the fact that ships sailing from infected ports were barred entry to several healthy ports on the Continent. As will be seen, this was a difficult issue for the Whigs in their attempts at reform, resulting in the alienation of the trading community and leading directly to the debate surrounding the Supplementary Act, with its measures designed to prevent the clandestine running of goods. Defoe's earliest letter on this issue, which directly challenged the provisions of Queen Anne's Act, was published on 12 August 1720 in the *Daily Post*. In this letter, he hinted at the need for tough import restrictions, suggesting that, in the seriously infected Marseilles, '[t]he Commerce is universally stopp'd', and indicating that goods were stored some leagues distant from the city so as to lessen the likelihood of infection.[17] In his letter to *Applebee's Weekly Journal* of 22 October 1720 Defoe intensified his attack on weak import restrictions by giving various examples of procedures in place on the Continent. By recounting numerous incidents of the strict enforcement of quarantine elsewhere, Defoe alerted his readers to the need for difficult and harsh law, despite its necessary incursion on individual liberties.[18]

On 25 January 1721, the Quarantine Act was passed by Parliament, apparently arousing little opposition and creating but minor, if any, controversy among merchants and common people. Besides its repeal of Queen Anne's Act in its entirety (legislation found to be 'defective', 'insufficient', and 'not adequate', according to recitals in the Quarantine Act), the legislation made two bold attempts to address the concerns described by Defoe in his essays and letters.

First, a method was prescribed by which ships arriving in England, having sailed from 'infected Places', were to be quarantined as the King might direct. Penalties were prescribed for a number of offences relating to possible breaches of quarantine, primarily directed towards the ship's master or others in control. For example, it was made a felony for a master to fail to report that he had on board a person infected with the plague, whether the master was aware of the infection or not. No one was permitted to leave a vessel committed to quarantine; anyone violating this prohibition would be compelled to return to the ship and forfeit £200. A term of imprisonment of six months was also provided for offenders of this provision, to be served, presumably, after the period of quarantine had run. Provision was made for the ship to be forfeited or, in some cases, burnt, and the master fined £200 for a variety of offences, including the failure to respond promptly to notice requiring the removal of the ship to quarantine. Very little attention was paid to the question of the illegal importation and possession of goods received from ships still liable to quarantine; the goods themselves would be committed according to the original quarantine orders, and the person found in possession was liable to a fine of £10.

In almost every respect, the Quarantine Act offered tighter measures to control the importation of goods than had Queen Anne's Act. This is especially clear in the removal of policing responsibilities from the master of a quarantined vessel, a provision that had placed the master in an obviously difficult position. The Quarantine Act was much more detailed than Queen Anne's Act had been, providing for penalties and forfeitures for specific offences to which only vague reference was made in the earlier legislation. It is evident from the recitals in the statute that the legislators felt the weaknesses in Queen Anne's Act were overcome in the Quarantine Act and that complaints about the ineffectiveness of the earlier statute, such as those offered up by Defoe, were answered by this new, more restrictive law.

Second, a series of powers was reserved to his Majesty in Council and his proper officers in the event that the plague were to establish itself onshore. These sections of the Quarantine Act had two primary thrusts. Powers were provided to construct separate facilities for the 'Entertainment of Persons infected', effectively allowing for the removal of persons from their homes to ships, lazarettos, or specially built houses of quarantine, and providing isolated places for the airing of once contaminated goods. Appointed officials were empowered to use all necessary force to compel committed persons to submit to quarantine and to ensure that those otherwise entering the quarantine area remained there as directed. The other principal focus of these sections of the Quarantine Act allowed for the isolation of individual homes, neighbourhoods, or towns, as required. Basically, powers were granted to dig trenches around infected areas so as to isolate the inhabitants more completely, and members of neighbouring parishes were to be conscripted to stand watch. Precedent for these measures could be seen in the way in which the French had responded to recent infections.

In the meantime, Defoe took a clear stand against the trading community, condemning their attempts to circumvent import restrictions, and thereby daring 'to venture the Welfare of the whole Kingdom, and the Lives of Men, Women and Children, for the wretched Gain of a private Man, and perhaps that Gain a Trifle.'[19] Strict and comprehensive laws, Defoe argued, were necessary to protect the populace against the plague, and equally strict laws were needed to punish those who would defy the quarantine for personal gain and those who would compromise the effectiveness of the law in its drafting.

> If I were to propose an equal and just Treatment of such Men, it should be, that if such a dreadful thing should ever happen among us, (as God forbid,) they who had concern'd themselves to obstruct the making needful Laws to prevent, or who had endeavour'd to leave out proper Clauses in such Laws, to make them more effectual; should be oblig'd, by the same Law, to stay in the City, and take their risk with the Poor, who are not able to fly, and have not wherewith to shift from one Country to another, as the Rich have.[20]

As we shall see, Walpole faced considerable opposition from the trading community, especially concerning the quarantine of imported goods to be provided for in the Supplementary Act. And even Defoe was sensitive to the complaints of the trading community relative to Walpole's proposals to stiffen restrictions on imports from infected locations. In fact, Defoe expressed considerable sympathy in particular for the losses that were bound to occur through the delays associated with the quarantine of shiploads of perishable goods. The following is from his letter to *Applebee's Weekly Journal* of 29 July 1721:

> The Damage of obliging Ships to Quarantine, is . . . very considerable to the Merchants; it spoils their Goods, and many sorts of Goods are perishable, and subject to decay in others. The Profit of the whole Voyage depends upon the Season of coming into the Market . . . some Times [the merchants are] quite disappointed, the profits of the voyage lost, and the Merchants perhaps ruin'd, by the ships lying for six Weeks for Product, as 'tis called. . . . Yet all this we chearfully submit to for the Reason of it; 'tis allow'd to be just, to be necessary, and what really ought to be done. But if one Villain can pass the Barriers set, . . . he may lodge the Plague among us, [and] the Merchants suffer all the Inconveniences for nothing.[21]

Defoe is clear here that the imposition of quarantine on ships arriving from infected ports is necessary for the public good, despite the likelihood that certain citizens—the merchants—would suffer as a result. But the important element of the law for Defoe was its effectiveness, without which the sacrifices made by the merchants could not be justified. The justice of the law, for Defoe, lay in the extent to which it protects the community as a whole despite the sufferings of the few, an early utilitarian view of the law unpalatable to the trading community that eventually assembled in opposition to Walpole. In Defoe's view, immediate action was required, and delay could prove disastrous.

> What then must we do? And what Remedy must we prepare for our Preservation, and to prevent the Confusions which threaten us in case of an Infection? I am not entering too far into such an Enquiry, because People will take more Alarm from such Things than is intended; but certainly it cannot be our Prudence entirely to neglect such a Thing, till it comes to the very Door, the consequence of which will be, that then we shall be all in Confusion and Distraction.[22]

While the Quarantine Act passed Parliament and was proclaimed into law with little fanfare, the trading community soon became concerned, especially with the prospect of perishable goods being liable to quarantine for lengthy periods in unfavourable weather. These concerns led to continued schemes to bring goods ashore in violation of quarantine. Because of the weakness in the provisions of the Quarantine Act related to the importation and possession of otherwise quarantined goods, the 'clandestine running of goods' was largely successful, leading Walpole to introduce the Supplementary Act into the House of Commons on 8 July 1721. While this legislation is relatively long, having a variety of aims apart from strengthening the Quarantine Act, for our purposes the relevant provisions are few. Recognition is made of the 'infamous and pernicious Practices' of illegal importation, despite clear law to the contrary and despite the threat to the health and lives of 'many thousands of his Majesty's innocent Subjects'. In Article X, the penalty for purchasing goods imported in defiance of quarantine was increased to £20 from £10. More severe penalties were provided in Article VI for those directly involved in the illegal importation of goods. Depending on the conditions under which the goods were run, persons convicted could be transported to America for seven years.

This legislation caused more difficulty for Walpole's administration than had the earlier Quarantine Act. The trading community rallied against it, enlisting the support of prominent opposition members in both the Commons and the House of Lords. Apparently seizing the opportunity to discredit the administration generally, the opposition Lords began a campaign to force repeal of the domestic provisions of the Quarantine Act (despite their failure to speak against those provisions when they were debated in the earlier session), as well as to prevent passage of the Supplementary Act. At this point the issue of civil liberties was raised, an issue that found considerable favour among London's municipal politicians, who were fearful of widespread civil unrest. The public at large, too, despite the failure of the opposition to have argued the point earlier, began to express its displeasure at the prospect of the potentially invasive provisions of the domestic aspects of the Quarantine Act. The model for these severe domestic provisions was found in France, and Walpole's parliamentary opposition capitalized on English contempt for what was perceived as French despotism by rallying the populace in defence of civil liberties.

At first, Walpole's government resisted the calls to reform the Quarantine Act. A petition lodged with the House of Lords by the City of London, asking that the City be heard on the adverse implications of the legislation for both trade and public order, was rejected by the Lords on 6 December 1721,[23] and many of the Opposition Lords were outraged that such an eminent body as London would be turned away on a matter of such importance. In a 'Dissentient' filed as part of the parliamentary record, these Lords stated their disgust with the rejection in unequivocal terms:

> Because the liberty of petitioning the king (much more that of petitioning either House of Parliament) is the birth-right of the free people of this realm, claimed by them, and confirmed to them, soon after the Revolution, in an Act, declaring the rights and liberties of the subject, and settling the succession of the crown; and whenever any remarkable check hath been given to the free exercise of this right, it hath always been attended with all consequences to the public.[24]

Again, the appeal was to civil liberties, an attempt to force Walpole to amend the legislation by reminding the government of the gains made since the Glorious Revolution of 1688 in the rights of the individual against the state.

A similar appeal was made by the same group of Lords when they introduced amendments to the Quarantine Act on 13 December 1721. These amendments were the initial attempt to repeal the domestic quarantine provisions of the legislation, but they were rejected in the House of Lords by a vote of 39–20.[25] Another 'Dissentient' was filed, and the Opposition Lords again appealed to concern for civil liberties as the grounds for their indignation. This record is especially interesting for its reference to King James's Act and its suggestion that the reformed legislation actually reversed libertarian allowances made in the Act of 1604. In the following excerpt, the Opposition Lords are speaking specifically of those aspects of the Quarantine Act related to the encirclement of towns and the removal of the infected to lazarettos and pesthouses:

> Because such powers as these are utterly unknown to our constitution, and repugnant, we conceive, to the lenity of our mild and free government; a tender regard to which was shown by the act of James I, which took care only to confine infected persons within their own houses, and to support them under their confinement, and lodged the execution of such powers solely in the civil magistrate; whereas the powers by us excepted against, as they are of a more extraordinary kind, so they will probably (and some of them must necessarily) be executed by military force; and the violent and inhuman methods which, on these occasions, may, as we conceive, be practised, will, we fear, rather draw down the infliction of a new judgment from Heaven, than contribute any way to remove that which shall then have befallen us.[26]

The Opposition Lords, in addition to drawing attention later in the 'Dissentient' to the parallels between the Quarantine Act and the 'arbitrary powers' exercised by the government of France,[27] argued forcefully against the Act as it was originally passed, appealing at once to the fear of military intervention, to humanitarian responsibilities during times of epidemic disease, to the fear of God and the idea of plague as punishment for sinful conduct, and to tradition in the shape of King James's Act. Despite such a persuasive attempt, the Opposition Lords were defeated, at least temporarily.

Faced with mounting criticism from the trading community, London's municipal council, and the general population, Walpole quickly introduced the Amending Act into Parliament. By its provisions, the offensive domestic restrictions and penalties of the Quarantine Act were repealed in their entirety. The Amending Act was marshalled through Parliament and passed on 29 January 1722. By 20 February 1722, the Supplementary Act, modified somewhat to appease the merchant community, had passed the Commons. Eventually it, too, was approved by the Lords, and finally received royal assent on 7 March.[28]

In the end, the attempt at reform of the domestic provisions of England's quarantine regime failed for the lack of support from the merchants and traders of London's moneyed classes. To appease the trading community's calls for fewer impediments to continued high levels of imports, and to answer those who, having been alerted by the trading community to the invasive nature of the reforms, were alarmed at the apparent incursions on civil liberties represented by tough innovations in domestic quarantine, Walpole was forced to dilute a regulatory initiative that was consistent with the vision of the public good embodied in the attempts at reform. Thus, the law of quarantine remained unaltered in any substantive aspect from the time of James I, the changes being minor enough to avoid offending commercial interests, and benign enough to avoid challenging the competing vision of the public good represented by those who maintained that a healthy, vibrant trading community was arguably the most certain route to sustained public health, perhaps especially during times of epidemic disease.[29]

It appears, then, that despite some considerable public alarm at the approach of yet another outbreak of bubonic plague, and despite at least some vociferous opinion that the age-old law of quarantine ought to be reformed so as to more certainly protect England's citizens, Robert Walpole was thwarted in his attempt to improve the law by the combined persuasive abilities of London's trading community and its civil libertarians. This failure is significant, for it represented the last major attempt at reform of the English law of domestic quarantine before the 1790s.[30]

Commerce and Quarantine in Upper Canada, 1832

When colonial legislatures of British North America confronted the possibility of epidemic disease at the end of the eighteenth century and early in the nineteenth, it was principally to this early English model, with its regime of domestic quarantine largely unchanged since the days of James I, that they looked for inspiration. The tenor of the English legislation on import and immigration restrictions was captured by the legislature of Lower Canada when it passed the first Canadian statute on quarantine in 1795.[31] New Brunswick followed with a modest statute in 1796, and then repealed and replaced it with two enactments in 1799. These were then amended in various respects throughout the period prior to mid-1831.[32] Later, as cholera approached the colonies in late 1831 and early 1832 on board vessels laden with immigrants, the House of Assembly in Nova Scotia passed statutes to regulate quarantine and help in the prevention of the spread of infectious diseases.[33] A revised statutory regime was put in place in New Brunswick[34] (replacing the earlier statutes of 1799, as amended) and Prince Edward Island[35] prior to the actual arrival of cholera in the Maritime colonies in mid-1832. Each of these later legislative initiatives was largely similar to the domestic quarantine inherited by Walpole's regime from King James's Act, and to the import restrictions and quarantine of immigrants secured by Walpole through the Supplementary Act, as it was amended from time to time during the eighteenth century.

When cholera made its first appearance in Upper Canada in mid-June 1832, it encountered little resistance from responsible regulatory authorities. Despite a fairly rigorous legal and administrative attempt to prevent the introduction and spread of the disease in both the United Kingdom and the other colonies of British North America, Upper Canada was virtually completely unprepared for the epidemic. It is not clear the extent to which the people of Upper Canada suffered for this unpreparedness, primarily because the measures used to combat the spread of the disease in other jurisdictions were either sadly ineffective or based on often conflicting medical opinion.[36] In any case, however, the executive government at York felt constrained in its ability to protect both the public health, through legislation similar to that in place in Lower Canada and the other colonies, and public economic well-being, through the encouragement and facilitation of continued high levels of emigration from the British Isles.

In the legal and administrative response to the cholera epidemic of 1832 in Upper Canada, we find the collision of two often incompatible visions of the public good, and the executive government had ultimate responsibility to see to the accomplishment of both. The cholera outbreak highlighted the tension between these two visions, and because of the extremely high level of public anxiety prevalent at the time, the response of the regulatory authority is particularly revealing. This was an administration in the throes of an emergency, when decisions as to the optimal employment of law in furtherance of policy objectives could be expected to have immediate impact on the lives of hundreds, perhaps thousands, of Upper Canadian settlers. Under this pressure, the government opted to resist the calls for a general quarantine statute, deciding instead to maintain the status quo by doing everything in its power to promote and maintain high levels of immigration into Upper Canada from the United Kingdom and Ireland. This course was taken despite the fact that immigration and disease were inextricably connected, both in the public mind and in the view of many experts. The reason for the official position, quite simply, was that immigration meant development for Upper Canada—the construction of roads, bridges, and canals; the clearing of vast tracts of unsettled land; the injection into the local economy of much-needed capital. Quarantine would interfere with all of this, and the government determined that such a threat to the economic health of the colony was too big a price to pay, even given the potential catastrophe represented by cholera.

When epidemic cholera became a real possibility with the arrival of news that it had begun to appear in England, the dilemma for decision-makers in the colonies became acute. Not only was immigration considered to be the source of many of the most dangerous diseases for the colonists, but the regulatory responses that might be expected—trade restrictions and immigration restrictions similar to those in place in the United Kingdom—were thought by many to be antithetical to commercial development. It was widely reported in the Upper Canadian press that immigration and trade restrictions in place in England were far more damaging to the public than cholera itself, and arguments such as these were used by many Upper Canadians to deny the value of legislation similar to that in place in Lower Canada. The public interest, apparently, was thought to suffer more by the interruption to trade and commerce than it did by the disease, and this position seemed to be supported by the suspension of domestic quarantine in England in February 1832, on the grounds 'that the interruption to trade would produce greater evils than the precautionary measures are likely to avert.'[37]

To be fair, there are alternative explanations for the hesitancy displayed by the executive government in responding to the threat of an epidemic in a manner similar to that displayed in neighbouring colonies. First, medical opinion was seriously split on the causes of cholera and on how it spread. An allegiance to one medical perspective or the other radically altered the way one understood the potential of law as a valuable regulatory force. 'Contagionists' believed that the disease was communicated through direct, person-to-person contact. This was the conventional, conservative understanding of the nature of epidemic disease, and it underlay the argument in favour of quarantine, the encirclement of towns, the isolation of the sick, and other standard regulatory measures employed in England and Continental Europe since the early seventeenth century. 'Anti-contagionists', on the other hand, considered the argument from contagion to be old-fashioned, suggesting instead that disease was spread through miasmatic emanations from the earth. For anti-contagionists, quarantine and related conventional legal responses to epidemic disease were misguided and futile. Disease was simply 'in the air', and isolating the sick could serve no purpose in preventing the inhalation of infectious vapours by the well.

It is possible that Lieutenant-Governor Colborne resisted the calls for a general quarantine in Upper Canada during the cholera outbreak of 1832 because of a sincere commitment to anti-contagionism. There is little question that the debate among medical professionals was current among Upper Canadians, and the two sides were well represented in both the conservative and reform presses. In the *Colonial Advocate* of 26 April 1832 a lengthy letter from London was published in which the two perspectives were given detailed treatment.[38] In the *Canadian Freeman* of 29 March 1832, the editor canvassed English opinion on the causes and communicability of disease, concentrating to some extent on the apparent lack of faith in the doctrine of contagion among many leading members of the English medical professions.[39] And the *Brockville Recorder*, quoting respected German experts, assured its readership that, despite the violent spread of cholera across Europe and into England, 'it is not of a contagious nature.'[40] There is little doubt, then, that relatively current opinion on the alternative medical explanations for the spread of cholera was available to Colborne, and it is therefore possible that he was sympathetic to the anti-contagionist view.

In fact, this possibility is borne out by certain correspondence from Colborne's office during the early weeks of the emergency. Despite the precautionary precedents available to him through the actions of the government at London and in the neighbouring colonies, Colborne insisted that the most expeditious way to respond to the crush of sick and indigent immigrants as they made their way up the St Lawrence during the summer of 1832 was to disperse them as quickly and widely throughout the province as possible. It would be foolhardy, he suggested, to confine large numbers of sick immigrants in cramped and unhealthy quarters; this would simply exacerbate their weakness, making them even more susceptible to miasma. In a letter from the Civil Secretary to one John Smyth of the Township of Bastard, Colborne's rejection of the call for a general quarantine of immigrants was relayed in the following terms:

> His Excellency desires me also to state that he has directed the agents for Emigrants stationed on the St Lawrence to use every exertion to forward the Emigrants to York whence they will be immediately dispersed over the Upper Districts, and to observe that your suggestion of keeping the Emigrants together in quarantine till the Cholera may disappear from the Country is a very injudicious proposal; and he can only call on you to recollect that the prevailing disease has been felt only with severity in crowded and unventilated and dirty towns; and that the many cases which have occurred at Prescott and on the St Lawrence may be attributed to the alarm and panic of the Inhabitants. . . . In a well regulated village there is no cause for alarm, and His Excellency trusts that you will lend your assistance in facilitating the conveyance of Emigrants through the Township in which you reside.[41]

The anti-contagionist elements of Colborne's response to the disease are obvious in this letter. He blames poor ventilation and panic for the extent of the epidemic, and hints strongly at the need for regulations in the area of cleanliness and air circulation as the preferred alternative to quarantine.

But this interpretation of Colborne's reluctance to employ conventional regulatory intervention in an attempt to stem the spread of the disease is inconsistent with much of the other historical evidence. For example, Colborne had been advised by his superiors in London and Quebec City as early as 30 June 1831 on the contagious nature of the disease, and had been provided with specific instructions on the best means to prevent the introduction of the disease into Upper Canada. Lord Goderich, Colonial Secretary in 1831, provided Colborne with copies of the King's speech on the threat inherent in the possible importation of cholera, and cautioned Colborne on the 'highly contagious nature' of the disease. Medical opinions, printed directions from the British House of Commons, recommendations on quarantine from respected English physicians, and the conclusion of the College of Physicians in England that the disease was in fact contagious—all were included with the Colonial Secretary's instructions.[42] Given that neighbouring colonies acceded to these instructions by enacting and implementing legislation consistent with their precautionary tone, it seems highly suspect to conclude that Colborne unilaterally decided to commit to the arguably more dangerous anti-contagionist position.

Another possible explanation for Colborne's reluctance to implement a general quarantine lies in the relative isolation of Upper Canada. During the 1830s the colony depended either on St Lawrence River traffic, or on the system of rivers and canals connecting New York City with the Niagara Peninsula, for the transportation of immigrants from the British Isles. In both cases, quarantine was already well established at the initial points of entry for immigrants, at Grosse Isle below the settlement at Quebec City and in New York City itself. Arguably, then, if quarantine were successful at all it would prevent the transmission of cholera to Upper Canada through Lower Canadian and American efforts. This seemed to be the attitude of Governor-General Aylmer, who, in a message to the Legislative Council of Lower Canada on 3 February 1832, said: 'The great influx of Emigrants into this Colony which . . . may be expected during the coming season, makes it necessary to adopt some precautionary measures calculated to avert the introduction of this dreadful malady into the Canadas.'[43] The efforts in Lower Canada, then, might help to protect the *entire* country west of Grosse Isle. And, of course, the other colonies of British North America did not enjoy the same isolation from first contact with immigrant ships as did Upper Canada, and that helps to explain their submission to the contagionist's point of view.

While at first this explanation has some logical appeal, it fails to recognize that the quarantine at Grosse Isle was an abject failure, not necessarily because of any weakness in the contagion argument that provided the theoretical foundation for the quarantine in the first place, but rather from problems of administrative and legal practices, especially in the areas of compliance and enforcement. It was well known in Upper Canada that Lower Canadian authorities were having serious difficulty in ensuring compliance with quarantine regulation, and weakness in this first line of defence against cholera could only spell trouble for Upper Canadian communities lying just west of the border with Lower Canada.[44] The threat to Upper Canada was felt by the public throughout the spring of 1832, and was realized when cholera erupted at Prescott in mid-June. In view of these difficulties, it is unlikely that the executive government at York simply depended on the quarantine at Grosse Isle to protect Upper Canada from the epidemic.

In any event, Lieutenant-Governor Colborne had considerable time to call the colonial legislature to York once it became obvious that the quarantine at Grosse Isle was inadequate to protect Upper Canada—if he had been interested at all in employing regulatory means consistent with the dominant medical view of the day, or if he had taken to heart his instructions from the Colonial Office of 1831, or if the initiatives in place in the other colonies of British North America had seemed of importance to him. In fact, the Board of Health at York, struggling badly during the height of the epidemic to find sufficient funds to care for the sick and to compel compliance with Board regulations that had no legislative foundation, actually petitioned Colborne to call the legislature to session to remedy these deficiencies.

> The Board cannot in justice to the public or itself, any longer delay in making known to Your Excellency its total inefficiency in carrying into effect any salutary regulations for the public good, not from any want of unanimity amongst its members, nor of concert on the part of the Magistracy of the district, but from the absence of any legal means to which it might resort to compel the observance of its ordinances and regulations.
> ... Thus without power and without funds the board is reluctantly compelled to address Your Excellency, praying Your Excellency (in the emergency in which this District and other parts of the Province are placed) will at as early a period as the Law will allow, summon the Legislature for the especial purpose of providing for the appointment of an efficient board with sufficient funds, and with power to act summarily and with promptness and decision, during the continuance of the dreadful malady which is now raging in this Province.[45]

In reply to this plea from the Board of Health of the most populous centre in the province, the Lieutenant-Governor simply wrote that he would consult the Executive Council on the matter, and that in the meantime the Board and the magistrates ought to use their persuasive powers to compel voluntary compliance with their regulations.[46] The next day, the York Board of Health resigned en masse,[47] and the legislature was not called to session until it was regularly scheduled to meet, at the end of October.

Given weaknesses in the argument that Colborne was a committed anti-contagionist, and in the possibility that he believed that quarantine at first points of entry would protect Upper Canada from the importation of cholera, an alternative explanation for his relative inaction in the face of the emergency must be found. Much of the historical evidence suggests that the threat posed by quarantine to continued high levels of immigration and commercial activity played a very significant role in Colborne's decision-making. The vision of the public good advanced by those who felt that quarantine was the responsible way to respond to the disease, those who preferred to sacrifice something of individual liberty to protect the public health to the greatest extent possible, was challenged by an alternative vision advanced by those who equated public welfare with economic well-being, and the possibility of economic well-being called for continued ambitious levels of immigration.[48]

To understand fully Colborne's dilemma, we must first recognize that it was the stated policy of the Colonial Office in London to encourage high levels of emigration to Canada and elsewhere, primarily to reduce the numbers of poor labourers in various parishes around the United Kingdom who depended to some extent on poor relief. A commission of prominent men was established in London for the purpose of collecting and disseminating general information on emigration to British possessions, and in Goderich's words, 'their chief object is to render their assistance for the emigration of such Persons, with their Families, as have been accustomed to earn their Subsistence by Manual Labour, such as Agriculturalists, Artizans, or Mechanics.'[49] While both Aylmer[50] and Colborne complained about the numbers of indigent emigrants arriving in 1831, we see in Colborne's correspondence on the issue the first suggestion of a possible solution:

> From the destitute state in which they arrived here, and the difficulty of forcing them into remote Townships from the Ports at which they disembarked, it is evident that a great burden will be thrown on industrious Settlers who can ill afford any disbursement, unless the local Government is authorized to direct the expenditure of the sums which may be collected for the temporary support of poor Emigrants. If they could be dispersed through the Province soon after their arrival they might readily find employment; but it is found impracticable to remove the numbers which reach Prescott and York.[51]

Goderich offered his opinion to both Aylmer and Colborne that some means must be found to deal with the large numbers of immigrants, but he noted that Colonial Office policy on the matter was not to be undermined. In a confidential circular to Aylmer of 10 September 1831, the contents of which were transmitted to Colborne on the same date, Goderich said: 'Legislation operating to check permanently present rate of emigration would be very objectionable. Question is not to be raised by Governor unless Legislature take initiative.'[52] In other words, the executives at Quebec City and York were to take some initiative to make the crush of immigration more manageable for the colonies, but that initiative was not to operate in a way detrimental to the overall thrust of emigration policy.

In 1832, then, Colborne was faced with apparently conflicting instructions from the Colonial Office, the first designed to prevent the introduction of cholera through the imposition of a general quarantine of immigrants, and the second designed to encourage and maintain high levels of immigration in support of Colonial Office policy. The best way to accomplish the latter was to disburse immigrants as quickly and inexpensively as possible throughout the colony, and, coincidentally, the same approach functioned nicely as an anticontagionist's response to the local call for a general quarantine. At the same time, Colborne's policy of moving immigrants quickly through the colony to their ultimate places of habitation and employment satisfied the calls for continued economic expansion and infrastructure development expressed by pamphleteers such as C. Sheriff.[53] All of this was confirmed by Colborne in his speech to open the colonial legislature in November 1831:

> [T]he industry and capital gained by the recent extensive emigration to this Province [will not be taken advantage of] without establishing a system for ensuring the effectual repair of Roads and Bridges, and the improvement of the principal communications of the back Townships.[54]

This, Colborne hoped, would prepare Upper Canada for the expected arrival of great numbers of the agricultural classes from the United Kingdom, which would in turn lead to the prosperous opening up of the back country. And in his speech to close the Legislature in January 1832, he again confirmed his commitment to this policy:

> The extent and fertility of the unoccupied lands, will continue to attract to this Country large portions of the redundant Population of the Parent State.—I am, therefore, persuaded, that on your return to your respective Counties, your influence may be usefully exerted in organizing societies for the purpose of affording the information to Emigrants which they so much require, at the Ports where they first disembark, and facilitating their dispersion in the Districts in which they may readily obtain employment.[55]

Given the impetus to foster continued high levels of immigration, both to satisfy Colonial Office imperatives and to realize the continued commercial and agricultural development of Upper Canada, it is not surprising that Colborne resisted calls for a general quarantine that would detract from the realization of these goals. Although early in the epidemic of 1832 Colborne did take some steps that, ostensibly at least, suggest a commitment to using legal and administrative measures to protect the public from the importation of the disease,[56] other events during the summer of 1832 suggest that his efforts were less than sincere in this respect. In particular, Colborne failed to support the Boards of Health in conflicts between the Boards and commercial interests, and this is confirmed by his reluctance to interfere with the ordinary flow of people and goods that might have resulted from an early recall of the legislature, as requested by the Board of Health at York.[57]

It is true that the executive government at York was placed in an extremely awkward spot by the arrival of cholera in Prescott in June 1832. Throughout British North America, other colonial legislatures had followed the directives of the Colonial Office by preparing themselves for the disease with the best regulatory means available. But John Colborne's administration was extremely slow to respond, perhaps out of faith in the possibility that Upper Canada would be protected through efforts in Lower Canada and New York, perhaps out of a genuine, albeit dangerous, commitment to anti-contagionism. But the preponderance of evidence suggests that it was a conflict between two competing visions of the public good, won in the end by the argument for commercial expansion supported by high levels of immigration, that compelled Colborne to resist calls for a general quarantine, to resist the temptation to recall the legislature to deal with the emergency, and to resist the urge to support local Boards of Health in their confrontation with commercial interests. The conclusion can only be that the executive's support of initiatives to protect the Upper Canadian settler from epidemic cholera was perfunctory and insincere, a token response to the call for a regulatory framework at once invasive and protective.

Conclusion

The regulatory response to epidemic disease in history cannot be understood completely without some recognition that influences other than strict considerations of public health have a great deal of impact on how the law develops. From the beginning of the state's acknowledgement of a regulatory responsibility for protection of public health, the ability to perform this responsibility adequately has been constrained by the need to balance competing policy interests, demonstrating an effectual transfer of regulatory power to places outside the legislative chamber. In the case of the law of quarantine, we have seen that, in both Robert Walpole's abortive attempt to reform the domestic law of quarantine in the 1720s and John Colborne's resistance to calls for a decisive regulatory initiative in the face of the cholera epidemic of 1832, competing interpretations of the public good had as much to do with the shape of legal regulation as did medical and other informed opinion on how disease is contracted and spread. It is not so simple as erecting a legal barrier around the sick or regulating the flow of goods and people into a frightened community. While measures such as these were perhaps the obvious choice, driven by a precautionary impulse if nothing else, and in most cases supported by legions of medical professionals and intellectuals, there can be no question that political and economic concerns undermined any real attempt at reform and innovation.

In both of the situations we have considered, however, it is relatively clear that commercial interests played a leading role in preventing reform of the law, and in preventing its employment in circumstances where it is only logical to have expected some aggressive regulatory response. If it were not for the lobbying of London's trading community, the strict domestic provisions of Walpole's Quarantine Act likely would never have captured the attention of the civil libertarians of the day. After all, those provisions had passed into law without so much as a murmur of protest until traders became alarmed at interference with the importation of goods. Similarly, in Upper Canada, there is simply no reasonable explanation for the inaction of the executive government, unless one allows for the influence of what were considered to be the economic imperatives of the day. In both cases, the opportunity was lost to champion the public health of the community over the commercial consequences of a difficult regulatory regime. And in both cases, the aging law of quarantine essentially remained in its early modern form, effectively reminding us of *la longue durée*, the time before medicine and law combined to release Western societies from the preoccupation with deadly epidemic disease.

It is true that we might point to the difficulties of controlling the global movement of goods in the early twenty-first century, and to the fresh threats to the public health that result, as yet further proof of the claims that the powers of the regulatory state of the twentieth century have been eroded substantially. The difficulties of the Canadian regulator with the quarantine of possibly contaminated blood products, referred to at the beginning of this chapter, function as substantiation of this position. Yet the conflicts today experienced by regulators in the area of public health are conflicts that have been suffered ever since the state assumed responsibility for the physical well-being of its citizenry in the early seventeenth century. The law of quarantine, with its restrictions on immigration and imports, is typical of the type of domestic regulation that has the potential to give the sovereignty of the state real substance in a shrinking global market, yet even here we see barriers to the efficacious employment of law that must often seem insurmountable to responsible officials. But rarely, even during historical moments when one might anticipate a forceful intervention by the state on behalf of its citizens, has the state been free to express its policy unencumbered by the demands of interstate commerce.

Notes

1. Details on the episode described in these paragraphs are taken from a series of articles in the *Ottawa Citizen* by Mark Kennedy: 'Canada Imposes Blood Quarantine: Supply May Be Contaminated with Sickness Linked to "Mad-Cow" Disease', 19 Dec. 1998, A1; 'Fate of Blood Product in Limbo: Users Demand Assessment of Product Linked to Brain Disorder', 23 Dec. 1998, A1; 'Quarantine Lifted on Blood Products: Policy Change Allows Release of Blood from CJD Victims', 26 Dec. 1998, A14; 'Blood Agency Fails Trust Test: CEO Admits More Could Have Been Done to Allay Fears After

Quarantine', 20 Jan. 1999, A6; 'Family Fears Another Blood-Borne Disaster', 27 Mar. 1999, A1; 'No Treatment, No Cure—Just Death', 28 Mar. 1999, A3; 'Plasma Donor Dies from Rare Brain Disease: Hemophiliacs Fear Blood Quarantine Lifted Prematurely', 28 Mar. 1999, A3; 'Hemophiliacs Demand Plasma Information', 13 Apr. 1999, A6. The references to scientists from Bayer Inc. are taken from the article of 27 Mar. 1999.
2. Mark Kennedy, 'Experts Fear Mad Cow Pandemic', *Ottawa Citizen*, 12 June 2001, A1.
3. Mark Kennedy, 'A Very Unhealthy Conflict of Interests', *Ottawa Citizen*, 1 Nov. 1999, A15.
4. 1 Jac. I c. 31. I will refer to this statute as 'King James's Act' in the balance of this chapter.
5. I borrow this term from Edward S. Golub, *The Limits of Medicine: How Science Shapes Our Hope for the Cure* (Chicago: University of Chicago Press, 1997), 32. He explains the term in the following way:

> The French historians who called themselves the *Annales* historians, in an attempt to understand what it was like to live in this 'world we have lost,' began to view the past from the peasants' perspective rather than following the traditional pursuit of studying only wars and treaties. They discovered the amazing fact that through most of history, little changed during the lifetime of one person, and they coined the phrase *la longue durée* to evoke the idea of the relative changelessness of conditions over a long time.

For a detailed discussion of regulatory sanitation reforms during the nineteenth century in the United Kingdom, see Anthony S. Wohl, *Endangered Lives: Public Health in Victorian Britain* (London: J.M. Dent & Sons, 1983).
6. Gabriel Kolko, *Railroads and Regulation 1877–1916* (Princeton, NJ: Princeton University Press, 1965).
7. Walpole's reforms resulted in three significant legislative enactments: 7 Geo. I c. 3, 8 Geo. I c. 10, and 8 Geo. I c. 18. In the balance of this chapter, the first will be referred to as the 'Quarantine Act', the second as the 'Amending Act', and the third as the 'Supplementary Act'.
8. An Act to establish Boards of Health, and to guard against the introduction of Malignant, Contagious and Infectious Diseases, in this Province, 3 Wm. IV c. 48 (U.C.), passed by the legislature on 13 Feb. 1833.
9. Paul Slack, *The Impact of Plague in Tudor and Stuart England* (London: Routledge & Kegan Paul, 1985), 61–2.
10. It should be noted that, as early as 1518, attempts had been made by Royal Proclamation to regulate activity during times of plague, and local councils and justices often used their powers in an attempt to control the disease. See ibid., 44–7. But King James's Act was certainly the first legislative attempt, the first systematic, comprehensive initiative, to deal with the plague. Considering that the statute was passed at the end of the most devastating plague to that point in English history, Slack (p. 200) is mistaken when, in referring to the prospect of legal reforms in the years prior to 1665, he writes:

Most obviously, none of them coincided with major epidemics of plague, either in London or in the country as a whole. In the immediate crisis caused by a great epidemic like that of 1563 or 1603, there was little opportunity to develop new policies. Rather, innovation occurred when relatively minor outbreaks of plague seemed to aggravate other pressing social problems, especially in London.

11. See ibid., 62, for a tabulation of plague years in 14 English towns between 1581 and 1666.
12. For a particularly vehement denunciation of the regulatory regime, see Thomas Cock, *Hygiene, or a Discourse upon Air* (London: Philem. Stephens, 1665). The literature of the Great Plague of 1665 and the years following is replete with similar examples.
13. William L. Payne, ed., *The Best of Defoe's Review* (New York: Columbia University Press, 1951), 219-20.
14. Ibid., 220.
15. Ibid., 148.
16. All of these letters, covering the period 22 Aug. 1719 to 28 Apr. 1722, are reproduced in William Lee, *Daniel Defoe: His Life and Recently Discovered Writings*, vol. 2 (Hildescheim: Georg Olms Verlagsbuchhandlung, 1968), 142, 265, 277-8, 281, 284-5, 291-2, 378-9, 399-401, 407-10, 427-30, 436-8, 449-51, 453-5, 464-5, 514-16.
17. Ibid., 265.
18. Ibid., 291.
19. Ibid., 409.
20. Ibid., 408.
21. Ibid., 409-10.
22. Ibid., 429.
23. William Cobbett, *Parliamentary History of England: From the Norman Conquest, in 1066, to the Year 1803* (London, 1806-1820), vol. 7, col. 930.
24. Ibid.
25. Ibid., col. 933.
26. Ibid., col. 934.
27. Ibid.
28. Despite remarks about the controversy surrounding the Supplementary Act and the Amending Act contained in Alfred James Henderson, *London and the National Government, 1721-1742* (Durham, NC: Duke University Press, 1945), 33-54, 138, 211-13, history has been remarkably quiet in considering the full implications of these statutes. See, for example, chapters 19-21 of vol. 1 of Robert Coxe's three-volume *Memoirs of the Life and Administration of Sir Robert Walpole, Earl of Orford* (London: T. Cadwell, Jun. and W. Davies, 1798), 126-58. The neglect of the quarantine legislation by historians can be explained by remembering that debates in Parliament during sessions in 1720, 1721, and 1722 were dominated by the 'South Sea Bubble' affair, an issue that threatened English public credit and jeopardized England's growing status as the world's premier trading nation. Coxe devotes virtually all of his discussion of the period to that question, applauding Walpole for steering England through especially difficult times. See especially pp. 147-55.

29. Perhaps oddly, Daniel Defoe himself raised this latter argument on the relationship between healthy trade and a healthy population in *A Journal of the Plague Year* (London: J.M. Dent/Everyman, 1994), first published in 1722, only months after Walpole's failed attempt to reform the law of domestic quarantine. See especially Defoe's reference (pp. 185–6) to the trade in corn and coal and to the difficulties experienced as a result of interruption in the free movement of those commodities.
30. While the law of domestic quarantine in the United Kingdom remained unchanged throughout the eighteenth century, the law with respect to import and immigration restrictions for the purposes of disease control was revamped twice before the end of the century. Those changes can be found in An Act to oblige Ships more effectually to perform their Quarantine; and for the better preventing the Plague being brought from foreign Ports into Great Britain or Ireland, or the Isles of Guernsey, Jersey, Alderney, Sark or Man, 26 Geo. II c. 6, and An Act more effectually to secure the Performance of Quarantine, and for amending several Laws relating to the Revenue of Customs, 28 Geo. III c. 34, both of which amended but did not repeal the Supplementary Act.
31. An Act to oblige ships and vessels coming from places infected with the plague or any pestilential fever or disease, to perform Quarantine, and prevent the communication thereof in this Province, 33 Geo. III c. 5 (L.C.). This statute dealt exclusively with restrictions on ships arriving from infected ports, and related matters. As cholera became more imminent in late 1831 and early 1832, the legislature introduced and then passed in February the statute 2 Wm. IV c. 16 (L.C.), by which standard provisions on domestic quarantine were provided.
32. The first statute was An Act to prevent bringing infectious distempers into the City of Saint John, 36 Geo. III c. 5 (N.B.) (1796). This Act was repealed and replaced by An Act to repeal an Act made and passed in the thirty-sixth year of His Majesty's reign, intituled 'An Act to prevent bringing infectious Distempers into the City of Saint John,' and to make more effectual provision for preventing the importation and spreading of such contagious Distempers, 39 Geo. III c. 9 (N.B.) (1799). It was passed concurrently with An Act to prevent the importation or spreading of Infectious Distempers within this Province, 39 Geo. III c. 8 (N.B.). Both statutes were amended several times prior to mid-1831.
33. An Act to prevent the spreading of Contagious Diseases, and for the performance of Quarantine, 2 Wm. IV c. 13 (N.S.), and An Act more effectually to provide against the introduction of Infectious or Contagious Diseases, and the spreading thereof in this Province, 2 Wm. IV c. 14 (N.S.), both passed 14 Apr. 1832.
34. There were four principal statutes in the revisions done in 1831 and 1832, prior to the actual arrival of cholera in the Maritime colonies. They were An Act to amend an Act, intituled 'An Act to repeal all the Acts now in force relative to the importation and spreading of infectious distempers in the City of Saint John, and to make more effectual provisions for preventing the same', 1 Wm. IV c. 35 (N.B.) (1831); An Act, to make more effectual provision for preventing the importation and spreading of infectious Distempers within the Towns and Settlements in the Counties of Charlotte and Northumberland, 1 Wm. IV c. 40 (N.B.); An Act to provide against the importation and spreading of distempers in the Counties of Westmorland, Gloucester and Kent, 2 Wm. IV c. 19 (N.B.) (1832); and An Act, to prevent the spreading of infectious or pestilential Distempers, 2 Wm. IV 2nd c. 5 (N.B.). These

statutes continued to be amended, repealed, and replaced throughout legislative sessions in 1833 and 1834.
35. An Act to prevent the importation and spreading of infectious diseases within this Island, 2 Wm. IV c. 13 (P.E.I.)
36. It is noteworthy that both New Brunswick and Prince Edward Island escaped the cholera infection in 1832, although it is not clear the extent to which that good fortune can be attributed to their quarantine regime.
37. *Upper Canada Herald*, 11 Apr. 1832, National Library of Canada (hereafter NLC).
38. *Colonial Advocate*, 26 Apr. 1832, NLC.
39. *Canadian Freeman*, 29 Mar. 1832, NLC.
40. *Brockville Recorder*, 9 Feb. 1832, NLC.
41. Civil Secretary to Smyth, 6 July 1832, National Archives of Canada (hereafter NAC), Governor-General's Office, Civil Secretary's Letterbooks, 1799-1841, RG7, G16C, vol. 26.
42. Goderich to Colborne, 30 June 1831, NAC, Upper Canada: Submissions to the Executive Council on State Matters, RG1, E3.
43. Aylmer's Message to the Legislative Council of Lower Canada, 3 Feb. 1832, NAC, Governor-General's Office, Miscellaneous Records, Governor-General's Speech and Message Book, 1831-3, RG7, G18, vol. 43.
44. Upper Canadian newspapers devoted considerable space to reports from Quebec City and Montreal on inefficiencies in the operation of Boards of Health in those communities and on enforcement of the quarantine at Grosse Isle. From the *Montreal Gazette*, for example: 'If we are asked why in Montréal this disease has been so fatal, we may with some safety reply, that the delay in the organization of the Board of Health, which prevented that complete purification of the city which is still wanted may be regarded as one; . . . that the careless treatment of some cases . . . have tended to swell the already large list of victims to this plague.' *Colonial Advocate*, 28 June 1832, NLC. See also the discussion of the inefficiencies of the Grosse Isle quarantine station in Geoffrey Bilson, *A Darkened House: Cholera in Nineteenth-Century Canada* (Toronto: University of Toronto Press, 1980), 9 ff.
45. Baldwin to Colborne, 9 Aug. 1832, NAC, Civil Secretary's Correspondence, Upper Canada Sundries, RG5, A1, vol. 120, 66646-55.
46. Civil Secretary to Baldwin, 10 Aug. 1832, NAC, RG7, G16C, vol. 27.
47. Baldwin to Colborne, 11 Aug. 1832, NAC, RG5, A1, vol. 120, 66671-2.
48. See, for example, the excerpt from a pamphlet on emigration written by C. Sheriff, in which Sheriff encourages government to press on with capital works, the improvement of roads, bridges, canals, waterways, thus to give immediate employment to the large numbers of labourers who form the bulk of the emigrant group, and also to make the country more attractive for those 'better classes to settle in the Colony that would otherwise leave it'. *Kingston Chronicle*, 2 July 1831, NLC. This position was later endorsed by the editor of the *Kingston Chronicle*, when he argued that, because of the natural richness of Upper Canada, emigrants of little education and means would be tempted to become dependent on the benevolence of the residents rather than turning to profitable employment. The chances of this occurring are reduced, so the argument went, when large capital projects offer employment to all who need it. *Kingston Chronicle*, 9 July 1831, NLC.
49. Goderich to Aylmer, 1 Aug. 1831, NAC, RG7, G1, no. 53.

50. Goderich to Aylmer, 3 Aug. 1831, NAC, RG7, G1, no. 54, in which Goderich chastises Aylmer for complaints about the state of emigrants on arrival at Quebec contained in an earlier letter from Aylmer to Goderich.
51. Colborne to Goderich, 5 Sept. 1831, NAC, Governor-General's Office, Letterbooks of Despatches to the Colonial Office, Upper Canada to Secretary of State, 1830-3, RG7, G12, vol. 18.
52. Confidential Circular, Goderich to Aylmer, 10 Sept. 1831, NAC, RG7, G1; Confidential Circular, Goderich to Colborne, 10 Sept. 1831, NAC, RG7, G1.
53. Sheriff, in *Kingston Chronicle*, 2 July 1831.
54. *Kingston Chronicle*, 26 Nov. 1831, NLC.
55. *Upper Canada Gazette*, Extraordinary Issue, 28 Jan. 1832, NLC.
56. On 20 June 1832 Colborne sent a circular to the various chairmen of the District Quarter Sessions, suggesting the formation of Boards of Health and allotting the sum of £500 to the use of each district for the construction of hospitals and other necessary expenses. Circular—Civil Secretary to Chairmen of District Quarter Sessions, 20 June 1832, NAC, RG7, G16C, vol. 26. The circular, however, was mainly perfunctory, for a number of reasons. First, there was no legislative authority to vest regulatory power in the Boards, such that their efforts were largely ignored by many in their jurisdictions. Second, many of the districts and towns, including Belleville, Kingston, and London, had already formed Boards of Health in advance of the instructions from Colborne. Magistrates of Belleville to Civil Secretary, 22 June 1832, NAC, RG5, A1, vols 116-17, 65779-81; John Kirby of Kingston to Civil Secretary, 23 June 1832, NAC, RG5, A1, vols 116-17, 65797-8; John B. Askin of Woodhouse, London District, to Civil Secretary, 25 June 1832, NAC, RG5, A1, vols 116-17, 65834-5. And third, the Civil Secretary failed to include instructions on the manner in which the funds might be accessed by the Boards, meaning many Board members were personally forced to guarantee payment of Board debts without assurances from Colborne's office that funds would be forthcoming without the usual audit of public accounts and without waiting many months for the legislature to approve payment. J.W. Macaulay of Kingston to Civil Secretary, 12 July 1832, NAC, RG5, A1, vol. 118, 66179-83.
57. Baldwin to Colborne, 9, 11 Aug. 1832; Civil Secretary to Baldwin, 10 Aug. 1832.

Chapter 15

Escaping Interventionism: Negotiating Regulation and Self-Governance in the Wake of the Indian Act

JANE DICKSON-GILMORE

If the purposive law of the past century has had as its primary goals the 'better allocation of resources, the protection of those who were marginalized and excluded from societal decision-making, and the provision of a more just distribution of wealth', as Peter Swan has suggested in Chapter 1 of this volume, then there is perhaps no better locus of critique of that law than the experience of First Nations within Canadian borders. Where historically colonial states existed in uneasy alliances with First Nations, when the need for those alliances diminished the move to a welfare approach and increasingly intrusive legal interventionism was swift. Successive legislation eroded modes of governance and regulation made vulnerable by the social, economic, and cultural dislocations caused by contact with newcomers, the fur trade, missionaries, and involvement in colonial wars imported from Europe. The modern consequences of legal interventionism for First Nations are readily apparent, and while Aboriginal communities exist at a range of locations on the socio-economic ladder and manifest variant levels of agency and activism in responding to the effects of interventionism, it is clear that none have escaped the 'culture of interventionism' and the creeping paternalism that is so much a part of it.

For many First Nations, the strategy by which they hope to eschew external legal intervention into their lives is the negotiation of self-determination or self-government; through self-government, it is argued, communities will achieve both the withdrawal of the state and its laws from the lives of its members and, by implication, a concomitant rise in individual autonomy. Like many Aboriginal communities, the self-government strategy of the Mohawks of Kahnawake, Quebec, has revolved around usurping the federal Indian Act, legislation that has long served as the leading edge of state legal interventionism into First Nations.

While the law had never sat comfortably on the backs of the Kahnawa'kehro:non, in 1988, following a federal police raid on contraband cigarette shops in the community, the Mohawk Council determined once and for all to end external regulation of their lives. Within days of the raid, the Council demanded that the federal government come to the table to negotiate a 'new relationship' between the two peoples. The negotiations were to focus on removal of the essential defining element of the old relationship—the Indian Act. Within one year negotiations between the Council's Intergovernmental Relations Team (IRT) and the federal government had commenced; by May 1999 their efforts had produced a Draft Umbrella Agreement (DUA) outlining the terms for removal of the Act and the creation of new legislation to replace it, the Canada/Kahnawake Intergovernmental Relations Act (CKIRA). The DUA, which the IRT expected to be open to public scrutiny within weeks of its completion, was released in Kahnawake only recently, as was the Sub-Agreement on Membership, which may prove to be one of the more contentious aspects of the DUA.[1] Taken together, these documents suggest the dawning of a new era in Canada's relationship with the Kahnawa'kehro:non and, by implication, the possibility that all Aboriginal peoples in Canada might one day follow Kahnawake's lead towards the realization of a degree of independence commensurate with their status as citizens of *First* Nations.

But is the independence anticipated by the DUA and the Sub-Agreement truly one unfettered by state intervention, and, if yes, does this necessarily lead to greater freedom of Aboriginal lives from legal intervention such as that implicit in the Indian Act? Critical scrutiny of the agreements suggests we would do well to be cautious, as the legal interventionism that has long circumscribed Aboriginal lives and actions has a range of origins and persists at a variety of levels. Thus, while successful negotiations may lead to a reduction in the specific form of state intervention implicit in the Indian Act, the intervention of other state laws will remain in the form of rights legislation, at least for a time, as will the potential for expansion of that layer of legal interventionism that originates with the Mohawk Council. The danger resides in the possibility that when all the talks are done, what will result is not less overall intervention but simply different forms of intervention, at least some of which will arise from new or unexpected origins; in essence, an exchange of state intervention for more local Mohawk intervention—and one wonders whether that exchange will be enough to quench the very real thirst for independence of individual Mohawks. Will that independence be more apparent than real? Although there can be little doubt that self-government will never approach the autonomy experienced by First Nations at the moment of the arrival of the newcomers, it seems clear that self-government should at least anticipate less outside interference in local business and less intervention in the lives of Aboriginal people. Is this what the DUA envisages? Will the escape from the strictures of the Indian Act lead to less intervention in Kahnawa'kehro:non lives?

There is little question that the Kahnawake people want the Indian Act out of their lives—a 1999 survey by the Council clearly documented this opinion, as slightly over 70 per cent of those polled voiced support for the IRT's efforts to usurp the Act.[2] Yet one wonders about the roots of this enthusiasm, given that a remarkable 75 per cent of those supporting the removal of the Act admitted that they have no knowledge of its contents.[3] This suggests that fervour for the removal of this fundamental intervention into Mohawk lives may have more to do with removing Canadian legal jurisdiction in general from the community and enhancing the freedoms of individual Kahnwa'kehro:non from state legal intervention, than with any particular quarrel with the quality of that intervention as evidenced in any specific aspect of the Act. That is, there would seem to be an assumption at work that removal of the Indian Act from the community will end the interventionism of the Canadian state and its laws more generally, leading to a better quality of law and life in Kahnawake under the 'Mohawk laws' and 'traditional government' intended to supplant the Canadian authorities. Yet the DUA, while clearly providing for the removal of some Indian Act provisions, nonetheless appears not only to preserve other parts of the Act but to assume explicitly the continued intervention into Kahnawa'kehro:non lives of other state laws, as well as a battery of new 'Mohawk laws' that have a clear echo in the state laws they succeed. Thus, the new powers that will fall to Kahnawake's local government, and that result from a significant reduction in *external* legal intervention, may well turn into much less reduction in legal intervention in Mohawk lives than was assumed to accompany the unseating of the Indian Act. That the source of at least a slightly greater part of that intervention is the Mohawk Council, rather than the Canadian state, may be significant at a philosophical level, but one wonders whether it will be so at more personal levels—especially with regard to the intimate and painful issue of membership. Self-government may lead to greater autonomy for the Mohawk Council, but will the enhanced production of local law and regulation, coupled with an admittedly more distant but remaining role of Canadian law, lead to greater freedom for individual Kahnawa'kehro:non?

The Beginnings of Rigorous Interventionism: Creating a Legacy

The Mohawks of Kahnawake are the most prominent modern community of a nation that was on the cutting edge of colonialism and the receiving end of much interventionism as the Canadian state evolved through the colonial period to its current status. The Mohawks, as the 'Keepers of the Eastern Door' of the famous Iroquois Confederacy, were among the first Iroquois to interact intensively with the Dutch, French, and British who coveted Aboriginal alliance, trade, and, later, land. At each turn of their shared history, the Mohawks and, after their 1667 relocation to Kentake (a precursor of modern Kahnawake), the Kahnawa'kehro:non actively resisted the intrusion of the laws and jurisdictions of the newcomer nations into their community and culture.

This resistance has persisted to the modern era, and while it has had limited success in ensuring the continued articulation of Mohawk traditions of regulation and governance, the philosophies and traditional ideologies that once animated community institutions remain a vital part of Kahnawa'kehro:non identity and activism. Mingling with centuries of interventionism and state paternalism, traditional philosophies and ideologies have fostered a unique and fascinating political culture in Kahnawake, a central feature of which is a strong ambivalence towards 'outside law' and governance.

Like many First Nations with whom the Iroquois shared this continent prior to the arrival of the newcomers, the Mohawk Iroquois regulated their communities through a detailed delineation of rights and duties. Although not committed to the written word, these 'laws' were nonetheless known and understood by all. International and interpersonal relationships were mediated by a robust network of cross-cutting ties and interdependencies holding at its core the extended family networks that were central to the exogamous, matrilineal Iroquoian clan system inherent to all the Iroquois Nations. These extended family networks, which among the Mohawk were identified with such totems as the Turtle, Bear, and Wolf, were the heart of law and regulation in Iroquoia: All individuals were members of clans, which in turn defined the communities that comprised the nations entering into international relationships through their collective political body, the Iroquois Confederacy. The Iroquois leaders who negotiated these relationships were put in place by the senior women of their respective clans, and the governance they practised at the international level was replicated in their role at the level of their individual nation and clan councils. At every level of governance their occupation was defined primarily by their membership in a clan and the obligations they owed to their extended families and the communities they together comprised.

Thus for the Iroquois people, philosophical traditions of governance and regulation depended on the pressures of clan to ensure that no individual existed free of the constraints and entitlements of blood ties. Laws were few but significant and well known; the oral constitution of the Iroquois Confederacy, or *Kaianerakowa*,[4] was recited regularly at government councils at many levels and was similarly clear in the minds of the people. Here, social control was self-control mediated by the ever-present clan structure; written laws were as irrelevant as they were unnecessary: the historic Mohawk Nation was independent, sovereign, and, in every sense of the word, self-determining. The Mohawk people of modern Kahnawake inherited these traditions and qualities, and the philosophy intrinsic to them continues to animate almost every part of their daily lives and stands as a constant foil to external legal intervention.

Although surviving records indicate alterations in Mohawk forms of governance and regulation over time, those enduring traditions suffered a serious blow with the Canadian government's passage in 1869 of legislation providing for the replacement of 'traditional' governments with elected band councils. This legislation was later revised and restated in the 1884 Indian Advancement Act.[5] Through this legal intervention, the Canadian state was able to achieve what it envisaged as a shift from 'irresponsible' traditional governmental forms to an allegedly more 'civilized' electoral system. It obtained this end through provisions in the law that permitted the federal Indian Department of the day to impose the council government system on any 'Indian band' declared by its resident 'Indian agent' to be ready and 'worthy' of council government, and the council system was imposed on many First Nations in Quebec and elsewhere as early as 1895.

In Kahnawake's case, the introduction of council government is a source of more than a little controversy historically and today. Local, modern reminiscences of the intrusion of council government refer to a group of Kahnawake families who have become known as 'The Fourteen', owing to their perceived role in supporting the introduction of the elective system. Two versions of this story appear to predominate. By one version, at a gathering of a large proportion of the community in a field to one side of the village, the resident agent purportedly traced a line in the soil and instructed all those Kahnawa'kehro:non supporting the introduction of an elected government to step over the line. Local lore informs that 14 families took the step, and they were deemed by the agent a sufficient majority to support government by council in Kahnawake. Modern versions of this historical event portray the descendants of those families as continuing to be viewed with some contempt by their peers, who perceive the act of their ancestors as constituting a crucial and life-altering betrayal of Kahnawake and a silent majority who rejected council government.[6] The second version of this event is linked with a number of petitions originating in Kahnawake and clearly supporting the intervention of the state to activate an elective system. Relevant primary reports provide evidence of a series of seven petitions sent from Kahnawake to Indian Department authorities in 1889, requesting that the community be allowed to begin holding elections prior to the passage of legislation legally empowering them to do so.[7] Clearly, not all Kahnawa'kehro:non were against the 'imposition' of an elective system, and although a connection between these Mohawks and the 'fourteen' is unclear, the latter have been associated with the petitions by many in Kahnawake.[8] On 5 March 1889, an Order-in-Council by the Canadian government brought the 'Iroquois Band Of Caughnawaga' under the terms of the Indian Advancement Act and into a system of governance by band council.[9]

Notwithstanding what appears to have been some support for the intervention of an elective system of government, there can be little doubt that the event was both painful and divisive for the Kahnawa'kehro:non. It also bears some relevance to the discussion of intervention central to this chapter. First, despite the 'amendments' to the 'old ways' of governance and regulation consciously made by the Kahnawa'kehro:non of the day and by their predecessors, those ways remained an important mode of local regulation in Kahnawake. In other words, the people of Kahnawake had a means of ordering their lives and relationships of their making and by their choice, and the Canadian state superimposed a state definition of 'appropriate' modes of regulation and governance on the community, albeit one putatively supported by at least some members of the community. Such interventionism not only reveals the legal imperialism typical of early First Nations-state relations in Canada, but also shows the reality that interventionism is rarely, as Purvis notes in Chapter 2 of this text, a simple act of replacement of one legal order with another. Rather, as subsequent records suggest, 'imposing' council government did not eradicate traditional government; according to some modern Kahnawa'kehro:non, it merely drove those traditions either underground or into the more apparently acceptable context of religion. Those who rejected council government thus became the objects not only of their chosen mode of regulation—the Longhouse—but also of a delegated arm of the Canadian state whose authority they could resist but were unlikely to unseat. In fact, as history shows, this 'over-interventionism' has proven remarkably enduring and intractable. The 1999 DUA that details Kahnawake's escape from state interventionism and, more pointedly, the Indian Act, while clearly embodying a degree of resistance, can be seen not only to fail to end interventionism, but to further it through the clear articulation of additional state and local government intervention into the lives of Kahnawa'kehro:non.

A second consequence of the interventionism of Canadian law into Kahnawake is the legacy of factionalism that the presence of council government nurtured and sustained. For while there can be little doubt that political tensions and factionalism had long existed in Mohawk communities, as they did and do in virtually all communities, none have proven so enduring in this and other First Nations as the conflict between those associated with council government systems and those preferring traditional modes of governance. Whether articulated as a split between 'progressives' and 'conservatives' or 'pro-government' and 'traditionalists', the persistence of friction between new and old forms of governance in most First Nations reveals both the fallacy of replacement in interventionism and the profound and intractable damage to the social and political cohesion of citizens it creates. Intervention not only often means more laws, but it also commonly means more conflict about those laws and about regulation and governance in general.

Escaping Interventionism? Negotiating a New Relationship between Canada and Kahnawake

In a twist of irony consistent with many in Kahnawake's past, the latest efforts by the Mohawk Council to redefine and improve the relationship with Canada find their origins in barricades that sprang into place on Kahnawake's borders on 1 June 1988 and that were, in many ways, little more than a new articulation of the long chronicle of division between Kahnawake and the newcomers. In this particular instalment, the blockades were the product of a dovetailing into violence of a protracted Kahnawake-Canada disagreement over the Mohawks' active interpretation of their Jay's Treaty rights to import and sell 'contraband cigarettes' and the Canadian government's apparent uncertainty over how to respond to that activity. That this disagreement had the potential for full-blown conflict had not gone unnoticed by at least some Mohawks. In January of 1988 the leaders of the largest traditional group in the community—and those most closely associated with the cigarette trade—had approached the governments of Quebec and Canada hoping to initiate talks in regard to the Mohawks' cross-border rights. The federal government apparently refused to participate in the talks, probably owing in equal measure to its desire not to appear to undermine the Mohawk Council as the legitimate local government in the community and a certain ambivalence in its own position on the trade. The province of Quebec agreed to join the negotiations, and its officials were actively working with the Longhouse towards an agreement on 'contraband' when the federal government executed a raid on Kahnawake that saw 200 Royal Canadian Mounted Police officers storm the community's borders and raid four cigarette shops.[10] It was an act that was both violent and designed to intimidate, and in its aftermath, Kahnawake leadership seized upon the raid as a spectacular act of interventionism demonstrating only too clearly the need to eliminate the presence of all outside law and forms of governance in their community. The Mohawk Council articulated that need in unambiguous terms to then Prime Minister Brian Mulroney:

> The issue that surrounds this incident is the sovereignty, the jurisdiction and the territorial integrity of the Mohawk Nation People of Kahnawake. . . . *The Federal government must agree to political level discussions to deal with the Federal/Kahnawake Mohawks relationship in the areas of jurisdiction, our sovereignty and our territorial integrity.* . . . Our relationship with the Federal government must be clarified.[11]

In essence, that clarification revolved around the role of the Indian Act in Kahnawake. Where previously the community had pushed the envelope of Indian Act powers and jurisdictions, and sometimes moved well beyond them, now the goal was nothing less than the total elimination of the Act from Kahnawake. The Canadian government agreed; in the minds of public officials, few communities were better suited or situated to go it alone—they also understood the reality, based on past experiences with the community, that Kahnawake has a tendency to do as it chooses regardless of outside opinions.

It seemed better to go along with the Mohawks and negotiate an independence compatible with both Kahnawake and the federal government's interests, rather than invite an inevitable conflict by denying that independence.

The Mohawk Council submitted its terms for the negotiations on 20 December 1988 in a document entitled 'A Framework Proposal for Negotiations Between Canada and the Mohawk Council of Kahnawake (on Behalf of the Mohawks of Kahnawake)'.[12] In this document the Council proposed the recognition of a significant degree of Kahnawa'kehro:non autonomy in a range of local government matters, all of which would be protected under the umbrella legislation, noted above, of the 'Canada/Kahnawake Intergovernmental Relations Act' (CKIRA). This legislation would define and guarantee the continuing relationship between Canada and Kahnawake in four categories. (1) Under the heading of those matters to be negotiated towards a 'Transitional Framework for Mohawk Government in Kahnawake' were included a number of important powers, some of which were already areas of substantial activity and authority by the Mohawk Council, including:

- legal authority, capacity, and structure of the Mohawk government;
- election procedures;
- duties and powers of the Mohawk government;
- procedures and guidelines for the Mohawk government, including accountability to the membership;
- functions and responsibilities of Mohawk institutions;
- provisions for amendment;
- nature of the traditional government and procedures for its implementation; and
- membership/citizenship.[13]

(2) In addition to these matters, the Council sought negotiation of those powers related to justice, wherein it was proposed that 'Canada recognize by legislation the legal capacity of Mohawk government to legislate, adjudicate and enforce laws for the community and territory of the Mohawks of Kahnawake'.[14] (3) There was also to be consideration and negotiation of the 'ongoing responsibility' of Canada to the Mohawks of Kahnawake within any agreed-upon financial arrangements, as well as over land management and control of the Mohawk territory within Kahnawake's borders. (4) Finally, negotiations would need to attend to a range of such other matters as taxation, social services, and family law.[15] If successful, these negotiations would lead to a considerable restriction of legal interventionism by the Canadian state and the rise of an independent, 'traditional' Mohawk government at Kahnawake.

The federal government agreed formally to initiate negotiation towards the removal of Kahnawake from the confines of the Indian Act, undertaking to define a new relationship that would not prejudice the Mohawks' treaty or Aboriginal rights, or existing or potential land claims. Ottawa also stipulated that this new relationship should not alter the existing federal-provincial division of powers and that where provincial areas of authority were likely to be affected by the negotiations, such as in those areas of justice articulated by the Mohawks, the involvement of the Quebec government would be required. This would prove to be a significant stumbling block, as the provincial government was understandably reluctant to involve itself in talks designed to alter relations of governance and regulation established in a Constitution it does not recognize.[16] Another hurdle resided in the minimum requirements of the federal government that any 'Canada/Kahnawake arrangements must be compatible with the established principles, jurisdictions and institutions of the government of Canada', including especially the Charter of Rights and Freedoms.[17] Clearly, legal interventionism would prove to be a hard habit to break.

When the Mohawk Council and the federal government reached an agreement on the parameters and process for their negotiations, they anticipated that an agreement-in-principle would be reached within 24 months of the appointment of their respective negotiators. That initial framework, discussions towards which were interrupted by a return to barricades in the Oka crisis of 1990, was signed by Grand Chief Joseph Tokwiro Norton, on behalf of the community of Kahnawake, and the Minister of Indian Affairs on 13 December 1991.[18] Those 24 months expanded into eight years, leading to the conclusion on 10 May 1999 of a 'Joint Final Draft' of the 'Umbrella Agreement with Respect to Canada/Kahnawake Intergovernmental Relations Act'.

The Draft Umbrella Agreement is a significant document on a number of fronts and, like the Indian Act it is putatively directed to replacing, leaves little of Kahnawake life untouched by legal intervention of some form or another. That this legal intervention appears to be at the hands of Mohawk government, rather than a Canadian one, is likely to affect Kahnawa'kehro:non reactions to it at least to some degree. Given this, it is perhaps surprising that the document was not released to the general public in Kahnawake until 2001.[19]

The DUA as it emerged in 1999 was expansive and, in places, somewhat unsettling in its content.[20] The DUA will form the basis of the proposed CKIRA, which, if passed, will confirm a form of Kahnawake governance defined in an as yet non-finalized 'Kahnawake Charter'. This Charter also achieved a draft form on 10 May 1999.[21] According to terms to be defined within the Charter, government in a post-Indian Act Kahnawake will be both 'democratic and accountable', and founded on 'the principles of Mohawk Government in Kahnawake based upon custom, traditions and traditional laws of the Iroquois Confederacy'.[22] Precisely what this will mean in a practical sense of things is not apparent in the document, although some idea of what 'customs, traditions and traditional laws' in modern Kahnawake might resemble will become apparent later in this chapter during a discussion of the controversies around membership and how the Sub-Agreement on Membership apparently proposes to respond to them. Membership or citizenship is one of the 28 matters that the DUA transfers to Kahnawake.[23] Most of these 28 jurisdictions will be phased in over time, consistent with the phasing out of those sections of the Indian Act incompatible with various elements of those jurisdictions.

The DUA and the CKIRA it anticipates offer an interesting example of the paradox of self-governance and the reduction of outside legal intervention in favour of the enhanced articulation of 'Mohawk law' and the achievement of greater autonomy for individual Mohawks. Reflection on the DUA and proposed CKIRA suggests strongly that the community's leaders have accepted both the external form of law and the inevitability of some measure of legal intervention, as it appears that the DUA not only does not eliminate the full articulation of the Indian Act in the community, but actually extends the interventionism implicit in much of the Act through replication or expansion of those aspects of the Act that the DUA and CKIRA appear to remove or reduce. Nowhere is this irony more apparent than in the context of the DUA and its Sub-Agreement on Membership. Where Kahnawa'kehro:non were once subject to state intervention and external definitions of 'Indian' and who qualifies as a 'member' of the 'Kahnawake Band', and later to this and to Mohawk Council definitions of 'Mohawks' and who qualifies for Council-administered services as a 'Mohawk', the DUA and Sub-Agreement appear to do little to ameliorate such interventions into Kahnawa'kehro:non definitions of self and may even further complicate them through the application of a complex interplay of Mohawk and Canadian charters, human rights regulations, and Mohawk custom. It is difficult to see how this constitutes a reduction in legal intervention in Kahnawa'kehro:non lives, even if it results from a putative reduction of state intervention in Council jurisdictions. Legal intervention tends to feel the same, whether it comes from an external state or a putatively community-friendly local government. This is especially so in the context of regulation of identity and those rights and entitlements that flow from particular qualities of identity. A brief look at the evolution of tensions around identity in Kahnawake reveals clearly that, while state legal intervention may have created much of the mess around membership, simply removing that intervention or passing authority for membership to local government will not lead to less intervention in Mohawk lives; indeed, less state intervention may do more harm than good.

Reordering Interventionism: The Draft Umbrella Agreement and the Challenge of Membership

Kahnawake's modern take on 'who belongs' has evolved somewhat from that witnessed historically among their Mohawk and Iroquois progenitors; it also differs significantly from the membership policies of the Department of Indian Affairs and Northern Development (DIAND), which have pertained to Kahnawake for most of its recent history. Although concerns over 'non-Mohawks' residing in Kahnawake had been evident consistently over much of the community's 300-year history, it was not until the 1980s that the community sought to control the presence of outsiders in their midst through control over the definition of insiders.[24] The catalyst for this initiative was the community's awareness that DIAND was contemplating changes to the Indian Act on the matter of Indian status and band membership. There was little support for additional interventionism through state definitions of 'Indians' and rights in Kahnawake, and the people were determined not to be pushed on this issue as they had been more than once in the past.

The Department's new position on status and rights emerged as part of its larger reaction to the international shaming the federal government had received at the hands of the United Nations Commission on Human Rights in the Lovelace decision. In that case, Canada's image as a proponent of human rights received a substantial blow when the federal government was admonished for the blatant gender discrimination implicit in section 12(1)(b) of the Indian Act. This section directed that Aboriginal women who chose to marry non-Aboriginal men would lose their Indian status and all rights and entitlements associated with that status, as would any children of such unions. The opposite did not hold true for Aboriginal men marrying non-Aboriginal women, as the latter assumed status by virtue of their husband's Indian status. The government's response to this humiliation was the proposal of Bill C-31. Passage of this bill would mean (and has meant) that DIAND would unilaterally reinstate such women and their children into their bands of origin, expecting councils to cope with the return of these people and their demands for recognition of membership rights. That this was to be achieved with federal funds, which were promised but often never materialized, added an additional pressure to already stressed reservation lands and band finances.

The proposed changes placed the community of Kahnawake in an interesting and highly complex position. Popular wisdom dictated that the community was simply without the resources to support an influx of returning band members. Modern traditional culture also suggested that such a return could not happen: consistent with the Two Row Wampum,[25] a person who had jumped from one canoe to the other would have to live with that choice. Thus a Mohawk woman who had lost her status and band membership by 'marrying outside' would have to stay outside, as would her children by that union. A return might be permitted if she were divorced or widowed, but no promises could be made either to her or to her children. Certainly the Indian Act was unfair when it stole the status of Mohawk women who married out, but these women had made a choice and would simply have to live with the consequences, amendments notwithstanding (the contradiction of the apparent ability of men to leap both ways also notwithstanding). Kahnawake could not be asked to pay, both monetarily and culturally, for Canada's efforts to improve its international image on gender equality and Aboriginal rights.

At the same time, the Indian Act and its inherent paternalism, patriarchy, and gender discrimination constituted an affront to Mohawk culture, which was traditionally and historically matrilineal and highly pro-woman in practice and politics. Thus the amendments could be seen, at least on a technical cultural level, as a positive development. That being said, however, those same amendments—and the 'membership' provisions of the Act generally—constituted a much larger affront to the enduring sovereignty and nationhood of the Mohawk people insofar as these constituted a further arrogation by the Canadian state of the fundamental right of the Mohawk Nation to define its own citizenry. In their eyes, the people of Kahnawake saw the impending amendments to the Indian Act as yet another theft by Canada of their nation's rights and powers.

The Council responded to this rather unique and conflicting set of circumstances by declaring its control over the articulation of Mohawk identity and associated rights a full three years before DIAND handed down its 'legislative permission' to do so. Transpiring as it did on the eve of the Bill C-31 changes to the Act, such a declaration need not have been a controversial one. The changes, which became law in 1985, created a context whereby band councils could create their own 'membership codes' that would act as a foil to those rights and entitlements associated with 'Indian status' and membership in a recognized 'Indian band'. This was a rather radical departure from prior practice. Pre-1985, Indian Affairs had maintained an 'Indian Registry' containing the names of all persons who met the Indian Act's criteria for Indian status, as well as individual 'band lists'. An apparent separation between the lists was, at that time, more theoretical than real: an individual was noted on the Registry as an Indian and as a member of a given band. After the passage of Bill C-31, the Indian Register and band lists became separate entities. Today, inclusion on a band list, which exists separately from the Registry, will be determined either by the federal department or by an individual band in those cases where the band in question has assumed formal control over its membership list via section10 of the current Indian Act. Where a band council has not assumed such control, s. 8 of the Indian Act requires the department to keep a list of all band members who fit the membership criteria defined in s. 6 of the Act. When a band assumes control over membership under s. 10, the department is no longer required to maintain that list since it has delegated statutory control over membership to that band. Thus the list created by the band council becomes the central document for purposes of membership and the allocation of benefits associated with membership, including rights of residence, education, land allotment, welfare, and the local franchise.

Consistent with its position of enduring Mohawk sovereignty and nationhood, the Council refused to consider a state-delegated control over 'membership' in the 'Kahnawake band'. Rather, it would take full control over defining those criteria to be met for anyone to be accepted as a Mohawk of Kahnawake and to enjoy any of the rights or privileges normally associated with 'membership'. This membership would now flow from designation as a Mohawk. It was not so much that the Mohawk Council rejected the concept of membership or bands or the Indian Act that defined and created these things; it was simply that such things are irrelevant to an independent nation, which by definition deals in matters of citizenship and the rights and responsibilities associated with it. The people of Kahnawake, as part of the larger, sovereign Mohawk Nation, retained as a fundamental right of citizenship the authority to define the contents of that citizenship and its concomitant rights.

Clearly, then, the Mohawk Council eschewed a delegated power over membership, preferring instead to reassert a perceived historic right of the Mohawk Nation to define its own identity and population. In essence, the Council chose to impose its own definition and regulation of membership on the community in defiance of the legal interventionism of the Canadian state. However, the Council also refused to ratify its membership policy within the terms for assuming control over membership defined within s. 10 of the Indian Act, which translated an act intended to reject state interventionism into Mohawk lives into one that not only magnified the degree of intervention but did so in a way that was both confusing and painful for many Mohawks. For example, the failure to request state sanction of the Council's membership code, while consistent with the Council's position of sovereignty, means that the 'Mohawk list' resulting from this position and currently maintained by the Mohawk Council has not officially superseded the DIAND list because it is not, strictly speaking, recognized by the department as resulting from a legal process. The consequence is that two lists exist of those persons deemed entitled to any rights and responsibilities associated with qualification as a Mohawk of Kahnawake—the band list kept by the department, wherein those rights and responsibilities arise from Indian status and membership in the Kahnawake band, and the list maintained by the Mohawk Council, wherein rights and responsibilities are linked with status as a Mohawk of Kahnawake. Given this, it is possible for an individual recognized as an 'Indian' and member of the Kahnawake band by DIAND to be denied any of the rights and benefits that historically flowed from such standing, as they are not defined as Mohawks by the Mohawk Council that actually administers those rights and benefits. It is thus possible to be an 'Indian' but not a 'Mohawk', or one might be a 'Mohawk' but not an 'Indian' and have access to membership-based rights while not necessarily a member.

The situation is confusing, to say the least, and has proven to be highly problematic and troubling for many people of Kahnawake as well.[26] In this case, interventionism by either government, Ottawa or Mohawk, seems to smell pretty much the same.

At least part of this controversy originates in the focus of the Kahnawake membership law on exclusion as opposed to inclusion. As a first task in determining who belonged in Kahnawake, the Council worked for the systematic removal of those non-Indian residents who resided in the community with no tangible connection to the people or culture. While this addressed one dimension of a historic problem, it did little to address the problem of those non-Indians whose presence on reserve territory was gained through their marriage or adoption into Mohawk families. The Council's response to the compromise of Mohawk culture perceived to reside in such unions and families was to render them illegal according to 'Mohawk law'. Thus, in a statement issued by the Council in 1981, a moratorium was placed on all 'mixed marriages' and on the adoption of non-Indians by Kahnawake Mohawks:

> As of that date [22 May 1982], any Indian man or woman who marries a non-Indian man or woman is not eligible for any of the following benefits that are derived from Kahnawake[:]. . .

- Band number
- Residency (to live in Kahnawake)
- Land allotment
- Housing assistance—loan or repair
- Welfare—in Kahnawake only
- Education—in Kahnawake only
- Voting privileges—in Kahnawake only
- Burial
- Medicines—in Kahnawake only
- Tax privileges—in Kahnawake only.[27]

The moratorium was expanded and complemented in 1984 with the passage of the Council's 'Mohawk Citizenship Law', which specified who could qualify as a 'Mohawk' for purposes of the aforementioned rights and services within the community.[28] Article 3 of the law specified the following definitions to determine rights and access to services within Kahnawake:

> *Mohawk*—any person, male or female, whose name appears on the present band and reinstatement lists and whose blood quantum is 50% and more shall comprise the Kahnawake Mohawk Registry.

Non-Indian—any person, male or female, whose name does not appear on the present band and reinstatement lists and whose blood quantum is less than 50%.[29]

The impact of the Council's definition of citizenship for the exercise of that most fundamental of citizenship rights, the franchise, became apparent at the next local government elections held in 1986. Having obtained ratification for its citizenship law at a public meeting of rather slim attendance (popular wisdom records attendance at about 50 people), the Council obtained approval for its specification of the local franchise as limited to anyone '18 years of age and over . . . who is presently on the Band List and the Mohawk Registry . . . excluding those persons who are non-Indian by birth and those persons who are affected by the Kahnawake Mohawk Citizenship Law'.[30] The result of this policy was the denial of the right to vote of 475 of the 1,266 persons who attended the polls, on grounds that their blood quantum had been judged inadequate, and thus they no longer possessed Mohawk citizenship or its concomitant right of enfranchisement.[31]

The franchise was not the only right restricted by the new citizenship law. Over the decade following the 1986 election controversy, the rules regulating citizenship in the Kahnawake community were used to deny residence rights to divorced or widowed Mohawk women who had been married to non-Indians but who wished to return to live in the community, and to prevent their children from participating in the local educational system. In other cases, Mohawk landholders found their livelihoods under threat as they were restricted from leasing privately held lands to non-Indians.[32] In a similar fashion, one lifelong resident of Kahnawake who applied to the Council to expand his automotive repair business was denied in that effort because, according to the Council, an investigation of his family history had revealed him to possess 'only a blood quantum of 25% Indian Blood'.[33] More recently, a Kahnawake man was denied the opportunity to run for the position of Grand Chief of the Mohawk Council because a similar investigation found that only 46.87 per cent of his blood was Mohawk. In a classic example of the Kahnawake sense of humour, the man was quoted in the local newspaper as remarking, 'How they get that number is beyond me. It is two percentage points lower than the last time I asked.'[34] Throughout this and other controversies, the Department of Indian Affairs and Northern Development tended towards the rather ironic policy position of nonintervention in Kahnawake's business. In doing so, the state preserved an optic of respect for local governance rights and, as will be seen, abandoned those who might have benefited from a little strategic intervention to protect their human rights.

Not everyone in Kahnawake was a fan of the blood quantum policy; voices of discontent had emerged with the moratorium in 1981. After all, as one

Kahnawa'kehro:non woman observed, 'we all have members in our family who have married non-Natives, whether it was with their knowledge or without their knowledge.'[35] A law that threatened many Mohawks on a personal level offended still others as an affront to Mohawk tradition, which was perceived as embracing difference and utterly unconcerned with such matters as blood quantum:[36]

> In our history we have accepted people who were non-Aboriginal. It is very clear in our current society. We have families and family lineages that go right back to non-Native great-grandparents or great-great grandparents. These families have been accepted into our community. Generally, people are accepted because they contribute to the community in some way and in some function that is not detrimental to the community. Historically, we adopted babies. We have adopted captives of one kind and another at different times to supplement the loss of our young men in war and disease. It is historically documented.[37]

By 1995, those initial few voices who protested against blood quantum had been joined by many others, and the Mohawk Council struck a Task Force on Membership for the purposes of reconsidering the membership policies. The Task Force was to be a 'non-political' vehicle for community consultation about membership, and was to condense the results of its consultations into a new draft policy on membership. The Task Force report, as released to the public, revealed how difficult it would be to keep politics out of the process of policy change, or to avoid legal intervention in the lives of Kahnwa'kehro:non. According to the editor of the Kahnawake newspaper, *The Eastern Door*, who was briefed by the Task Force on their consultations with the community, 'they told me basically that the majority of people they met with have a big problem with the whole issue of blood quantum.'[38] When the draft membership law was released to the community, it proposed the replacement of the blood quantum requirement with the concept of a 'clear, recognizable lineage', whereby a prospective Mohawk citizen and member of the Kahnawake band would possess, among other capacities, 'at least two grandparents who are Mohawk or Native'.[39] It is difficult to see precisely how lineage may be determined without some consideration of 'blood'—a criticism levied against DIAND when it adopted the policy of lineage sometime prior to Kahnawake's discussions around it.

This draft law apparently remains unratified, and the blood quantum requirement continues to be a central and hotly disputed aspect of Kahnawake's criteria for Mohawk citizenship and band membership. The most recent entry into the debate has been a committee of Elders, who released in April 1998 a 'Declaration on Kanienke:haka Membership of Kahnawa:ke'. By this document, 'Kanienke:haka membership' required the demonstration of four things: knowledge of the Kanienke:haka language, Kanienke:haka lineage back to three generations, respect for Mother Earth, and identification with a clan.[40] The Elders added to these requirements a directive that:

> any Kanienke:haka who marries a non-Native is to leave the Kanienke:haka along with any children resulting from this union and such persons will be deprived of certain privileges such as residency rights, land ownership and political participation. . . . Persons who fall under this customary exclusion may return if widowed and furthermore the children of this union may return and be recognized as full members of the Kanienke:haka community if they marry a recognized member of the Kanienke:haka of Kahnawa:ke.[41]

At the present time, then, the realities of qualification for membership in the Kahnawake Mohawk community and any eligibility for benefits stemming from membership depend on a determination by the Councils' Membership Office of an acceptable amount of 'Mohawk blood'. As noted above, this policy has attracted more than a little controversy and caused more than a little heartache within Kahnawa'kehro:non families. It has also been rejected by the Canadian Human Rights Commission, which in 1998 determined that the policy discriminates on the basis of race when it denies rights and services to 'Indians' deemed not to be Mohawks. In the case of *Peter and Trudy Jacobs and the Canadian Human Rights Commission and Mohawk Council of Kahnawake*, a Human Rights Tribunal found that not only was the Council engaging in discrimination, but that DIAND was indirectly condoning that discrimination by refusing to come to the aid of persons it defined as 'status Indians' and members of the Kahnawake band, but who failed to meet the Council's apparently non-legal membership code:

> Clearly the Minister [of Indian Affairs and Northern Development] has turned over considerable responsibility to the MCK [Mohawk Council of Kahnawake] with respect to services covered in its [Funding] Arrangement and does not interfere with the decisions made by the MCK. . . . It was also clear that DIAND is aware of the current membership criteria being applied at Kahnawake and nonetheless has taken a 'hands-off' approach. This is so notwithstanding that Peter [Jacobs] is a status Indian within the meaning of the Indian Act who resides in a Territory and is therefore entitled to the benefits of programs funded by DIAND under the Act. While one can well understand the politics of the Kahnawake situation, in our view DIAND has abrogated its statutory responsibility by refusing to assist Peter and his family. . . . In satisfying its political requirements, the federal government has become a party to the resulting discriminatory practice.[42]

When it comes to the negotiation of membership in Kahnawake, it would appear that both in its pure legal interventionist form and in its retraction of intervention and apparent support of a highly regressive policy on the part of the Mohawk Council, the Canadian state failed to demonstrate any interest in legal intervention as a means of protection for marginalized or disempowered people (so much for the 'primary goals of purposive law' noted at the opening of this chapter). In the same moment that the government appeared to empower local government by withdrawing from discussions over membership, it also abandoned individuals deserving and expectant of state support through the negation of any state protection of individual human rights. While arguments could be—and have been—made that respect for 'Mohawk politics' or 'tradition' required that the state remove itself from legal intervention into putatively traditional approaches to defining citizenship or membership, when the impact of that withdrawal on individuals abandoned both by local government and by the state is considered, and that abandonment is considered in terms of a long history of state intervention in definitions of 'Indian' and 'membership', the actions of the state seem less consistent with a principled political correctness than with a desire to withdraw from a mess of its own making, collateral damage notwithstanding.

The DUA explicitly states in s. 85 that the 'relationship between Kahnawake laws and the *Canadian Human Rights Act* will be addressed in the Umbrella Final Agreement.'[43] What this means in the absence of the latter is unclear, but it is worthy of some speculation. The response to the membership conundrum in the Sub-Agreement on Membership seems to offer limited insight. Part I of the Sub-Agreement articulates the right of the Council to 'maintain its membership regime and institutions for the purpose of controlling and regulating all matters of Kahnawake membership on behalf of all the Mohawks of Kahnawake', and that this will be done 'pursuant to the Kahnawake Law on Membership.'[44] It is further provided that some provisions of the Indian Act will continue to apply, but that 'in the event of a conflict between a provision of a Kahnawake law, which provision is in relation to membership, and a provision of a federal law, which provision is in relation to First Nation membership, the provision of the Kahnawake law will prevail.'[45] This qualification is perhaps not surprising given that the Sub-Agreement's Canada clause confirms that 'Any person registered or entitled to be registered as an Indian under the *Indian Act* will continue to be registered as an Indian under the *Indian Act*.'[46] This would seem simply to replicate the rather unsatisfactory standoff that led to the Jacobs case. Here, while the Sub-Agreement clearly states an expectation that 'Indians' will continue to exist in Kahnawake, it also qualifies that the Mohawk laws relevant to membership will be the primary authority in the community and, moreover, states that these laws will be paramount over federal laws (such as the Indian Act) in the case of a conflict. The safety net for those, like the Jacobs family, who become entwined in such conflicts will be the Canadian Charter (which the Sub-Agreement notes must be applied mindful of Mohawk custom, a matter over which there is some disagreement in Kahnawake, especially in regard to membership and such concomitant matters as adoption and marriage), and a yet to be confirmed Kahnawake Charter that might be expected to contain the same override provisions as the Sub-Agreement on Membership.

What does all this mean for the alleged 'end to intervention' assumed to reside in a return to self-determination through self-government? The DUA and the Sub-Agreement appear to create a context in which the Mohawk Council has full control over membership, which is definitely consistent with a reduction of state intervention as it withdraws fully and legally from this issue. At the same time, that control would seem to be fettered by the application of the Canadian Charter of Rights and Freedoms as well as a proposed Kahnawake Charter, not to mention by the continued presence of individuals who might not qualify as 'Mohawks', but who will continue to qualify as 'Indians' according to DIAND and, one must assume, an Indian Act that putatively no longer applies to Kahnawake. In this regard, it would seem that some provision has been made for Indians who fail to qualify as Mohawks—suggesting a degree of continued state intervention—and that this intervention will lessen as non-Mohawk Indians become a smaller and ultimately non-existent component of the Kahnawake population. What might happen to the children of these people is unclear, but once again one must assume that if their parents did not qualify as Mohawks, they can hardly inherit an identity or rights as Kahnawa'kehro:non. Clearly, membership in Kahnawake is unlikely to become less contentious or less regulated in a self-governing Mohawk community. Indeed, what we see here seems less a reduction in interventionism than a proliferation of regulation through the reordering of legal relationships between the Canadian state and Mohawk government, and between both of these entities and individual Kahnawa'kehro:non.

As the Jacobs case suggests, ending a legacy of state intervention may not be as simple as the mere withdrawing of that intervention in favour of a similar intervention by local government. Rather, it seems that, in the realm of self-government, a carefully orchestrated intervention is required in some contexts to oversee state withdrawal in favour of community, while the state might retain a role in supporting individuals in other situations. The remarkable intervention that Kahnawake now seeks to eschew took a century (at least) to reach its current state, and it will take time to change this situation. The Mohawks recognize this, and have chosen to move to greater independence over time; at some points in this transition it may be necessary to have more, rather than less, intervention to ensure that the rights of individuals are respected both in the transition and in the achievement of self-government. The challenge of self-government for the state might well prove to be balancing when to retreat and when to intervene; for the Mohawk Council, the challenge may reside in its ability to resist the pull to create laws to fill any regulatory gaps left by the withdrawal of state intervention. After all, if one of the primary motivations towards self-government has been the desire to increase the rights and freedoms of Kahnawa'kehro:non by rejecting the over-regulation of their lives implicit in the Indian Act, how can this be achieved if a withdrawal of outside law is met simply with its replacement by local law? Traditional government was founded on ensuring the maximum freedom of the individual within the constraints of clan and community; while clan has largely been eroded by time and state intervention, community persists. That persistence has clearly been in spite of state intervention, and hopefully the persistence and prosperity of a self-governing Kahnawake will not have to be in spite of local intervention.

Conclusion: Can Intervention Assist Independence?

As the discussion of the controversies around membership indicate, the sword of interventionism can cut more than one way. While there can be little doubt that an initial act of state intervention appropriated the right of First Nations like Kahnawake to define their own peoples and rights of citizenship, and thus paved the way for the Jacobs case, at the same time the intervention of one arm of the state, a Canadian Human Rights Tribunal, ensured the respect of individual rights in the face of a highly divisive legal intervention of the Mohawk Council into the lives of the Jacobs family. The state created a context for the abuse of individual rights when it acted purposively to undermine Aboriginal cultures, destabilize Aboriginal communities, and erode traditional political, social, and economic structures, all of which provided clear guidelines on how to protect and prioritize communities by ensuring rights and respect for individuals. State legal intervention sowed the seeds of the modern conundrum around membership in Kahnawake, and there are few easy avenues through which the state can respect the rights of self-government of communities by withdrawing state regulations and laws, and at the same time maintain some degree of intervention in order to limit the collateral damage of self-government, such as was experienced by the Jacobs family. In this way it is possible to understand the importance of expanding, for a time, the degree of intervention in Kahnwa'kehro:non lives. The challenging question is when to withdraw entirely, and whether full withdrawal of the Canadian state from Kahnawake's business is a realistic or possible outcome of self-government. Canada and Kahnawake are very intimate neighbours; there is probably no way our fates can ever be completely distinct, and our laws must find a mutually acceptable means of coexistence. If the DUA and the Sub-Agreement on Membership are any indication, for at least some time that coexistence will lead to an enhanced level of legal intervention by state and local government. If this acts to preserve human rights in the face of self-government rights, such expansionism may be permissible.

The key, then, becomes when to rein in intervention. The latter is, as noted earlier, a hard habit to break. Too often in Canada's history has intervention been justified by its putatively positive goals—the 'better allocation of resources, the protection of those who were marginalized and excluded from societal decision-making, and the provision of a more just distribution of wealth'. Indeed, to a fair extent the same sentiments were enlisted to 'justify' such interventions as residential schools and reserves. If we have learned anything from this history and our mistakes, we should know where the extremes reside and how to avoid them, yet this arguably does little to assist us to find a positive middle ground. How, then, can we balance the good and bad of intervention to achieve the just society that may lie somewhere in that middle ground? The answer to this question is, at best, a work in progress. One important part of that work is currently underway in Kahnawake; how it will look in the end, and the degree of intervention it will produce, will remain to be seen.

Notes

This chapter is a revised and truncated version of a paper originally written for the Law Commission of Canada through its Legal Dimensions Initiative 1999, 'Perspectives on Legislation'; it is reproduced here with the permission of the Commission. The original paper was published by the Commission in its *Perspectives on Legislation: Essays from the 1999 Legal Dimensions Initiative* under the title 'As Good As Our Word: Reconciling Kahnawake Mohawk and Canadian Legal Culture in the Negotiation and Articulation of Self-Determination'. The author would like to acknowledge and thank the Commission for its support of this project.

1. The DUA and the Sub-Agreement, which are the focus of this chapter, are the DUA given to me in May of 1999, within days of its completion, and the Sub-Agreement that I am informed became public only a short time ago and was faxed to me by a member of the community. I am drawing on these documents with the understanding that they have not been altered in any significant way since I received them; that some changes may have occurred that are either not public in Canada, or only in Kahnawake, is possible, and I apologize if I have in any way misrepresented the situation anticipated by the DUA owing to changes in it of which I am not aware.
2. 'A total of 70.1 percent of the people surveyed by the [Mohawk Council of Kahnawake Intergovernmental Relations Team] opinion surveyors said they want to move away from the Indian Act as the foundation for Kahnawake governance. Support for elimination of the Indian Act was found higher among youth and adults (76%) than among elders (59.4%)—although support for removal of the Act from Kahnawake was highest (75%) among those who reported no knowledge of the

Act's contents. Still, a clear majority of the people (69%) who said they were familiar with the powers outlined in the Indian Act endorsed moving away from the Act as the basis for Kahnawake governance. A clear majority of Kahnawa'kehro:non (84%) support establishing traditional government in Kahnawake. Those who oppose moving towards traditional government express concerns about what such a move would involve. They want further clarification about how a Traditional form of government would operate. Supporters of a governmental change say they want to unify political factions within the community. They called for a clear action plan that would guide a transition to Traditional government.' See the Kahnawake Web site at: < http://www.kahnawake.com/ckr/99survey/sep99-survey_p4.htm >.
3. Ibid.
4. The law and regulation that the clans observed and administered were retained orally, recorded at most through mnemonic devices, such as wampum belts, which represented the mutual understandings and commitments contained in that unwritten law. Legislation and regulation resided in the spoken word and were by all accounts entirely as effective as (if not more so than) the written documents that have come to dominate modern Canadian society. Little of that oral law was transcribed and has thus survived in its original form in modern Iroquois societies. The *Kaianerakowa*, or Great Law of Peace, contains the constitution that defined the terms by which the Seneca, Cayuga, Onondaga, Oneida, Mohawk, and, after 1722, Tuscarora came together as the famed Iroquois Confederacy sometime between AD 1400 and AD 1600. The Great Law was a complex entity, and for purposes here may be understood as consisting of two essential components or layers. The first was a substratum that held the defining principles—the spirit of the Iroquois Great Law—upon which the confederated Iroquois Nations would base their government and international relations. From this philosophical base sprang the second layer of the Peace, which detailed in roughly 176 articles or 'wampums' constitutional matters ranging from the mechanics of the process of confederation through to the appointment, tenure, and jurisdiction of government officials, and included specific instructions relating to inter- and intra-Confederacy relations. See Paul A.W. Wallace, *The Great Law of Peace* (Philadelphia: University of Pennsylvania Press, 1946).
5. S.C. 1884, c. 28, 47 Vic. The first formal Indian Act would not be passed until 1876. In this, its initial incarnation, the Act simply consolidated all earlier 'Indian legislation'—and did it so badly that it required amendment less than a year later.
6. Interview with K. Deer, Kahnawake, 21 Feb. 1996.
7. Canada, Department of Indian Affairs and Northern Development, File #E 4218-D6 (vol.1), 1/1980–10/1980 Council Elections—Montreal District, 'Order-in-Council, PC 466, Privy Council, Canada, Government House, Ottawa. Tuesday 5th day of March 1889'.
8. Interview with J. Deom, L. Norton, and L. Delormier, 20 Feb. 1996; C. Deom, 19 Feb. 1996. Such a linkage is possible, insofar as the petitions appear to be grouped by clan, including Onrwaskerewake (small bear), Roineniotronon (stone), Rotiskirewakekowa (big bear), and Ratiniaten (turtle). Canada, Department of Indian Affairs and Northern Development, File #E4218-D6, vol. 2, 11/80–11/83, Council—Elections Montreal District, Seven Petitions forwarded by A. Brosseau, Agent, Caughnawaga, to Superintendent General, Indian Affairs, Ottawa, 16 Jan.

1889. That being said, there are far fewer than 14 of those extended family groupings identified in the petitions, although that does not mean that only those clans identified had members who signed them, or that those acting as translators or who committed the names to writing accurately identified all those who wished to be signatories or their proper clan affiliation. There is also some question regarding whether the signatories who made their marks on the petitions knew, in fact, what they were signing. Two informants who translated the Kanienkehaka (Mohawk) paragraph introducing each petition advised that the English translation accompanying the petitions to Ottawa, and that translated the Kanienkehaka phrases into a desire for the elective system, may not have been entirely accurate. By their understanding of the phraseology (and my grasp of the Kanienkehaka language is far too rudimentary to support or challenge that understanding), a more accurate translation would have depicted a desire by the signatories to have their chiefs made into councillors, which may not be equated with an invitation for the imposition of a band council government. Interview with C. Deom and L. Francis, Kahnawake, 19 Feb. 1996.

9. Canada, Department of Indian Affairs and Northern Development, File #E4218-D6, vol. 1, Order-in-Council, PC 466, 5 Mar. 1889.
10. 'The Sovereignty Position of the Kanienkehaka (Mohawk Nation), Kahnawake Territory of the Haudenosaunee Six Nations Iroquois Confederacy', 15 June 1988 (Ottawa: Department of Indian Affairs and Northern Development, File #E4200-070, vol. 2 7/88-12/88, Band Management—Mohawks of Kahnawake, 2.
11. Letter from Joseph Norton to Prime Minister Brian Mulroney, 6 June 1988. Copy of original in researcher's possession.
12. Annex B, 'Discussion Paper, Canada/Kahnawake Relations' (18 May 1990), in 'Canada/Kahnawake Relations, An Agreement on an Agenda and Process for the Negotiation of a New Relationship Between The Mohawks of Kahnawake and Canada', 13 Dec. 1991.
13. Ibid., 4.
14. Ibid.
15. Ibid., 5.
16. Interview with Arnold Goodleaf, Mohawk Council of Kahnawake Intergovernmental Relations Team, Kahnawake, 21 May 1999. Ultimately, the Mohawk Council's Intergovernmental Relations Team engaged in simultaneous bilateral talks with both levels of government, permitting it the ability to accommodate or avoid federal-provincial constitutional conflicts and reach two separate agreements that, taken together, describe the terms of Kahnawake's 'new relationship' with Canada. For a brief discussion of the provincial agreement, see Dickson-Gilmore, 'As Good As Our Word'.
17. Annex D, 'Parameters for Canada/Kahnawake Relations Negotiations', in *Canada/Kahnawake Relations. Agreement on an Agenda and Process for the Negotiation of a New Relationship Between The Mohawks of Kahnawake and Canada*, 13 Dec. 1991, 1.1.
18. The Mohawk Council of Kahnawake articulated its agenda and process in two resolutions: no. 138/1989-90 (17 Apr. 1990) and no. 22/1991-92 (5 Aug. 1991); the latter resolution also grants Grand Chief Joe Tokwiro Norton the right to sign the 'Canada/Kahnawake Relations Negotiation Agreement' on behalf of the Mohawk

people of Kahnawake. At the time of their passage, the resolutions and the negotiations generally were said to have emerged through an extensive public education campaign launched by the Council in Kahnawake, via meetings, circulars, and local newspaper articles and radio spots. Despite this apparent 'media blitz', reports at this earlier phase of the process suggested that many Kahnawa'kehro:non did not feel a part of the consultation process and were generally troubled by what they perceived to be the profound distance between the negotiations, the Council, and themselves. A more recent opinion survey taken by the Council in 1999 suggests that enthusiasm for the process and the negotiators has grown substantially. See the Intergovernmental Relations Team report of the survey at < http://www.kahnawake.com/ckr/99survey/sep99-survey_pt4.htm > . See also The Mohawk Council of Kahnawake and Canada, *Agreement on an Agenda and Process for the Negotiation of a New Relationship Between the Mohawks of Kahnawake and Canada*, 13 Dec. 1991. Also, interview with K. Deer, Editor, *The Eastern Door*, Kahnawake, 21 May 1999.

19. As of 22 May 1999, the document had not been released to the community and was released to me by a senior negotiator on the negotiating team. At that time, I was granted permission to discuss the agreement owing to the intention of the Council to begin full community consultations on the draft within a couple of weeks of its completion. That two weeks stretched out to almost two years, and the draft agreement has only recently, to my knowledge, become widely available to the community. Most of the discussion contained in this paper is based on the document I received in 1999, supplemented with what seems to be a more succinct version of the agreement made available through the Intergovernmental Relations Team's Internet site (see note 18, above).

20. All information regarding the DUA is taken from the original draft agreement, which is dated 10 May 1999; discussion of the clauses on membership is supplemented by the recently released 'Draft Sub-Agreement with Respect to Membership', which I received on 19 June 2001.

21. I was unable to obtain access to a copy of the draft Kahnawake Charter.

22. Clause 6, 'Kahnawake Governance', DUA, 5.

23. Clause 9, 'Jurisdiction', DUA, 6.

24. The discussion of the recent history of Kahnawake's struggles with membership is taken from my paper '"More Mohawk than My Blood": Citzenship, Membership and the Struggle over Identity in Kahnawake', *Canadian Issues* 21 ('Aboriginal Peoples in Canada: Futures and Identities') (1999): 44–62. Paper originally presented at the annual meeting of the Association for Canadian Studies and Law and Society Association meetings, Ottawa, June 1998.

25. The Two Row Wampum is an important constitutional principle for the Iroquois people, which speaks to a 'separate but equal' relationship with external states. By the Two Row, the Kahnawa'kehro:non and Canada are envisaged as travelling together down the river of life, their people, cultures, traditions, and laws residing separately in their respective canoes, neither imposing nor infringing upon the other. While the Canadian government seems quite fond of enlisting the metaphor of the Two Row in its public relations (the Penner Report on Self-Government, released in October 1983, displayed the Two Row Wampum belt prominently on its back cover, lauding it as a symbol of a new state relationship with First Nations),

the government's historic and current affection for interventionism indicates clearly that it has had little concept of the true meaning of this principle.
26. For a full discussion of one example of those troubles, see Dickson-Gilmore, '"More Mohawk Than My Blood"'; *Decision of Tribunal in the matter of Peter and Trudy Jacobs and the Canadian Human Rights Commission and Mohawk Council of Kahnawake* (1998).
27. Canada, Department of Indian Affairs and Northern Development, File #E4200-070, vol. 1, 1975—30/6/88, Band Management—Mohawks of Kahnawake, Summary prepared for the Minister of Indian Affairs, re Actions of the Mohawk Council of Kahnawake, Annex C: Mohawk Council of Kahnawake Moratorium.
28. Canada, Department of Indian Affairs and Northern Development, File #E4218-070, vol.1, 1/86-3/88, Council Elections—Kahnawake Band. Mohawk Citizenship Law, undated.
29. Ibid., 2.
30. Canada, Department of Indian and Northern Affairs, File #E4218070, vol. 1, 1/86-3/88, Council Elections—Kahnawake Band, Letter from Davis Rice, Chief, Mohawk Council of Kahnawake, to Gaetan Pilon, Lands, Revenues and Estates, Department of Indian and Northern Affairs, 18 Mar. 1987.
31. Canada, Department of Indian Affairs and Northern Development, File #E4218-2-070, vol. 3, 4/87-6/88, Council Elections—Appeals—Caughnawaga Band, Letter and materials from R.B. Kohls, Director General, Lands, Revenues and Trusts (Operations), to R.A. Hodgkinson, Director, Statutory Requirements, re Kahnawake Band Election Appeal; 'Norton re-elected Grand Chief, 475 denied right to vote', *Chateauguay Sun*, undated; 'Caughnawaga council vote illegal says Indian refused right to vote', *Montreal Gazette*, undated. (Undated papers from DIAND files.)
32. Canada, Department of Indian Affairs and Northern Development, File #E4200-070, vol. 1, 1975—30/6/88, Band Management—Mohawks of Kahnawake, Summary prepared for the Minister of Indian Affairs, re Actions of the Mohawk Council of Kahnawake, Annex C: Mohawk Council of Kahnawake Moratorium, 2.
33. Ibid.
34. K. Deer, 'Carl Curotte challenges election regulations', *The Eastern Door* 5, 17 (1996): 4.
35. Elena Mayo Diabo, Testimony before the Canadian Human Rights Commission Tribunal, in the matter of Peter and Trudy Jacobs and the Canadian Human Rights Commission and the Mohawk Council of Kahnawake, 11 Dec. 1995, Montreal.
36. Mark McComber, Testimony before the Canadian Human Rights Commission Tribunal, in the matter of Peter and Trudy Jacobs and the Canadian Human Rights Commission and the Mohawk Council of Kahnawake, 16 Sept. 1997, Montreal.
37. Kenneth Atsenhaienton Deer, Testimony before the Canadian Human Rights Commission Tribunal, in the matter of Peter and Trudy Jacobs and the Canadian Human Rights Commission and the Mohawk Council of Kahnawake, 15 Sept. 1997, Montreal.
38. Ibid.
39. Kenneth Deer, 'New Draft Membership Law Ready', *The Eastern Door* 5, 25 (1996): 1, 10.
40. 'Elders come to consensus on membership', *The Eastern Door* 7, 13 (24 Apr. 1998): 7.

41. Ibid.
42. Decision of Tribunal in the matter of Peter and Trudy Jacobs (complainants) and the Canadian Human Rights Commission, and the Mohawk Council of Kahnawake (respondents), 11 Mar. 1998, 36.
43. DUA, s. 85, 17.
44. Draft Sub-Agreement on Membership, Part I, ss. 3–4, 2.
45. Ibid., Part IV, s. 8; Part VIII, s. 13, 3.
46. Ibid., Part II, s. 5, 2.

Index

Abel, Richard L., 233
Abella, Rosalie, 195-6
Aboriginal justice, 12, 212, 214-15
Aboriginal peoples/First Nations, 136-7, 317-22, 334, 336. *See also* indigenous peoples; Iroquois; Mohawks
Aboriginal rights, 8, 325, 328
Aboriginal self-government, 8, 12, 22, 212, 317-19, 335-7
Aboriginal women, 320, 327-8, 330-3
accountability, 7, 13-14, 17-18, 38, 87, 124-5, 127, 129-32, 135, 137, 145-6, 148-51, 154-5, 157, 162, 164-5, 191, 214, 233, 235, 242, 246, 273, 324-5
Act of Settlement (United Kingdom), 83
Adler, Peter S., 208
administrative law, 78
ADR. *See* alternative dispute resolution
Africa, 226
agency, 66-7, 147, 152, 317
alternative dispute resolution (ADR), 19-20, 60-1, 68, 206, 208, 216-17, 226, 230-2
alternative justice, 204-18
Amending Act (England), 301
American Bar Association, 232
anti-trust law, 242, 245
arbitration, 60, 180, 184, 194-5, 197, 208, 226
Arbitration Act (Ontario), 230
Arthur Young and Company, 191
Asia, 244

Asia-Pacific Economic Cooperation Summit (Vancouver), 92
Atkinson, Logan, 22
Automobile Association of America, 245
autopoiesis, 104-5, 116
Aylmer, Matthew Whitworth-Aylmer, 5th Baron, 306, 308
Ayres, Ian, 17, 98, 107-9, 114-16

band councils, 321-2, 328-9
Bayer Inc., 288-9
Bennett, Colin, 273
Berger, Thomas, 136
Berle, Adolfe A., 148
Better Business Bureau, 245
Bhopal chemical disaster, 245
bills (Canada): C-31, 327-8; C-36, 92; C-42, 92
Black, Julia, 9, 126
Blackstone, Sir William, 174
blood products, 288-9, 311
Borden, Robert, 191
bovine spongiform encephalopathy, 288
Boyle, James, 267
Braithwaite, John, 17, 98, 107-9, 114-16
breaches of the peace, 86, 213
breach of contract, 157, 161
British Columbia Justice Reform Committee Report, 226
British North America Act, 81
broadcasting, 265, 271, 278-82

Broadcasting Act (Canada), 280
bubonic plague, 292–302
Bush, Robert A. Baruch, 208

Campbell, R. Lynn, 18
Canada, 12, 21–2, 36, 71, 86–90, 92–3, 97–8, 101–3, 109–16, 149, 164, 172, 174–5, 179, 184, 214, 223, 225–6, 229, 232, 245–54, 264–5, 270–4, 276–82, 288–90, 317–37
Canada/Kahnawake Intergovernmental Relations Act (CKIRA), 318, 324, 326
Canada Labour Code, 184
Canadian Association of Chiefs of Police, 276–7
Canadian Automobile Association, 245
Canadian Bar Association, 232
Canadian Broadcasting Corporation, 278, 280
Canadian Charter of Rights and Freedoms, 55, 60, 73, 81–2, 90, 151, 204, 251, 325, 335
Canadian Chemical Producers' Association (CCPA), 245–54
Canadian Eco-Management and Audit System, 111–15
Canadian Environmental Assessment Act, 230
Canadian Environmental Protection Act, 249–50
Canadian federal civil service, 18–19, 188–99
Canadian Food Inspection Agency, 289
Canadian Human Rights Act, 197, 334
Canadian Human Rights Commission, 197–8, 333
Canadian Human Rights Tribunal, 198, 333, 336
Canadian Postal Employees Association, 194
Canadian Radio-television and Telecommunications Commission (CRTC), 271, 279–82
Canadian rebellions (1837-8), 86, 88
Canadian Standards Association, 273–4
capitalism, 2, 4–5, 18, 28, 32–4, 44, 100, 123, 133, 171, 175, 189, 191, 271
CCPA (Canadian Chemical Producers' Association), 245–54
Chemical Manufacturers Association, 128
children, 63–4, 68, 70, 181, 217, 241, 244, 327, 331, 333, 335
China, 226
cholera, 292, 302–10
Chornenki, Genevieve A., 209
Christie, Nils, 212
circle sentencing, 214–15
citizenship, 2, 8, 12–13, 48, 58–9, 73, 79, 90–1, 136; Mohawk, 22, 324, 326–36
civility, 71, 91
Civil Justice Review (Ontario), 231
civil law, 216
civil rights, 55, 66, 71–2
civil servants, 18–19, 188–99
Civil Service Act (Canada), 191
Civil Service Association of Canada, 194
Civil Service Commission (Canada), 191
Civil Service Federation of Canada, 194
CJD (Creutzfeld-Jakob Disease), 288–9
CKIRA. See Canada/Kahnawake Intergovernmental Relations Act
Cobbett, William, 82–4
Cohen, Joshua, 131
Colborne, Sir John, 304–10
Cold War, 91
collective agreements, 19, 174, 179–81, 184, 196
collective bargaining, 101, 149, 172, 174, 177–81, 184, 190, 192, 194–5, 197
Colonial Office, 307–10
Commission on Human Rights (United Nations), 327
Competition Act (Canada), 250
Competition Bureau (Canada), 250
competition law, 242, 245
constitutional law, 81–2
consumer rights, 64
Consumers' Association (United Kingdom), 254
contract(s), 33, 38, 144–5, 150, 156–7, 161, 164, 242, 249, 253, 255, 266–7, 269; of employment, 172–8, 180, 183–4; freedom of, 5–6, 175; law,

18-19, 172-3
Cook, R.F., 228, 234
copyright, 267
corporations, regulation of, 18, 106, 109-15, 144-65
courts, 18-20, 23, 42, 60, 69, 80-7, 90, 105, 145, 147, 150-8, 160-2, 164-5, 173-6, 180, 183-4, 199, 204-6, 208-12, 215-18, 226-7, 231, 234-5, 240, 242-3, 248-9, 254, 270. *See also* Federal Court of Canada; judges; Supreme Court of Canada
Creutzfeld-Jakob Disease (CJD), 288-9
Criminal Code (Canada), 93, 152
criminal justice, 12, 55, 60, 213-15, 234
criminal law, 56, 79, 86, 204, 212, 216, 265, 271
critical legal studies movement, 81
Crown corporations, 112
CRTC (Canadian Radio-television and Telecommunications Commission), 271, 279-82
Cuba, 92
cyberspace, 21, 44, 264-82. *See also* Internet

Daubney Report, 226
Dean, Mitchell, 11, 42, 89
Defence of the Realm Act (United Kingdom), 87
Defoe, Daniel, 296-9
delegalization, 20, 99, 204-5, 212
democracy, 9-10, 13-14, 17, 58, 65-6, 71, 74, 102, 121-4, 126-7, 129-38, 147, 172, 178-9, 185, 264, 325
Denmark, 296
Department of Finance (Canada), 198
Department of Health (Canada), 289
Department of Indian Affairs and Northern Development (DIAND) (Canada), 327-9, 31-3, 335. *See also* Indian Department
deregulation, 2, 7, 9, 15, 28, 39, 41-2, 49, 60, 98, 110-11, 121, 126, 173, 279, 289
destabilization rights, 68
Dewey, John, 126, 131-2

DIAND. *See* Department of Indian Affairs and Northern Development
Dickson-Gilmore, Jane, 22
Dillon, Michael, 91
directors and officers, corporate, 18, 106, 110, 144-65
discrimination, 134, 182; gender, 71, 327-8; racial, 71-3, 134, 333
dispute resolution, 19-20, 58, 78, 80, 100, 204, 207-11, 216, 223-36, 244, 282. *See also* alternative dispute resolution; alternative justice
divorce, 56, 59, 73, 205, 210, 216-17, 327
Divorce Act (Canada), 230
domestic violence, 70, 215
Draft Umbrella Agreement (DUA) (Kahnawake), 318-19, 322, 325-6, 334-6
due diligence, 153, 247-8, 253
Durham, Lord, 88
Durkheim, Émile, 57

Eastern Door, The (newspaper), 332
Eco-Management and Audit Scheme (EMAS) (European Union), 21, 102-3, 109, 112, 254
electronic commerce (e-commerce), 21, 272, 274, 277-8
Electronic Frontier Foundation, 277
EMAS. *See* Eco-Management and Audit Scheme
Emergencies Act (Canada), 87
Emergency Planning and Right to Know Act (United States), 128
Emergency Powers (Defence) Act (United Kingdom), 87
Emond, Paul, 208
employers and employees, 6, 18, 57-9, 144-5, 149-51, 153, 171-85
employment, 4, 6, 18-19, 37, 57, 171-85
employment equity, 196. *See also* pay equity
Employment Equity Act (Canada), 196
employment standards legislation, 177, 181-4
empowerment, 14, 65-70, 172, 178,

181–2, 206–14, 224–5, 228–9, 234, 255, 274–5, 282
encryption, 265, 267, 271, 274–8, 281–2
England, 174, 226, 244, 290, 292–304
environment, 10, 15, 17–18, 45–7, 97–116, 122–38, 165, 216–17, 230, 240–1, 245–55
environmental justice, 17, 123, 134–8
environmental law, 21, 98, 101–3, 106–7, 109–11, 121–2, 125
Environmental Protection Agency (EPA), 127–8
equality, 2, 4–5, 7, 14, 17, 36, 55, 71–3, 135–8, 182, 198, 270, 328
Espionage Act (United States), 87
Europe, 4, 34–5, 244, 295, 304
European Directive on the Protection of Personal Data, 273
European Economic Community, 102
European Union, 21, 23, 43, 254
Evans, John Munroe, 198
executive enabling legislation, 16, 87–8

families, 35, 42, 56, 63–4, 67, 70, 73, 156, 215, 217, 226, 330–3. *See also* divorce
family law, 204, 216–17, 324
Federal Court of Canada, 198
feminists, 45, 55, 215
Fenians, 88
fiduciary duties, 18, 157–60, 165
First Nations. *See* Aboriginal peoples/First Nations; indigenous peoples
Fish, Stanley, 62
Folger, Joseph P., 208
food safety, 289
formal law, 2, 4–5, 8, 60–1, 100–2, 107, 233, 292
Foss v. Harbottle, 155
Foucault, Michel, 6, 10–11, 13, 15–16, 31–2, 40, 57–8, 62–4, 66–7, 72, 85, 88–9, 91
Fox, A., 175
France, 300–1
free riders, 21, 242, 247–9, 273
French Revolution, 83
Friends of the Earth, 254

Fung, A., 127, 129–30, 132, 137–8

Galanter, Marc, 209, 216
gender discrimination, 71, 327–8
General Agreement on Tariffs and Trade (GATT), 44, 92
Germany, 6
Gibbons, Llewellyn Joseph, 267–8
globalization, 1–2, 7–8, 15, 43–9, 65, 79, 89, 92, 163, 268, 289–90
Glorious Revolution, 83
Goderich, Lord, 305–6, 308
Goldsmith, Jack L., 269
Gordon, C., 40
governance: corporate, 18, 144–65; of cyberspace and the Internet, 21, 264–82; definition of, 10–11, 62, 145; economic, 16; in employer-employee relations, 18–19, 171–85; environmental, 17, 122–3, 125–33, 135–8; in Kahnawake, 320–3, 325–6, 331; law as, 54, 57–8, 62–3, 72; liberal, 63–5; of mediation, 224–5, 230, 235–6; multiple sites of, 3, 6, 10–12, 15, 32, 46–7, 49; social, 16, 29, 31, 34, 37, 63; sociology of, 11, 13, 15–16. *See also* self-governance
governmentality, 6, 10–11, 16, 31–2, 47, 62, 65, 89, 91
Great Britain/United Kingdom, 22, 42, 82, 86–7, 288, 290, 303–4, 306, 308–9. *See also* England
Great Depression, 35, 149
Greenpeace, 254
Grosse Isle, 306–7
guilds, 244–5

Habeas Corpus Act (United Kingdom), 83
Habermas, Jürgen, 57–8, 68, 100, 104
Hale, Robert L., 4–6
Handler, Joel, 68
Hardy, Trotter, 269
Harrington, Christine B., 217, 234
Hayek, F.A., 59
Heller, Hermann, 4–5, 12

Helms-Burton Act (United States), 92
Hobbes, Thomas, 88
Howell's English State Trials, 82-4
Howell, Thomas Bayly, 83
Howell, Thomas Jones, 83
Huber, Peter, 264-5
human rights, 67, 73, 149, 165, 181, 216, 278, 326, 331, 334, 337
Hunt, Alan, 11, 16, 40, 90
Hydro One Networks, 112
Hydro-Québec, 112

IMF (International Monetary Fund), 44, 49
immigration, 1-2, 12, 22, 47-8, 88, 291, 302-11
Immigration Act (Canada), 88
Incitement to Disaffection Act (United Kingdom), 87
Income Tax Act (Canada), 114, 160
incorporation, 145
Indian Act (Canada), 22, 88, 317-19, 322-9, 333-6
Indian Advancement Act (Canada), 321
Indian Department (Canada), 321. *See also* Department of Indian Affairs and Northern Development
indigenous peoples, 45. *See also* Aboriginal peoples/First Nations
Industrial Revolution, 245
Industry Canada, 271
information and communications technologies, 44, 46. *See also* cyberspace; Internet
information-based (performance-based) regulation (IR), 17, 122, 125-7, 129, 132, 135, 138
Information Highway Advisory Council, 270
IntellectualCapital.com, 275
intellectual property, 267, 271
interest groups, 9, 70, 101, 114, 250
Intergovernmental Relations Team (IRT) (Kahnawake), 318-19
International Labour Organization, 171; convention 100, 197
international law, 268

International Monetary Fund (IMF), 44, 49
International Organization for Standardization (ISO) 14000 series, 102-3, 109, 112, 249, 254
Internet, 1-2, 21, 44, 46, 92, 264-8, 270-1, 275, 279-82; broadcasting, 265, 271, 280-2. *See also* cyberspace
interventionist law, 1-2, 5, 9, 14, 16. *See also* legal interventionism
IR. *See* information-based (performance-based) regulation
Ireland, 303
Iroquois, 320, 326. *See also* Mohawks
Iroquois Confederacy, 319-20, 325
IRT (Intergovernmental Relations Team) (Kahnawake), 318-19
ISO 14000. *See* International Organization for Standardization (ISO) 14000 series

Jacobs, Peter and Trudy, 333-6
Japanese Canadians, 88
Japanese law, 226
Jay's Treaty, 323
Jessop, Bob, 189
Joint Union/Management Initiative (Treasury Board), 198
judges, 4, 55, 81-2, 86, 90, 92, 208, 210, 216-17, 227, 229, 234, 269. *See also* courts
Judicial Committee of the Privy Council, 81
juridification, 54, 58, 68, 70, 100, 102, 173, 205
Jutzi, Samuel, 289

Kahnawake, 22, 317-37
Kahnawake Charter, 325, 335
Kahnawake Mohawk Citizenship Law, 331
Karkkainen, B., 127, 129-30, 132, 137-8
Kentake, 319
Keynesianism, 36-7
King James's Act (England), 293-4, 301-2
Kirchheimer, Otto, 5-6
Kolko, Gabriel, 291
KPMG, 110

labour laws, 172
Lac Minerals v. International Corona, 159
La Forest, Gérard V., 159
Laskin, Bora, 162
law(s): administrative, 78; of agency, 152; anti-trust, 242, 245; civil, 216; competition, 242, 245; constitutional, 81-2; contract, 18-19, 172-3; corporation, 156; criminal, 56, 79, 86, 204, 212, 216, 265, 271; crisis of, 4-6, 10, 14-16, 19-20, 55-61; decentring of, 15-16; employment, 177; environmental, 21, 98, 101-3, 106-7, 109-11, 121-2, 125; family, 204, 216-17, 324; formal, 2, 4-5, 8, 60-1, 100-2, 107, 233, 292; as governance, 54, 57-8, 62-3, 72; as an instrument of social control, 97-8; international, 268; interventionist, 1-2, 5, 9, 14, 16 (*See also* legal interventionism); Japanese, 226; in Kahnawake, 317-26, 330-2, 334; labour, 172; legitimacy of, 4-5; material, 4, 8; nuisance, 101-2; private, 5, 7, 265; procedural, 8-9, 150; proceduralization of, 8, 17, 58, 61, 68, 70, 107; property, 173; public, 2, 5, 7, 11, 80; of quarantine, 22, 290-303, 310-11; reflexive/responsive, 9, 16, 57, 60-2, 72, 98, 100-4, 106-9, 111-12, 114-16, 125-6; reform, 16, 70-1, 82, 121, 226; regulatory, 29, 36, 60; role within the state, 1-4, 7-8, 11, 13-16; rule of, 1, 5, 8, 78-85, 93, 180; school, 68; social, 54, 58-9; state, 13, 23, 218, 266, 317-19; state security/national security, 15-16, 78-93; substantive, 65, 100-2, 106-7, 115-16, 150; as a vehicle of social policy, 54-74, 121
lawyers, 20, 60-1, 69, 86, 164, 211-12, 216, 218, 223, 227, 230-3, 235-6
legal formalism, 60
legal interventionism, 2-4, 9, 22, 59, 317-27, 329-30, 332, 334-7. *See also* interventionist law
legalism, 4, 6, 62, 65, 70-4, 234
legal pluralism, 11, 16, 60-1

legal realism, 3-7, 11, 81
legal subjectification, 16, 66
Lessig, Lawrence, 267
Libel Act (United Kingdom), 86
liberalism, 3-4, 6, 8-9, 37, 39-40, 63-4, 67, 89, 207, 272-3. *See also* neo-liberalism
liberal state(s), 2-3, 35, 38, 43, 80, 188-91, 199, 212. *See also* neo-liberal state
litigation, expansion of, 69
Lobban, Michael, 86
London, 300-2, 308, 311
Love Canal chemical disaster, 128, 245
Lovelace decision (*Lovelace v. Canada*), 327
Lower Canada, 302-4, 306, 310
Luhmann, Niklas, 104-5

McCormick, John P., 1, 7
Macdonald, Sir John A., 88
McGuinty, Dalton, 56
Mackenzie Valley Pipeline Inquiry (Berger Inquiry), 136-7
Mac Neil, Michael, 18, 21, 92, 199, 275
Mahon, Rianne, 189
Majone, G., 32
Marseilles, 297
Marshall, T.H., 2, 58
Martin, Diane, 215
Marx, Karl, 190
Massachusetts Toxic Use Reduction Act (TURA), 127-9
material law, 4, 8
Means, Gardiner C., 148
mediation, 19-20, 60, 205-6, 208-12, 215-18, 223-36
mens rea offences, 153
Merry, Sally Engle, 217-18
minority shareholders, 18, 147-8, 151, 154-5, 165
Model Code for the Protection of Personal Information (Canadian Standards Association), 273-4
Mohawk Council (Kahnawake), 317-19, 322-4, 326, 328-36

Mohawks, 22, 317-37
Moonkin, Robert H., 209-10
Mulroney, Brian, 323

Nader, Laura, 206
National Policy (Macdonald), 88
national security. *See* state security/national security law(s)
National Security Act (United States), 87
Native peoples. *See* Aboriginal peoples/First Nations; indigenous peoples
negligence, 63, 69, 213-14, 247-9
neo-liberalism, 2, 7, 13, 31, 37-43, 48-9, 121, 138, 273, 281
neo-liberal state, 207
Neumann, Franz, 5-6
New Brunswick, 302
New Deal, 7
New York City, 306, 310
Nonet, P., 57
non-governmental organizations (NGOs), 13, 46, 240-3, 250, 253-5
non-state/non-governmental actors, 3, 5, 7, 13-16, 49, 64, 124-5, 241, 251
North America, 4, 19, 34-5, 122, 124, 226, 228, 230, 254
North American Free Trade Agreement, 43
Northern Ireland Civil Authorities (Special Powers) Act, 87
Northern Ireland Emergencies Provisions Act, 87
Northwest Rebellion, 88
Norton, Joseph Tokwiro, 32
Nova Scotia, 302
nuisance law, 101-2

occupational/worker health and safety, 18, 149, 179, 181-4, 241, 245, 255
October Crisis, 88
Offe, Claus, 12
Official Secrets Act (United Kingdom and Canada), 87
Oka crisis (1990), 325

Olson, Mancur, 244
Ontario, 56, 182, 210, 226, 230-1. *See also* Upper Canada
Ontario Insurance Commission, 230
Ontario Labour Relations Board, 182-3
Ontario Mandatory Mediation Program, 210-11
oppression remedies, 150, 155-8

pay equity, 19, 58, 70, 196-8
Pentland, H. Clare, 174
performance-based regulation. *See* information-based (performance-based) regulation
Personal Information Protection and Electronic Documents Act (Canada), 273-4
Picard, Cheryl, 20, 211
piercing the corporate veil, 157, 161-2
Pirie, Andrew, 232
Pitkin, Hannah, 138
plague, 292-302
Polanyi, K., 41
police/policing, 33-4, 46, 48, 63, 72-3, 81, 85-91, 214, 271
political rights, 66
political trials, 16, 79, 82. *See also* state trials
Politics of Informal Justice, The (Abel), 233
politics, regulation of, 15-17, 29, 78-93
popular justice, 233-4
Poulantzas, Nicos, 190
Pound, Roscoe, 56
pragmatism, 122, 126, 130-2, 135, 137
Prevention of Terrorism (Temporary Provisions) Act (United Kingdom), 87
Prince Edward Island, 302
privacy, 21, 207, 265, 267, 271-8, 281-2
Privacy Commissioner (Canada), 274
private law, 5, 7, 265
privatization, 15, 28, 41-2, 189-90, 282
proceduralism, 8-10, 13-15, 17, 58, 61; thick, 9-10, 13-14, 17, 126, 173; thin, 9, 17, 126. *See also* law(s), proceduralization of

procedural law, 8-9, 150
Project XL (Environmental Protection Agency Program for Regulatory Excellence), 127, 129
property, 33, 38, 87, 217, 266, 269; freedom of, 39; intellectual, 267, 271; law, 173; rights, 5, 66, 99, 101, 121, 124, 127, 147, 172, 179
Protection of Personal Information, The (Canadian government discussion paper), 272-4
prudentialism, 63-4
PSAC (Public Service Alliance of Canada), 19, 194-8
public health, 22, 60, 288-311
public law, 2, 5, 7, 11, 80
public/private distinction, 5-6, 11
Public Service Alliance of Canada (PSAC), 19, 194-8
Public Service Employment Act (Canada), 195
Public Service Staff Relations Act (Canada), 194-5, 197
Purvis, Trevor, 15-16, 40, 90, 207, 225, 236, 268, 273, 281, 322

Quakers, 226
quarantine, 22, 288-311
Quarantine Act (England), 297-301, 311
Quebec, 88, 321, 323, 325. *See also* Lower Canada
Queen Anne's Act (England), 294-7

racial discrimination, 71-3, 134, 333
railroads, 291
RCMP (Royal Canadian Mounted Police), 88, 92
reasonable care, 247-8
reflexive regulation, 17, 137
reflexive/responsive law, 9, 16, 57, 60-2, 72, 98, 100-4, 106-9, 111-12, 114-16, 125-6
Reform Act (United Kingdom), 86
regulation: of broadcasting, 278-81; of civil servants, 19, 188-99; of corporations, 18, 106, 109-15, 144-65; crisis of, 2-4, 7-11, 15, 28, 39; of cyberspace and the Internet, 21, 265-71, 279-82; decentring of, 14-23; economic, 2, 12, 16, 29-38, 40, 49; of employment, 18, 172-85; of encryption, 265, 271, 277-8, 281; environmental, 10, 15, 17-18, 97-116, 121-38, 240-1, 245-55; of food safety, 289; informal, 18; information-based (performance-based) (IR), 17, 122, 125-7, 129, 132, 135, 138; in Kahnawake, 318-22, 325-6, 329, 335-6; of mediation, 223-36; moral, 85, 89; multiple sites of, 10, 12-13, 15-16, 31, 47; of politics, 15-17, 29, 78-93; of privacy, 265, 272-4, 281; private, 21, 240-55; of public health, 22, 288-311; of railroads, 291; reflexive, 17, 137; responsive, 17, 98, 107-9; social, 2, 12, 16, 28-37, 42-3, 46, 49, 78, 85, 89, 97, 207; of telecommunications, 278-9. *See also* deregulation; self-regulation
regulatory law, 29, 36, 60
Reidenberg, Joel R., 267
religion, 243-4
responsibilization, 40-1, 48, 63
Responsible Care program (CCPA), 243, 245-53
responsive regulation, 17, 98, 107-9
Revenue Canada, 160
rights, 8, 11-12, 16, 18, 58-60, 65-8, 73, 81-2, 91, 121, 138, 182-3, 204, 213, 216, 223-4, 233-4, 264, 268, 318, 320, 327, 329-31, 335-7; Aboriginal, 8, 325, 328; citizenship, 8, 48; civil, 55, 66, 71-2; consumer, 64; corporate voting, 147; destabilization, 68; environmental, 123, 136-7; group, 91; human, 67, 73, 149, 165, 181, 216, 278, 326, 331, 334, 337; individual, 91, 227, 335-6; minority, 91; political, 66; property, 5, 66, 99, 101, 121, 124, 127, 147, 172, 179; victim's, 56, 215
Rise and Decline of Nations, The (Olson), 244
Rock, Alan, 289

Roehl, J.A., 228, 234
Roosevelt, Franklin Delano, 7
Rose, Nikolas, 11, 49
Rottleuthner, H., 39
Royal Canadian Mounted Police (RCMP), 88, 92
rule of law, 1, 5, 8, 78–85, 93, 180
R. v. Sault Ste Marie, 153

Sabel, Charles, 17, 122–3, 126–38
Salomon v. Salomon, 161
Santos, B. de Sousa, 42, 49
Sargent, Neil, 19–20
Saunders, Ron, 20, 211
Scandinavia, 36
Schmitt, Carl, 4, 6
Schneider, David, 17
school law, 68
Seaborn Panel, 114
securities legislation, 149
sedition, 79, 82, 86, 88, 93
self-governance, 14, 64–5, 68, 125, 207, 266–7, 326
self-regulation, 3, 14–15, 17, 20–1, 61–2, 68–9, 89, 99, 101, 106–9, 111, 114–15, 122, 173, 241, 249–53, 264, 266, 272–4, 280, 282
Selznick, P., 57
Seveso chemical disaster, 245
Shapiro, Andrew L., 268, 281
shareholders, 18, 144–51, 154–9, 161–2, 164–5
Sheppard, D., 228
Sheriff, C., 309
Smith (Alien Registration) Act (United States), 87
smokers, 69
Smyth, John, 305
Snider, Laureen, 215
social control, 10, 13, 28, 30–1, 59, 97–8, 106, 109, 233–4, 320
social law, 54, 58–9
Spinner, Jeff, 71
spongiform encephalopathy, 288
Standards Council of Canada, 103
state: crisis of the, 1, 3, 7, 22, 39, 289–90; decentring of the, 11–15, 19–20, 22–3, 29, 42–3, 46, 48, 207, 265; regulatory, 9, 14–15, 29–33, 36–7, 43, 45, 47, 49, 290, 311; role of law within the, 1–4, 7–8, 11, 13–16
state law(s), 13, 23, 218, 266, 317–19
Statement of Responsible Care and Guiding Principles (CCPA), 245–6
state security/national security law(s), 15–16, 78–93
state trials, 82–5. *See also* political trials
strict liability offences, 153, 247–9
strikes, 19, 149, 178–9, 195–6
Stuart, Barry, 214
Stychin, Carl, 90
Sub-Agreement on Membership (Kahnawake), 318, 326, 334–6
substantive law, 65, 100–2, 106–7, 115–16, 150
Summit of the Americas (Quebec City), 92
Supplementary Act (England), 296, 298–301, 303
Supreme Court of Canada, 153, 159, 176–7
surveillance, 16, 41, 65, 69–70, 79, 85–8, 92, 267, 271
Swan, Peter, 17, 207, 289–90, 317
Sweden, 296
Sydenham, Lord, 88
systems theory, 13, 17, 99, 103–7

terrorist attacks (11 September 2001), 79, 92, 274
Teubner, Gunther, 9, 13, 17, 68, 98–108, 110–11, 114–16, 125, 253–4
Thompson, Edward, 83–4
Toronto Stock Exchange, 162–4
tort(s), 144, 157, 161, 247–8, 253, 272
Toxic Release Inventory (TRI), 127–9
Toxic Use Reduction Act (TURA) (Massachusetts), 127–9
Trachtman, Joel P., 269
treason, 79, 82–3, 86, 88, 93
Treason Act (United Kingdom), 83, 86
Treasury Board (Canada), 195–8

Treaty of Versailles, 171
TRI (Toxic Release Inventory), 127-9
TURA (Massachusetts Toxic Use Reduction Act), 127-9
Two Row Wampum, 327

Unger, Roberto, 68, 123
unions, 59, 70, 101, 172, 174, 178-81, 183-4, 190, 194-8
United Kingdom. *See* Great Britain/United Kingdom
United Nations, 327
United States, 6-7, 36, 69, 87-8, 92, 122, 126-7, 229-30, 232-3, 255, 288-9, 291, 306
Universal Declaration of Human Rights, 73
unlawful assembly, 86
Upper Canada, 22, 292, 303-11

Venice, 244

Walpole, Robert, 292, 295, 298-303, 310-11
War Measures Act (Canada), 87-8
Warskett, Rosemary, 19
Wassenaar Arrangement, 277-8
Watering Cove (Newfoundland), 101-2
Webb, Kernaghan, 21, 273
Weber, Max, 3-4, 188, 190, 195
Weimar Constitution, 6
Weimar Republic, 5
welfare, 28, 31, 36, 38, 41-2, 59-60, 70, 189, 193, 207

welfare state(s), 1-4, 7-8, 13-14, 28, 34, 36-41, 54, 59-60, 97-8, 100, 102, 115-16, 121, 163, 190
welfarist liberalism, 35
Western Europe, 4
Where Were the Directors? (Toronto Stock Exchange), 163
White Paper on Employment and Income (Canada), 193
Winnipeg General Strike, 88
Wisebrod, Dov, 266
women, 55; Aboriginal, 320, 327-8, 330-3; in the Canadian federal civil service, 190-4, 196-8; male violence against, 215
worker health and safety. *See* occupational/worker health and safety
Working Group on Directors' Liability (Industry Canada), 163-4
World Bank, 44, 49
World Trade Organization, 23, 44
World War I, 87
World Wildlife Fund, 254
Wright, Barry, 16
wrongful/unjust dismissal, 174-7, 180-1, 183-4
Wynne, Brian, 135, 138

York (Upper Canada) Board of Health, 307, 310
Young, Iris Marion, 134
Yukon, 230

Zehr, Howard, 212-14
Zuber Inquiry, 226